SECOND EDITION

Introduction to Probability and Its Applications

RICHARD L. SCHEAFFER

University of Florida

Duxbury Press

An Imprint of Wadsworth Publishing Company

I(T)P™ An International Thomson Publishing Company

Belmont • Albany • Bonn • Boston • Cincinnati • Detroit • London • Madrid • Melbourne
Mexico City • New York • Paris • San Francisco • Singapore • Tokyo • Toronto • Washington

Editorial Assistants   *Michelle O'Donnell, Janis Brown*
Production   *The Book Company*
Print Buyer   *Randy Hurst*
Permissions Editor   *Peggy Meehan*
Copy Editor   *Steven Gray*
Text Design   *Cloyce Wall*
Cover Design   *Craig Hanson*
Illustrations   *Interactive Composition Corporation*
Compositor   *Interactive Composition Corporation*
Printer   *Quebecor Printing/Fairfield*

Printed on acid-free recycled paper.

Printed in the United States of America
1 2 3 4 5 6 7 8 9 10—01 00 99 98 97 96 95

For more information, contact Wadsworth Publishing Company:

Wadsworth Publishing Company
10 Davis Drive
Belmont, California 94002, USA

International Thomson Publishing Europe
Berkshire House 168-173
High Holborn
London, WC1V 7AA, England

Thomas Nelson Australia
102 Dodds Street
South Melbourne 3205
Victoria, Australia

Nelson Canada
1120 Birchmount Road
Scarborough, Ontario
Canada M1K 5G4

International Thomson Editores
Campos Eliseos 385, Piso 7
Col. Polanco
11560 México D.F. México

International Thomson Publishing GmbH
Königswinterer Strasse 418
53227 Bonn, Germany

International Thomson Publishing Asia
221 Henderson Road
#05-10 Henderson Building
Singapore 0315

International Thomson Publishing Japan
Hirakawacho Kyowa Building, 3F
2-2-1 Hirakawacho
Chiyoda-ku, Tokyo 102, Japan

Library of Congress Cataloging-in-Publication Data

Scheaffer, Richard L.
    Introduction to probability and its applications / Richard L. Scheaffer. – 2nd ed.
        p.   cm.
    Includes bibliographical references and index.
    ISBN 0-534-23790-8
    1. Probabilities.   I. Title.
QA273.S357   1994
519.2– dc20   94-38064
                                                                  CIP

This textbook is designed to be covered in a one-semester or one-quarter course in probability for students who have a solid background in basic differential and integral calculus. Probability is presented as an area of mathematics that has many and varied applications to the solutions of real-world problems. Therefore, an axiomatic development of the theory is given, including a thorough discussion of random variables and the commonly used probability distributions. This is mixed with discussions of practical uses of probability, some involving real data. The text should provide a sound background for those students going on to more advanced courses in probability or statistics and, at the same time, provide a working knowledge of probability for those who must apply it in engineering or the sciences.

Many exercises are provided, most for practice in the calculation of probabilities, but some for allowing the student to extend the theory. It has been this author's experience that many probability problems appear deceptively easy, and students must be encouraged to practice the calculations in a variety of settings in order to really master the subject. Thus, students should work through most of the problems at the end of each section, along with a sampling of those from the supplementary set at the end of each chapter.

In this age of technology, much of the applied work in probability is done with the aid of a computer, especially in cases involving nonstandard distributions. Thus, activities for the computer are presented and guidelines are given for generating probability distributions by use of random numbers. These sections should be especially useful for those who must simulate the properties of complex systems. Except for those sections designated for the computer, knowledge of computing is not required for completing the text.

Attention should be directed to a few of the major changes for the second edition. Applications of a more advanced nature, particularly in the area of stochastic processes, have been removed from within the chapters and placed in a new Chapter 8. This provides greater flexibility in the use of the book as these applications can be woven into earlier chapters, used at the end of the course for review and extension, or skipped entirely. New or expanded sections include Section 2.7 on odds, odds ratios and relative risk; Section 3.11 in which applications of Markov chains have been added; Section 4.6 in which much has been added on the use of the normal distribution as a model for data (including Q-Q plots); Section 4.9 on reliability; and Section 6.5 on an important use of moment-generating functions.

I extend my thanks to the following individuals for providing valuable suggestions for improvements to be made in the second edition: Richard Bolstein, George Mason University; Paul T. Holmes, Clemson University; and Joseph J. Walker, Georgia State University.

I also extend thanks to Chris Franklin of the University of Georgia for her insightful comments and for writing the solutions manual, and to Nancy Watson for her superb work with the word processing.

Richard L. Scheaffer

1 Probability in the World Around Us 1

2 Probability 6

8 Extended Applications of Probability   327

Probability in the World Around Us

1.1 Why Study Probability?

We live in an information society. We are confronted—in fact, inundated—with quantitative information at all levels of endeavor. Charts, graphs, rates, percentages, averages, forecasts, and trend lines are an inescapable part of our everyday lives. They affect our decisions on health, citizenship, parenthood, jobs, financial concerns, and many other important matters. Today, an informed person must have some facility for dealing with data and making intelligent decisions based on quantitative arguments that involve uncertainty or chance.

We live in a scientific age. We are confronted with arguments that demand logical, scientific reasoning, even if we are not trained scientists. We must be able to make our way successfully through a maze of reported "facts," in order to separate credible conclusions from specious ones. We must be able to weigh intelligently such issues as the evidence on the causes of cancer, the effects of pollutants on the environment, and the likely results of a limited nuclear war.

We live amidst burgeoning technology. We are confronted with a job market that demands scientific and technological skills; and students must be trained to deal with the tools of this technology productively, efficiently, and correctly. Much of this new technology is concerned with information processing and dissemination, and proper use of this technology requires probabilistic skills. These skills are in demand in engineering, business, and computer science for jobs involving quality control, reliability, product development and testing, market research, business management, data management, and economic forecasting, to name just a few.

Few results in the natural or social sciences are known absolutely. Most results are reported in terms of chances or probabilities: the chance of rain tomorrow, the chance of your getting home from school or work safely, the chance of your living past sixty years of age, the chance of contracting (or recovering from) a certain disease, the chance of inheriting a certain trait, the chance of your annual income's exceeding $30,000 in two years, the chance of winning an election. Students must obtain some knowledge of probability and must be able to tie this concept to real scientific investigations if they are to understand science and the world around them.

This book provides an introduction to probability that is both mathematical, in the sense that a theory is developed from axioms, and practical, in the sense that applications to real-world problems are discussed. The material is designed to provide a strong basis in probability for students who may go on to deeper studies in statistics, mathematics, engineering, or the physical and biological sciences; at the same time, it should provide a basis for practical decision making in the face of uncertainty.

1.2 Deterministic and Probabilistic Models

1.2.1 Modeling Reality

It is essential that we grasp the difference between theory and reality. Theories are ideas proposed to explain phenomena in the real world. As such, they are approximations or models of reality. Theories are presented in verbal form in some (less quantitative) fields and as mathematical relationships in others. Thus, a theory of social change might be expressed verbally in sociology, whereas the theory of heat transfer is presented in a precise and deterministic mathematical manner in physics. Neither gives an accurate and unerring explanation for real life, however. Slight variations from the mathematically expected can be observed in heat-transfer phenomena and in other areas of physics. The deviations cannot be blamed solely on the measuring instruments (the explanation that one often hears); they are due in part to a lack of agreement between theory and reality. Anyone who believes that physical scientists now completely understand the wonders of this world need only look at history to find a contradiction. Theories assumed to be the "final" explanation for nature have been superseded in rapid succession during the past century.

In this text, we shall develop certain theoretical models of reality. We shall attempt to explain the motivation behind such a development and the uses of the resulting models. At the outset, we shall discuss the nature and importance of model building in the real world, to convey a clear idea of the meaning of the term *model* and of the types of models generally encountered.

1.2.2 Deterministic Models

Suppose that we wish to measure the area covered by a lake that, for all practical purposes, has a circular shoreline. Since we know that the area A is given by $A = \pi r^2$, where r is the radius, we attempt to measure the radius (perhaps by averaging a number of measurements taken at various points), and then we substitute the value obtained into the formula. The formula $A = \pi r^2$, as used here, constitutes a *deterministic model*. It is *deterministic* because, once the radius is known, the area is assumed to be known. It is a *model* of reality because the true border of the lake has some irregularities and therefore does not form a true circle. Even though the planar object in question is not exactly a circle, the model identifies a useful relationship between the area and the radius, which makes approximate measurements of area easy to calculate. Of course, the model becomes poorer and poorer as the shape of the figure deviates more and more from that of a circle until, eventually, it ceases to be of value and a new model must take over.

Another deterministic model is Ohm's Law, $I = E/R$, which states that electric current is directly proportional to the voltage and inversely proportional to the resistance in a circuit. Once the voltage and the resistance are known, the current can be determined. If we investigated many circuits with identical voltages and resistances, we might find that the current measurements differed by small amounts from circuit to circuit, owing to inaccuracies in the measuring equipment or other uncontrolled influences. Nevertheless, any such discrepancies are negligible, and Ohm's Law thus provides a useful deterministic model of reality.

1.2.3   Probabilistic Models

Contrast the two preceding situations with the problem of tossing a balanced coin and observing the upper face. No matter how many measurements we may make on the coin before it is tossed, we cannot predict with absolute accuracy whether the coin will come up heads or tails. However, it is reasonable to assume that, if many identical tosses are made, approximately one-half will result in outcomes of heads; that is, we cannot predict the outcome of the next toss, but we can predict what will happen in the long run. We sometimes convey this long-run information by saying that the "chance" or "probability" of heads on a single toss is 1/2. This probability statement is actually a formulation of a *probabilistic model* of reality. Probabilistic models are useful in describing experiments that give rise to random, or chance, outcomes. In some situations, such as the tossing of an unbalanced coin, preliminary experimentation must be conducted before realistic probabilities can be assigned to the outcomes; but it is possible to construct fairly accurate probabilistic models for many real-world phenomena. Such models are useful in varied applications, such as in describing the movement of particles in physics (Brownian motion) and in predicting the profits for a corporation during some future quarter.

1.3 Applications of Probability

We shall now consider two uses of probability theory. Both involve an underlying probabilistic model, but the first hypothesizes a model and then uses this model for practical purposes, whereas the second deals with the more basic question of whether the hypothesized model is in fact a correct one.

Suppose that we attempt to model the random behavior of the arrival times and lengths of service for patients at a medical clinic. Such a mathematical function would be useful in describing the physical layout of the building and in helping us determine how many physicians are needed to service the facility. Thus, this use of probability assumes that the probabilistic model is known and offers a good characterization of the real system. The model is then employed to enable us to infer the behavior of one or more variables. The inferences will be correct—or nearly correct—if the assumptions that governed construction of the model were correct.

The problem of choosing the correct model introduces the second use of probability theory, and this use reverses the reasoning procedure just described. Assume that we do not know the probabilistic mechanism governing the behavior of arrival and service times at the clinic. We might then observe an operating clinic and acquire a sample of arrival and service times. Based on the sample data, inferences can be drawn about the nature of the underlying probabilistic mechanism—a type of application known as *Statistical inference*. This book deals mostly with problems of the first type; but on occasion, it makes use of data as a basis for model formulation. Ideally, readers will go on to take a formal course in statistical inference later in their academic studies.

Consider the problem of replacing the light bulbs in a particular socket in a factory. A bulb is to be replaced either at failure or at a specific age T, whichever comes first. Suppose that the cost c_1 of replacing a failed bulb is greater than the cost c_2 of replacing a bulb at age T. This may be true because in-service failures disrupt the factory, whereas scheduled replacements do not. A simple probabilistic model enables us to conclude that the average replacement cost C_a per unit time, in the long run, is approximately

$$C_a = \frac{1}{\mu}[c_1(\text{Probability of an in-service failure})$$

$$+ c_2(\text{Probability of a planned replacement})]$$

where μ denotes the average service time per bulb. The average cost is a function of T; and if μ and the indicated probabilities can be obtained from the model, a value of T can be chosen to minimize this function. Problems of this type are discussed more fully in Chapter 8.

Biological populations are often characterized by birth rates, death rates, and a probabilistic model that relates the size of the population at a given time to these rates. One simple model allows us to show that a population has a high probability of becoming extinct even if the birth and death rates are equal. Only if the birth rate exceeds the death rate might the population exist indefinitely.

Again referring to biological populations, models have been developed to explain the diffusion of a population across a geographic area. One such model concludes that the square root of the area covered by a population is linearly related to the length of time the population has been in existence. This relationship has been shown to hold reasonably well for many varieties of plants and animals.

Probabilistic models like those mentioned give scientists a wealth of information for explaining and controlling natural phenomena. Much of this information is intuitively clear, such as the fact that connecting identical components in series reduces the system's expected life length compared to that of a single component, whereas parallel connections increase the system's expected life length. But many results of probabilistic models offer new insights into natural phenomena—such as the fact that, if a person has a 50:50 chance of winning on any one trial of a gambling game, the excess of wins over losses will tend to stay either positive or negative for long periods of time, given that the trials are independent. (That is, the difference between number of wins and number of losses does not fluctuate rapidly from positive to negative.)

1.4   A Brief Historical Note

The study of probability has its origins in games of chance, which have been played throughout recorded history and, no doubt, during prehistoric times as well. Game boards have been found in excavations of Egyptian tombs, and gaming was so popular in Roman times that laws had to be passed to regulate it. Despite this long history of games of chance, formal study and development of probability theory did not take root until the middle of the seventeenth century.

It is generally agreed that a major impetus to the formal study of probability was provided by the Chevalier de Méré when he posed a gambling question to the famous mathematician Blaise Pascal (1623–1662). The question was along the following lines. To win a particular game of chance, a gambler must throw a 6 with a die; he has eight throws in which to do it. If he has no success on the first three throws, and the game is thereupon ended prematurely, how much of the stake is rightfully his? Pascal cast this problem in probabilistic terms and engaged in extensive correspondence with another French mathematician, Pierre de Fermat (1608–1665), about its solution. This correspondence began the formal mathematical development of probability theory.

Scientists in the eighteenth century (such as James Bernoulli and Abraham de Moivre) continued to develop probability theory and recognized its usefulness in solving important problems in science. The normal curve was introduced during this period. Their work was carried forward by Carl Friedrich Gauss and Pierre de Laplace in the nineteenth century, when the use of probability in data analysis emerged as a forerunner of modern statistics. In the twentieth century, probability theory became a major branch of mathematical research; and applications of the theory have spread to virtually every corner of scientific research.

1.5   A Look Ahead

This text is concerned with the theory and applications of probability as a model of reality. We shall postulate theoretical frequency distributions for populations and develop a theory of probability in a precise mathematical manner. The net result will be a theoretical or mathematical model for acquiring and utilizing information in real life. It will not be an exact representation of nature, but this should not disturb us. Like other theories, its utility should be gauged by its ability to assist us in understanding nature and in solving problems in the real world. Such is the role of the theory of heat transfer, the theory of strengths of materials, and other models of nature.

Probability

2.1 Understanding Randomness: An Intuitive Notion of Probability

2.1.1 Randomness with Known Structure

At the start of a football game, a balanced coin is flipped into the air to decide which team will receive the ball first. What is the chance that the coin will land heads up? Most of us would say that this chance or probability is 0.5, or something very close to that. But, what is the meaning of this number 0.5? If the coin is tossed 10 times, will it come up heads exactly 5 times? Upon deeper reflection, we recognize that 10 tosses need not result in exactly 5 heads; but in repeated flipping, the coin should land heads up approximately one-half of the time. From there on, the reasoning begins to get more fuzzy. Will 50 tosses result in exactly 25 heads? Will 1000 tosses result in exactly 500 heads? Not necessarily, but the fraction of heads should be close to one-half after "many" flips of the coin. So the 0.5 is regarded as a **long-run** or **limiting relative frequency** as the number of flips gets large. John Kerrich, a mathematician interned in Denmark during World War II, actually flipped a coin 10,000 times, keeping a careful tally of the number of heads. After 10 tosses, he had 4 heads, a relative frequency of 0.4; after 100 tosses he had 44 heads (0.44); after 1000 tosses he had 502 heads (0.502); and after 10,000 tosses he had 5067 heads (0.5067). The relative frequency of heads remained very close to 0.5 after 1000 tosses, although the actual figure at 10,000 was slightly farther from 0.5 than was the figure at 1000.

In the long run, Kerrich obtained a relative frequency of heads close to 0.5. For that reason, the number 0.5 can be called the **probability** of obtaining a head on the toss of a balanced coin. Another way of expressing this result is to say that Kerrich **expected** to see about 5000 heads among the outcomes of his 10,000 tosses. He actually came close to his expectations, and so have others who have repeated the coin-tossing study. This idea of a stabilizing relative frequency after many trials lies at the heart of random behavior; another example that illustrates the point comes from studying properties of random digits.

A table of random digits (such as Table 1 in the Appendix) or random digits generated by computer are produced according to the following model. Think of ten equal-sized chips, numbered from 0 through 9, with one number per chip, and thoroughly mixed in a box. Without looking, someone reaches into the box and pulls out a chip, recording the number on the chip. That process constitutes a single draw of a random digit. Putting the chip back into the box, mixing the chips, and then drawing another chip produces a second random digit. A random number table is the result of hundreds of such draws, each from the same group of thoroughly mixed chips.

What is the probability of selecting a digit that is a multiple of 3 from a random number table? We can find this probability, at least approximately, by selecting digits from a random number table and counting the number of multiples of 3 (3's, 6's, and 9's). Figure 2.1 shows the results of two different attempts at doing this with 100 digits each. The relative frequencies are recorded as a function of the number of digits selected.

F I G U R E **2.1** Proportion of multiples of 3.

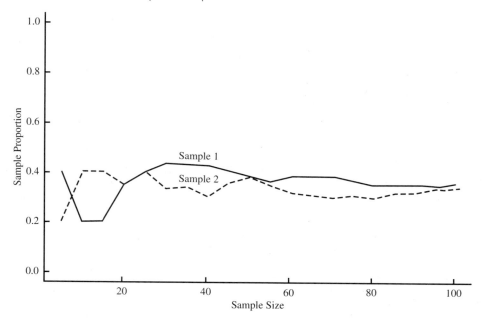

The graph in Figure 2.1 has two important features. First, both sample paths oscillate greatly for small numbers of observations and then settle down around a value close to 0.32. Second, the variation between the two sample fractions is quite large for small sample sizes and quite small for the larger sample sizes. From these data, we can approximate the probability of selecting a multiple of 3 from a random number table as 0.32, recognizing that the approximation tends to get better as the sample size increases. It should not be surprising to see about 32 multiples of 3 in a selection of 100 random digits. (What would you have suggested as the probability of seeing a 3, 6, or 9 in one draw before looking at the data?)

2.1.2 Randomness with Unknown Structure

For the two examples discussed in the preceding subsection, we knew what the resulting long-run relative frequency should be. Now, we will consider an example for which the result is less obvious before the data are collected. A standard paper cup, with the open end slightly larger than the closed end, was tossed in the air and allowed to land on the floor. The goal was to approximate the probability of the cup's landing on the open end. (You might want to generate your own data here before looking at the results presented next.) After 100 tosses on each of two trials, the sample paths looked like those shown in Figure 2.2.

F I G U R E **2.2**   Proportions of cup landings.

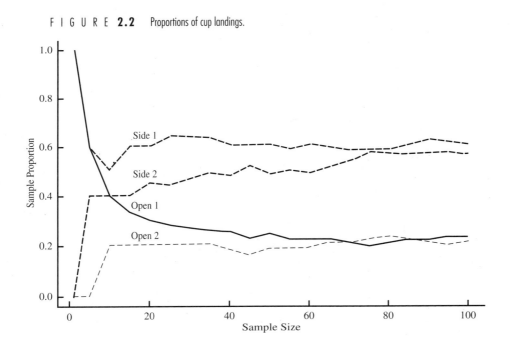

Notice that the pattern observed in the other examples occurs here, too; both sample paths seem to stabilize around 0.2, and the variability between the paths is much greater for small sample sizes than for large ones. We can now say that a tossed cup's probability of landing on its open end is approximately 0.2; in 100 tosses of this cup, we would expect to see it land on its open end about 20 times. Figure 2.2 also includes two sample paths for the outcome "landed on side," which seem to stabilize at slightly less than 0.6. Thus, the basic notion of probability as a long-run relative frequency works here just as well as it did in the cases for which we had a good theoretical idea about what the relative frequency should be!

Notice that the approximate probabilities for the cup toss depend on the dimensions of the cup, the material from which the cup is made, and the manner of tossing the cup. Another experiment with a different cup should generate different approximate probabilities.

2.1.3 Sampling a Finite Universe

In the preceding subsection, there was no limit to the number of times the coin could have been tossed, the random digits could have been selected, or the paper cup could have been tossed. The set of possible sample values in such cases is both infinite and conceptual; that is, it doesn't exist on a list somewhere. (A list of random digits may exist, but it could always be made larger.) Now, however, let's consider a jar of differently colored beads that sits on my desk. I plan to take samples of various sizes from this jar, with a goal of estimating the proportion of yellow beads in the jar. After mixing the beads, I take a sample of 5, then 5 more (to make a total of 10), and so on, up to a total of 100 beads. The beads sampled are **not** returned to the jar before the next group is selected. Will the relative frequency of yellow beads stabilize here, as it did for the infinite conceptual universes sampled earlier? Figure 2.3 shows two actual sample paths for this type of sampling from a jar containing more than 600 beads. The same properties observed earlier in the infinite cases do, indeed, hold here as well. The probability of randomly sampling a yellow bead is about 0.2; a sample of 100 beads can be expected to contain about 20 yellows (or equivalently, the jar can be expected to contain more than 120 yellows). Actually, the jar contains 19.6% yellow beads, so the sampling probabilities came very close to the true proportion.

In general, as long as the selection mechanism remains random and consistent for all selections, the sample proportion for a certain specified event will eventually stabilize at a specific value that can be called the **probability** for the event in question.

If n is the number of trials of an experiment (such as the number of flips of a coin), one might define the probability of an event E (such as observing an outcome of heads) by

$$P(E) = \lim_{n \to \infty} \frac{\text{Number of times } E \text{ occurs}}{n}$$

But will this limit always converge? If so, can we ever determine what it will converge to, without actually conducting the experiment many times? For these and other reasons, this is not an acceptable mathematical definition of probability, although it is a property that should hold in some sense. Another definition of probability must be found that allows such a limiting result to be proved as a consequence. This will be done in Section 2.3.

2.2 A Brief Review of Set Notation

Before embarking on a formal discussion of probability, let's go over the set notation we will use. Suppose that we have a set S consisting of points labeled 1, 2, 3, and 4.

F I G U R E **2.3** Proportions of yellow beads.

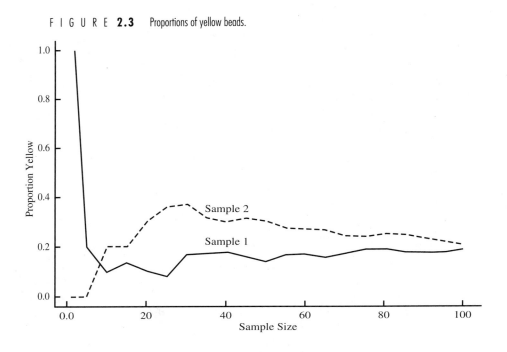

We denote this by $S = \{1, \ 2, \ 3, \ 4\}$. If $A = \{1, \ 2\}$ and $B = \{2, \ 3, \ 4\}$, then A and B are subsets of S, denoted by $A \subset S$ and $B \subset S$ (B is "contained in" S). We denote the fact that 2 is an element of A by $2 \in A$. The union of A and B is the set consisting of all points that are in A or B or both. This is denoted by $A \cup B = \{1, \ 2, \ 3, \ 4\}$. If $C = \{4\}$, then $A \cup C = \{1, \ 2, \ 4\}$. The intersection of two sets A and B is the set consisting of all points that are in both A and B, denoted by $A \cap B$, or merely AB. For the example $A \cap B = AB = \{2\}$ and $AC = \emptyset$, where $\emptyset$ denotes the null set (the set consisting of no points).

The complement of A with respect to S is the set of all points in S that are not in A, denoted by $\bar{A}$. For the specific sets just given, $\bar{A} = \{3, \ 4\}$. Two sets are said to be mutually exclusive or disjoint if they have no points in common, as is the case with A and C.

Venn diagrams can be used to portray graphically the concepts of union, intersection, complement, and disjoint sets (see Figure 2.4).

We can easily see from Figure 2.4 that

$$\bar{A} \cup A = S$$

for any set A. Other important relationships among events are the *distributive laws*:

$$A(B \cup C) = AB \cup AC$$
$$A \cup (BC) = (A \cup B)(A \cup C)$$

F I G U R E **2.4**
Venn diagrams of set relations.

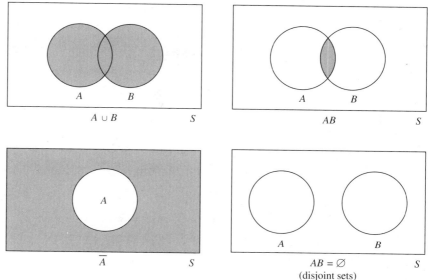

and *DeMorgan's Laws*:

$$\overline{A \cup B} = \bar{A}\bar{B} \quad \text{or} \quad \overline{\left(\bigcup_{i=1}^{n} A_i\right)} = \bigcap_{i=1}^{n} \overline{A_i},$$

$$\overline{AB} = \bar{A} \cup \bar{B} \quad \text{or} \quad \overline{\left(\bigcap_{i=1}^{n} A_i\right)} = \bigcup_{i=1}^{n} \overline{A_i}$$

It is important to be able to relate descriptions of sets to their symbolic notation, using the symbols given here, and to list or count correctly the elements in sets of interest. The following example illustrates this point.

E X A M P L E   **2.1**   Twenty electric motors are pulled from an assembly line and inspected for defects. Eleven of the motors are free of defects, eight have defects on their exterior finish, and three have defects in their assembly and will not run. Let A denote the set of motors having assembly defects, and let F denote the set having defects on their finish. Using A and F, write a symbolic notation for each of the following sets; then identify the number of motors in each set.

1   The set of motors having both types of defects.

2   The set of motors having at least one type of defect.

3   The set of motors having no defects.

4   The set of motors having exactly one type of defect.

Solution   **1**   The motors with both types of defects must be in A and in F; therefore, this event can be written AF. Since only nine motors have defects, whereas A contains three motors and F contains eight motors, two motors must be in AF (see Figure 2.5).

F I G U R E **2.5**
Venn diagram for Example 2.1
(numbers of motors shown for
each set).

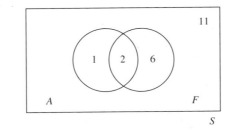

2 The motors having at least one type of defect must have either an assembly defect or a finish defect. Hence, this set can be written $A \cup F$. Because eleven motors have no defect, nine must have at least one defect.

3 The set of motors having no defects is the complement of the set of those having at least one defect; it is written $\overline{A \cup F} = \bar{A}\bar{F}$ (by DeMorgan's Law). Clearly, eleven motors fall into this set.

4 The set of motors having exactly one type of defect must either be in A but not in F or in F but not in A. This set can be written $A\bar{F} \cup \bar{A}F$; and seven motors fall into the set.

■

Exercises

2.1 Of 25 microcomputers available in a supply room, 10 have circuit boards for a printer, 5 have circuit boards for a modem, and 13 have neither board. Using P to denote those that have printer boards and M to denote those that have modem boards, symbolically denote the following sets, and identify the number of microcomputers in each set.

a The set of all microcomputers that have both boards.

b The set of all microcomputers that have neither board.

c The set of all microcomputers that have printer boards only.

d The set of all microcomputers that have exactly one of the boards.

2.2 Five applicants (Jim, Don, Mary, Sue, and Nancy) are available for two identical jobs. A supervisor selects two applicants to fill these jobs.

a List all possible ways in which the jobs can be filled; that is, list all possible selections of two applicants from the five.

b Let A denote the set of selections containing at least one male. How many elements are in A?

    c   Let B denote the set of selections containing *exactly* one male. How many elements are in B?

    d   Write the set containing two females in terms of A and B.

    e   List the elements in $\bar{A}$, AB, $A \cup B$, and $\overline{AB}$.

2.3   Use Venn diagrams to verify the distributive laws.

2.4   Use Venn diagrams to verify DeMorgan's Laws.

2.3 Definition of Probability

Probability, as we have seen, requires three elements: a target population (conceptual or real) from which observable outcomes are obtained; meaningful categorizations of these outcomes; and a random mechanism for generating outcomes. Each outcome in a conceptually infinite population of coin tosses can be categorized as "heads" or "tails." If the tosses are random, a meaningful statement can be made about the probability of "heads" on the next toss.

To change the setting a bit, suppose that a regular six-sided die is tossed onto a table and the number on the upper face is observed. This is a probabilistic situation, since the number that occurs on the upper face cannot be determined in advance. We shall analyze the components of the situation and arrive at a definition of probability that permits us to model mathematically what happens in die tosses, as well as in many similar situations.

First we might toss the die several times, to collect data on possible outcomes. This data-generating phase, called a *random experiment,* allows us to see the nature of the possible outcomes, which we can list in a **sample space.**

DEFINITION   **2.1**   A **sample space** S is a set that includes all possible outcomes for a random experiment, listed in a mutually exclusive and exhaustive manner. ■

The phrase *mutually exclusive* means that the elements of the set do not overlap, and the term *exhaustive* means that the list contains all possible outcomes.

For the die toss, we could identify a sample space of

$$S_1 = \{1, \ 2, \ 3, \ 4, \ 5, \ 6\}$$

where the integers indicate the possible numbers of dots on the upper face, or of

$$S_2 = \{\text{even, odd}\}$$

Both S_1 and S_2 satisfy Definition 2.1, but S_1 seems the better choice because it maintains a higher level of detail. S_2 has three possible upper-face outcomes in each listed element, whereas S_1 has only one possible outcome per element.

As another example, suppose that a nurse is measuring the height of a patient. (This measurement process constitutes the experiment.) A sample space could be listed as

$$S_3 = \{1, \ 2, \ 3, \ \ldots, \ 50, \ 51, \ 52, \ \ldots, \ 70, \ 71, \ 72, \ \ldots\}$$

if the height is rounded to the closest integer number of inches. On the other hand, an appropriate sample space could be

$$S_4 = \{x \mid x > 0\}$$

which is read "the set of all real numbers x such that $x > 0$." Whether S_3 or S_4 should be used in a particular problem depends on the nature of the measurement process. If decimals are to be used, we need S_4. If only integers are to be used, S_3 will suffice. The point is that sample spaces for a particular experiment are not unique and must be selected to provide all pertinent information for a given situation.

Let us go back to our first example, the toss of a die. Suppose that player A can have first turn at a board game if he or she rolls a 6. Therefore, the *event* "roll a 6" is important to that player. Other possible events of interest in the die-tossing experiment are "roll an even number," "roll a number greater than 4," and so on.

DEFINITION   **2.2**   An **event** is any subset of a sample space.   ∎

Definition 2.2 holds as stated for any sample space that has a finite or countable number of elements. Some subsets must be ruled out if a sample space covers a continuum of real numbers, as S_4 (given earlier) does, but any subset likely to occur in practice can be called an *event*.

From this definition of *events*, we see that an event is a collection of elements from the sample space. For the die-tossing experiment, we can see how this works by establishing the following definitions:

   A is "an even number."

   B is "an odd number."

   C is "a number greater than 4."

   E_1 is "observe a 1."

   E_i is "observe an integer i."

Then, if $S = \{1, 2, 3, 4, 5, 6\}$,

$$A = \{2, 4, 6\}$$
$$B = \{1, 3, 5\}$$
$$C = \{5, 6\}$$
$$E_1 = \{1\}$$
$$E_i = \{i\}, \qquad i = 1, 2, 3, 4, 5, 6$$

Sample spaces and events often can be conveniently displayed in Venn diagrams. Some events for the die-tossing experiment are shown in Figure 2.6. For an experiment, we now know how to establish a sample space and how to list appropriate events. The next step is to define a probability for these events. We have already seen that the intuitive idea of probability is related to relative frequency of occurrence. When tossed, a regular die should come up an even number about 1/2 of the time

FIGURE **2.6**
Venn diagram for a die toss.

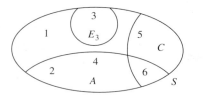

and a 3 about 1/6 of the time. All probabilities should be fractions between 0 and 1, inclusive. One of the integers 1, 2, 3, 4, 5, or 6 must occur every time the die is tossed, so the total probability associated with the sample space must be 1. In repeated tosses of the die, if a 1 occurs 1/6 of the time, then a 1 or 2 must occur $1/6 + 1/6 = 1/3$ of the time. Because relative frequencies for mutually exclusive events can be added, so must the associated probabilities. These considerations lead to the following definition.

DEFINITION   **2.3**

Suppose that a random experiment has associated with it a sample space S. Let A and B denote any two mutually exclusive events in S. A **probability** is a numerically valued function that assigns a number $P(A)$ to every event A so that the following axioms hold:

1 $P(A) \geq 0$.

2 $P(S) = 1$.

3   If $A_1, A_2, \ldots$ is a sequence of mutually exclusive events (that is, a sequence in which $A_i A_j = \emptyset$ for any $i \neq j$), then

$$P\left(\bigcup_{i=1}^{\infty} A_i\right) = \sum_{i=1}^{\infty} P(A_i)$$

From axiom 3, it follows that, if A and B are mutually exclusive events,

$$P(A \cup B) = P(A) + P(B)$$

This result is similar to the addition of relative frequencies in the die-tossing example discussed earlier.

Two elementary properties of probability now follow immediately. First, if $A \subset B$, then $P(A) \leq P(B)$. To see this, write

$$B = A \cup \bar{A}B$$

so that

$$P(B) = P(A \cup \bar{A}B) = P(A) + P(\bar{A}B)$$

Because $P(\bar{A}B) \geq 0$ by axiom 1, it follows that $P(A) \leq P(B)$. In particular, because $A \subset S$ for any event A and because $P(S) = 1$, then $P(A) \leq 1$.

Second, we can show that $P(\emptyset) = 0$. Because S and $\emptyset$ are disjoint with $S \cup \emptyset = S$,

$$1 = P(S) = P(S \cup \emptyset) = P(S) + P(\emptyset)$$

The definition of probability tells us only the axioms such a function must obey; it does not tell us what numbers to assign to specific events. The actual assignment of numbers usually comes about from empirical evidence or from careful thought about the experiment. If a die is balanced, we could toss it a few times to see whether the upper faces all seem equally likely to occur. Or we could simply assume that this result would be obtained and assign a probability of 1/6 to each of the six elements in S; that is, $P(E_i) = 1/6$, $i = 1, 2, \ldots, 6$. Once we have done this, the model is complete; by axiom 3, we can now find the probability of any event. For example, for the events defined on page 14 and in Figure 2.6,

$$
\begin{aligned}
P(A) &= P(E_2 \cup E_4 \cup E_6) \\
&= P(E_2) + P(E_4) + P(E_6) \\
&= \frac{1}{6} + \frac{1}{6} + \frac{1}{6} = \frac{1}{2}
\end{aligned}
$$

and

$$
\begin{aligned}
P(C) &= P(E_5 \cup E_6) \\
&= P(E_5) + P(E_6) \\
&= \frac{1}{6} + \frac{1}{6} = \frac{1}{3}
\end{aligned}
$$

Definition 2.3, in combination with the actual assignment of probabilities to events, provides a probabilistic model for an experiment. If $P(E_i) = 1/6$ is used in the die-tossing experiment, we can assess the suitability of the model by examining how closely the long-run relative frequencies for each outcome match the numbers predicted by the theory underlying the model. If the die is balanced, the model should be quite accurate in telling us what we can expect to happen. If the die is not balanced, the model will fit poorly with the actual data obtained, and other probabilities should be substituted for the $P(E_i)$. Throughout the remainder of this book, we shall develop many specific models based on this underlying definition of probability and discuss practical situations in which they work well. None is perfect, but many are adequate for describing real-world probabilistic phenomena.

EXAMPLE 2.2 A purchasing clerk wants to order supplies from one of three possible vendors, which are numbered 1, 2, and 3. The supplies offered by all three vendors are equal with respect to quality and price, so the clerk writes each number on a piece of paper, mixes the papers, and blindly selects one number. The order is then placed with the vendor whose number is selected. Let E_i denote the event that vendor i is selected $(i = 1, 2, 3)$, let B denote the event that vendor 1 or 3 is selected, and let C denote the event that vendor 1 is *not* selected. Find the probabilities of events E_i, B, and C.

Solution Events E_1, E_2, and E_3 correspond to the elements of S because they represent all the "single possible outcomes." Thus, if we assign appropriate probabilities to these events, the probability of any other event can easily be found.

Because one number is picked at random from the three numbers available, it should seem intuitively reasonable to assign a probability of 1/3 to each E_i:

$$P(E_1) = P(E_2) = P(E_3) = \frac{1}{3}$$

In other words, we find no reason to suspect that one number has a greater likelihood of being selected than any of the others. Now,

$$B = E_1 \cup E_3$$

and by axiom 3 of Definition 2.3,

$$P(B) = P(E_1 \cup E_3) = P(E_1) + P(E_3) = \frac{1}{3} + \frac{1}{3} = \frac{2}{3}$$

Similarly,

$$C = E_2 \cup E_3$$

and therefore,

$$P(C) = P(E_2) + P(E_3) = \frac{2}{3}$$

Notice that different probability models could have been selected for the sample space connected with this experiment, but only this model is reasonable under the assumption that the vendors are all equally likely to be selected. The terms *blindly* and *at random* are interpreted as imposing equal probabilities on the finite number of points in the sample space. ∎

The examples discussed so far have assigned equal probabilities to the elements of a sample space, but this is not always the case. If you have one quarter and one penny in your pocket and you pull out the first coin you touch, the quarter may have a higher probability of being chosen because of its larger size.

Often the probabilities assigned to events are based on experimental evidence or observational studies that yield relative frequency data on the events of interest. The data provide only approximations to the true probabilities, but these approximations often are quite good and usually are the only information we have on the events of interest.

E X A M P L E 2.3 A serious quality-of-life issue in today's world is the threat of HIV infection and AIDS. Good decisions on how to battle this threat must come from sound data, but reliable and representative data on such a personal issue are difficult to obtain. Improvements are being made in data collection, however, as is evidenced by a recent study reported in the November 13, 1992, issue of *Science* (J. Catania et al., "**Prevalence of AIDS-related Risk Factors and Condom Use in the United States**"). In this study, a random sample of 2673 residents of the United States between the ages of 18 and 75 was selected by random digit dialing (essentially, randomly dialing of telephone numbers). A larger sample of 8263 residents was randomly selected from high-risk cities. A summary of the sample data is given in Table 2.1.

After a national advertising campaign, a company hires a new worker.

1 What is the probability that the worker falls into the "risky partner" category?

2 What is the probability that the worker falls into at least one of the risk groups?

3 If the firm hires 1000 workers, how many can be expected to be at risk if the 1000 come from the population at large?

4 If the firm hires 1000 workers, how many can be expected to be at risk if the 1000 come from high-risk cities?

Solution **1** Solutions to practical problems of this type always involve assumptions. To answer parts 1 and 2 with the data given, we must assume that the new worker is randomly selected from the national population, which implies that

$$P(\text{Risky partner}) = 0.032$$

2 Let's label the seven risk groups, in the tabled order, $E_1, E_2, \ldots, E_7$. Then, the event of being in **at least one** of the groups can be written as the union of these seven:

$$E_1 \cup E_2 \cup \cdots \cup E_7$$

The event "at least one" is the same as the event "E_1 or E_2 or E_3 or $\cdots$ or E_7." Since these seven groups are listed in mutually exclusive fashion,

$$P(E_1 \cup E_2 \cup \cdots \cup E_7) = P(E_1) + P(E_3) + \cdots + P(E_7)$$
$$= 0.070 + 0.032 + 0.023 + 0.017 + 0.000 + 0.002 + 0.007$$
$$= 0.151$$

T A B L E **2.1**
Prevalence of HIV-related Risk Groups Among Adult Heterosexuals: National and High-risk Cities Samples

	National		High-risk Cities	
	%	*n*	%	*n*
Multiple partner (over past 12 months)	7.0	170	9.5	651
Risky partner	3.2	76	3.7	258
Transfusion recipient	2.3	55	2.1	144
Multiple partner and risky partner	1.7	41	3.0	209
Multiple partner and transfusion recipient	0.0	1	0.3	20
Risky partner and transfusion recipient	0.2	4	0.3	19
All others	0.7	16	0.7	51
No risk	84.9	2045	80.4	5539

Note: The number of respondents for each risk group is indicated by *n*.

Source: J. Catania et al., "Prevalence of AIDS-related Risk Factors and Condom Use in the United States," *Science*, November 13, 1992.

Note that the mutual exclusive property is essential here; otherwise, we could not simply add the probabilities.

3   We must assume that all 1000 new hires are randomly selected from the national population (or a subpopulation of the same structure). Then, 15.1% of the 1000, or 151 workers, can be expected to be at risk.

4   We assume that the 1000 workers are randomly selected from high-risk cities. Then, 19.6% of these, or 196 workers, can be expected to be at risk.

Do you think that the assumptions necessary to answer the probability questions are reasonable here?   ▪

Exercises

2.5   A vehicle arriving at an intersection can turn left, turn right, or continue straight ahead. An experiment consists of observing the movement of one vehicle at this intersection.

a   List the elements of a sample space.

b   Attach probabilities to these elements if all possible outcomes are equally likely.

c   Find the probability that the vehicle turns, under the probabilistic model of part b.

2.6   A manufacturing company has two retail outlets. It is known that 30% of all potential customers buy products from outlet 1 alone, 50% buy from outlet 2 alone, 10% buy from both 1 and 2, and 10% buy from neither. Let A denote the event that a potential customer, randomly chosen, buys from outlet 1, and let B denote the event that the customer buys from outlet 2. Find the following probabilities.

a $P(A)$ 　　　　　　　　　　　　　 **b** $P(A \cup B)$

c $P(\bar{B})$ 　　　　　　　　　　　　 **d** $P(AB)$

e $P(\bar{A} \cup \bar{B})$ 　　　　　　　　　 **f** $P(\bar{A}\bar{B})$

g $P(\overline{A \cup B})$

2.7   Among donors at a blood center, 1 in 3 gave O^+ blood, 1 in 15 gave O^-, 1 in 3 gave A^+, and 1 in 16 gave A^-. What is the probability that the first person who shows up tomorrow to donate blood has the following blood type?

a   type O^+ 　　　　　　　　　　　 **b**   type O

c   type A 　　　　　　　　　　　　 **d**   either type A^+ or type O^+

2.8   Information on modes of transportation for coal leaving the Appalachian region is shown in the accompanying chart. If coal arriving at a certain power plant comes from this region, find the probability that it was transported out of the region by the following means.

a   by truck to rail

b   by water only

c   at least partly by truck

d   at least partly by rail

e   by modes not involving water (assume that *other* does not involve water)

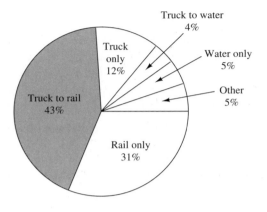

Truck to water
4%

Truck only
12%

Water only
5%

Truck to rail
43%

Other
5%

Rail only
31%

Source: G. Elmes, "Coal Transportation: An Undervalued Aspect of Energy Modeling," *Transportation Research*, 18A, no. 1 1984: 19. Used by permission.

2.9 Hydraulic assemblies for landing gear produced by an aircraft rework facility are inspected for defects. History shows that 8% have defects in the shafts alone, 6% have defects in the bushings alone, and 2% have defects in both the shafts and the bushings. If one such assembly is randomly chosen, find the probability that it has the following characteristics.

a a bushing defect

b a shaft or bushing defect

c only one of the two types of defects

d no defects in shafts or bushings

2.10 Of the 2.7 million engineers in the United States (as of 1988) 95.5% are male. In addition, 90.6% are white, 1.6% are black, 5.6% are Asian, and 0.4% are Native American. Of the 0.7 million computer specialists, 69.1% are male, 88.3% are white, 3.7% are black, 6.6% are Asian, and 0.1% are Native American.

a Construct a meaningful table to compare numbers of engineers and computer scientists by gender.

b Construct a meaningful table to compare numbers of engineers and computer specialists by racial group.

Suppose that 1000 engineers are interviewed in a national pool on engineering education.

c How many would you expect to be female?

d How many would you expect to be Asian?
In a random poll of computer specialists, it is desired to get responses from approximately 50 blacks.

e How many computer specialists should be polled?

2.4 Counting Rules Useful in Probability

Let us now look at the die-tossing experiment from a slightly different perspective. Because the six outcomes should be equally likely for a balanced die, the probability of A, an even number, is

$$P(A) = \frac{3}{6} = \frac{\text{Number of outcomes favorable to } A}{\text{Total number of equally likely outcomes}}$$

This "definition" of probability will work for any experiment that results in a finite sample space with *equally likely* outcomes. Thus, it is important to be able to count the number of possible outcomes for an experiment. Unfortunately, the number of outcomes for an experiment can easily become quite large, and counting them is difficult unless one knows a few counting rules. Four such rules are presented as theorems in this section.

Suppose that a quality control inspector examines two manufactured items selected from a production line. Item 1 can be defective or nondefective, as can item 2. How many outcomes are possible for this experiment? In this case, it is easy to list them. Using D_i to denote that the ith item is defective and N_i to denote that the ith item is not defective, the possible outcomes are

$$D_1 D_2, \quad D_1 N_2, \quad N_1 D_2, \quad N_1 N_2$$

These four outcomes can be placed on a two-way table, as in Figure 2.7, to help clarify that the four outcomes arise from the fact that the first item has two possible outcomes and the second item has two possible outcomes—and hence, the experiment of looking at both items has $2 \times 2 = 4$ outcomes. This is an example of the multiplication principle, given as Theorem 2.1.

FIGURE **2.7**

Possible outcomes for inspecting two items (D_i denotes that the ith item is defective; N_i denotes that the ith item is not defective).

	Second Item	
	D_2	N_2
D_1	D_1D_2	D_1N_2
N_1	N_1D_2	N_1N_2

First Item

THEOREM **2.1**

If the first task of an experiment can result in n_1 possible outcomes and, for each such outcome, the second task can result in n_2 possible outcomes, then there are $n_1 n_2$ possible outcomes for the two tasks together. ▪

The multiplication principle extends to more tasks in a sequence. If, for example, three items were inspected and each of these could be defective or not defective, there would be $2 \times 2 \times 2 = 8$ possible outcomes.

Tree diagrams are also helpful in verifying the multiplication principle and for listing outcomes of experiments. Suppose that a firm is deciding where to build two new plants—one in the east and one in the west. Four eastern cities and two western cities are possible locations. Thus, there are $n_1 n_2 = 4(2) = 8$ possibilities for locating the two plants. Figure 2.8 lists these possibilities on a tree diagram.

The multiplication principle (Theorem 2.1) helps only in identifying the number of elements in a sample space for an experiment. We must still assign probabilities to these elements to complete our probabilistic model. This is done in Example 2.4 for the site selection problem.

E X A M P L E   **2.4** In connection with the firm that plans to build two new plants, the eight possible locations are as shown in Figure 2.8. If all eight choices are equally likely (that is, if one of the pairs of cities is selected at random), find the probability that city E is selected.

Solution City E can get selected in four different ways, because four possible eastern cities may be paired with it. Thus,

$$(E \text{ gets selected}) = (AE) \cup (BE) \cup (CE) \cup (DE)$$

Each of the eight outcomes has a probability of 1/8, because the eight events are assumed to be equally likely. Because these eight events are mutually exclusive,

$$P(E \text{ gets selected}) = P(AE) + P(BE) + P(CE) + P(DE)$$
$$= \frac{1}{8} + \frac{1}{8} + \frac{1}{8} + \frac{1}{8} = \frac{1}{2}$$

■

F I G U R E **2.8**
Possible locations for two plants (A, B, C, D denote eastern cities; E, F denote western cities).

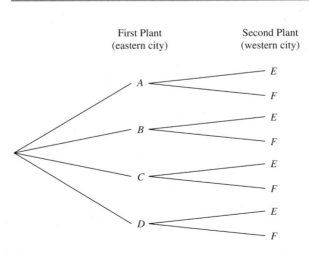

First Plant
(eastern city)

Second Plant
(western city)

E X A M P L E   **2.5**   Five motors (numbered 1 through 5) are available for use, and motor number 2 is defective. Motors 1 and 2 come from supplier I, and motors 3, 4, and 5 come from supplier II. Suppose that two motors are randomly selected for use on a particular day. Let A denote the event that the defective motor is selected, and let B denote the event that at least one motor comes from supplier I. Find $P(A)$ and $P(B)$.

Solution   We can see from the tree diagram in Figure 2.9 that this experiment has 20 possible outcomes, which agrees with our calculation using the multiplication rule. In other words, there are 20 events of the form $\{1,\ 2\}$, $\{1,\ 3\}$, and so forth. Since the motors are randomly selected, each of the 20 outcomes has a probability of 1/20. Thus,

$$P(A) = P(\{1,\ 2\} \cup \{2,\ 1\} \cup \{2,\ 3\} \cup \{2,\ 4\} \cup \{2,\ 5\} \cup \{3,\ 2\} \cup \{4,\ 2\} \cup \{5,\ 2\})$$
$$= \frac{8}{20} = 0.4$$

because the probability of the union is equal to the sum of the probabilities of the events in the union.

F I G U R E   **2.9**   Outcomes for experiment of Example 2.5.

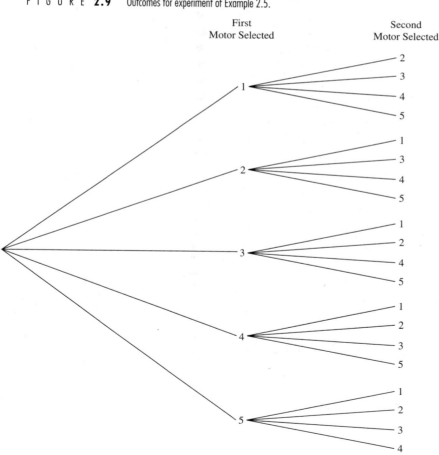

The reader can see that B contains 14 of the 20 outcomes and, hence, that

$$P(B) = \frac{14}{20} = 0.7$$

∎

The multiplication rule is often used to develop other counting rules. Suppose that, from among three pilots, a crew of two is to be selected to form a pilot-copilot team. To count the number of ways this can be done, observe that the pilot's seat can be filled in three ways and the copilot's in two ways (once the pilot has been selected), so there are $3 \times 2 = 6$ ways of forming the team. This is an example of a *permutation*, for which a general result is given in Theorem 2.2.

THEOREM 2.2

The number of ordered arrangements or permutations P_r of r objects selected from n distinct objects $(r \leq n)$ is given by

$$P_r^n = n(n-1) \cdots (n-r+1) = \frac{n!}{(n-r)!}$$

Proof   The basic idea of a permutation can be thought of as a process of filling r slots in a line, with one object in each slot, by drawing these objects one at a time from a pool of n distinct objects. The first slot can be filled in n ways, but the second can only be filled in $(n-1)$ ways after the first is filled. Thus, by the multiplication rule, the first two slots can be filled in $n(n-1)$ ways. Extending this reasoning to r slots, the number of ways of filling all r slots is

$$n(n-1) \cdots (n-r+1) = \frac{n!}{(n-r)!} = P_r^n$$

Hence, the theorem is proved.   ∎

Examples 2.6 and 2.7 illustrate the use of Theorem 2.2.

EXAMPLE 2.6   From among ten employees, three are to be selected to travel to three out-of-town plants, A, B, and C, one to each plant. Since the plants are located in different cities, the order in which the employees are assigned to the plants is an important consideration. In how many ways can the assignments be made?

Solution   Because order is important, the number of possible distinct assignments is

$$P_3^{10} = \frac{10!}{7!} = 10(9)(8) = 720$$

In other words, there are ten choices for plant A, but then only nine for plant B, and eight for plant C. This gives a total of $10(9)(8)$ ways of assigning employees to the plants. ∎

EXAMPLE **2.7** An assembly operation in a manufacturing plant involves four steps, which can be performed in any order. The manufacturer wishes to compare experimentally the assembly times for each possible ordering of the steps. How many orderings will this experiment involve?

Solution The number of orderings is the permutation of $n = 4$ things taken $r = 4$ at a time. (All steps must be accomplished each time.) This turns out to be

$$P_4^4 = \frac{4!}{0!} = 4! = 4(3)(2)(1) = 24$$

because, by definition, $0! = 1$. (In fact, $P_r^r = r!$ for any integer r.) ∎

Sometimes order is not important, and we are interested only in identifying the number of subsets of a certain size that can be selected from a given set.

THEOREM **2.3** The number of distinct subsets or combinations of size r that can be selected from n distinct objects $(r \leq n)$ is given by

$$\binom{n}{r} = \frac{n!}{r!(n-r)!}$$

Proof The number of ordered subsets of size r, selected from n distinct objects, is given by P_r^n. The number of unordered subsets of size r is denoted by $\binom{n}{r}$. Since any particular set of r objects can be ordered among themselves in $P_r^r = r!$ ways, it follows that

$$\binom{n}{r} r! = P_r^n$$

$$\binom{n}{r} = \frac{1}{r!} P_r^n$$

$$= \frac{n!}{r!(n-r)!}$$

∎

E X A M P L E   **2.8**   In the manufacturing setting described in Example 2.6, suppose that three employees are to be selected from among the ten available to go to the same plant. In how many ways can this selection be made?

Solution   Here, order is not important; we merely want to know how many subsets of size $r = 3$ can be selected from $n = 10$ people. The result is

$$\binom{10}{3} = \frac{10!}{3!7!} = \frac{10(9)(8)}{1(2)(3)} = 120$$

∎

E X A M P L E   **2.9**   Refer to Example 2.8. If two of the ten employees are women and the other eight are men, what is the probability that exactly one woman will get selected among the three?

Solution   We have seen that there are $\binom{10}{3} = 120$ ways to select three employees from the ten.

Similarly, there are $\binom{2}{1} = 2$ ways to select one woman from the two available, and

$\binom{8}{2} = 28$ ways to select two men from the eight available. If selections are made at random (that is, if subsets of three employees are equally likely to be chosen), then the probability of selecting exactly one woman is

$$\frac{\binom{2}{1}\binom{8}{2}}{\binom{10}{3}} = \frac{2(28)}{120} = \frac{7}{15}$$

∎

E X A M P L E   **2.10**   Five applicants for a job are ranked according to ability, with applicant number 1 being best, applicant number 2 second best, and so on. These rankings are unknown to an employer, who simply hires two applicants at random. What is the probability that this employer hires exactly one of the two best applicants?

Solution   The number of possible outcomes for the process of selecting two applicants from five is

$$\binom{5}{2} = \frac{5!}{2!3!} = 10$$

If exactly one of the two best is selected, the selection can be done in

$$\binom{2}{1} = \frac{2!}{1!1!} = 2$$

ways. The other selected applicant must come from among the three lowest-ranking applicants, which can be done in

$$\binom{3}{1} = \frac{3!}{1!2!} = 3$$

ways. Thus, the event of interest (hiring one of the two best applicants) can come about in $2 \cdot 3 = 6$ ways. The probability of this event is therefore $6/10 = 0.6$.   ∎

THEOREM **2.4**

The number of ways of partitioning n distinct objects into k groups containing n_1, n_2, ..., n_k objects, respectively, is

$$\frac{n!}{n_1! \, n_2! \cdots n_k!}$$

where

$$\sum_{i=1}^{k} n_i = n$$

Proof

The partitioning of n objects into k groups can be done by first selecting a subset of size n_1 from the n objects, then selecting a subset of size n_2 from the $n - n_1$ objects that remain, and so on until all groups are filled. The number of ways of doing this is

$$\binom{n}{n_1} = \binom{n - n_1}{n_2} \cdots \binom{n - n_1 - \cdots - n_{k-1}}{n_k}$$

$$= \left(\frac{n!}{n_1!(n - n_1)!} \right) \left(\frac{(n - n_1)!}{n_2!(n - n_1 - n_2)!} \right) \cdots \left(\frac{(n - n_1 - \cdots - n_{k-1})!}{n_k! \, 0!} \right)$$

$$= \frac{n!}{n_1! \, n_2! \cdots n_k!}$$

∎

EXAMPLE **2.11**

Suppose that ten employees are to be divided among three job assignments, with three employees going to job I, four to job II, and three to job III. In how many ways can the job assignments be made?

Solution

This problem involves partitioning the $n = 10$ employees into groups of size $n_1 = 3$, $n_2 = 4$, and $n_3 = 3$, which can be accomplished in

$$\frac{n!}{n_1! \, n_2! \, n_3!} = \frac{10!}{3! \, 4! \, 3!} = \frac{10(9)(8)(7)(6)(5)}{3(2)(1)(3)(2)(1)} = 4200$$

ways. (Notice the large number of ways this task can be accomplished!) ▪

EXAMPLE **2.12** In the setting of Example 2.11, suppose that three employees of a certain ethnic group all get assigned to job I. Assuming that they are the only employees among the ten under consideration who belong to this ethnic group, what is the probability of this happening under a random assignment of employees to jobs?

Solution We have seen in Example 2.11 that there are 4200 ways of assigning the ten workers to the three jobs. The event of interest assigns three specified employees to job I. It remains for us to determine how many ways the other seven employees can be assigned to jobs II and III, which is

$$\frac{7!}{4!\,3!} = \frac{7(6)(5)}{3(2)(1)} = 35$$

Thus, the chance of randomly assigning three specified workers to job I is

$$\frac{35}{4200} = \frac{1}{120}$$

which is very small, indeed! ▪

One interesting application of the counting rules we have just developed is in assessing the probabilities of runs of like items in a sequence of random events. Let's examine this idea in terms of monitoring the quality of an industrial process in which manufactured items are produced at a regular rate throughout the day. Periodically, items are inspected; a D is recorded for a defective item, and a G is recorded for a good (nondefective) item. A typical daily record might look like this:

G D G G D G G G G D D G

for twelve inspected items.

If the D's bunch together noticeably, one might conclude that the rate of producing defective items is not constant across the day. For example, a record such as

G G G G G G G G G D D D D

might suggest that workers get careless toward the end of the day or that the machinery used in the process got badly out of tune.

A key difference between these patterns involves the number of runs of like items. The first shows three runs of D's and four runs of G's, for a total of seven runs. The second shows one G-run and one D-run, for a total of two runs. If the rate of producing defects remains constant during the day, all possible arrangements of m D's and n G's, in a total of $(m + n)$ trials, are equally likely. There are

$$\binom{m + n}{m} = \binom{m + n}{n}$$

such arrangements.

Suppose that the number of D-runs and the number of G-runs both equal k. The number of ways dividing m D's into k groups can be seen by considering the placement of $(k-1)$ bars into the spaces between runs of D's:

$$\text{D} \mid \text{DD} \mid \text{DD} \cdots \text{DDD} \mid \text{DD}$$

This can be done in

$$\binom{m-1}{k-1}$$

ways. Similarly, n G's can be divided into k groups in

$$\binom{n-1}{k-1}$$

ways. Since both of these divisions must occur together, there are a total of

$$\binom{m-1}{k-1}\binom{n-1}{k-1}$$

ways of producing k D-runs and k G-runs. When these two sets are merged, we could begin with either a D or a G; so the probability of getting $2k$ runs becomes

$$P(2k \text{ runs}) = \frac{2\binom{m-1}{k-1}\binom{n-1}{k-1}}{\binom{m+n}{m}}$$

The number of D-runs can be one more (or one less) than the number of G-runs. To get $(k+1)$ D-runs and k G-runs, we must begin the sequence with a D and end it with a D. To get k D-runs and $(k+1)$ G-runs, we must begin and end the sequence with a G. Thus,

$$P(2k+1 \text{ runs}) = \frac{\binom{m-1}{k}\binom{n-1}{k-1} + \binom{m-1}{k-1}\binom{n-1}{k}}{\binom{m+n}{m}}$$

E X A M P L E **2.13** A football team produced a record of 8 wins and 4 losses over its season.

1   What is the probability that these wins and losses occurred in only two runs?

2   A sportswriter, observing that the team's four losses came in the last four games of the season, remarked that the team had "lost its ability to win late in the season." Was the sportswriter correct?

Solution   1   The probability of having only two runs is given by

$$P(2 \text{ runs}) = \frac{2\binom{3}{0}\binom{7}{0}}{\binom{12}{4}} = \frac{2}{495}$$

assuming that all arrangements of 8 wins and 4 losses are equally likely. (What does this assumption say about the team's ability to win?)

The probability of having one run of 8 wins followed by one run of 4 losses, assuming equally likely arrangements, is

$$\frac{\binom{3}{0}\binom{7}{0}}{\binom{12}{4}} = \frac{1}{495}$$

(Why do we not need to multiply by 2 in this case?)

2 If the team had a constant ability to win across the season, the chance of all four of its losses coming at the end of the season would be small, indeed! Thus, the sportswriter could be correct; but what other explanations are possible?

∎

Exercises

2.11 An experiment consists of observing two vehicles in succession as they move through the intersection of two streets.

a List the possible outcomes, assuming that each vehicle can go straight, turn right, or turn left.

b Assuming that the outcomes are equally likely, find the probability that at least one vehicle turns left. (Would this assumption always be reasonable?)

c Assuming that the outcomes are equally likely, find the probability that at most one vehicle makes a turn.

2.12 A commercial building is designed with two entrances: entrance I and entrance II. Two customers arrive (separately) and enter the building.

a List the elements of a sample space for this observational experiment.

b Assuming that all elements in part (a) are equally likely, find the probability that both customers use door I; that both customers use the same door.

2.13 A corporation has two construction contracts that are to be assigned to one or more of three firms bidding for them. (It is possible for one firm to receive both contracts.)

a List the possible outcomes for the assignment of contracts to the firms.

b If all outcomes are equally likely, find the probability that both contracts will go to the same firm.

c Under the assumptions of part (b), find the probability that one specific firm—say, firm I—will get at least one contract.

2.14 Among five portable generators produced by an assembly line in one day, two are defective. If two generators are selected for sale, find the probability that both will be nondefective. (Assume that the two selected for sale are chosen in such a way that every possible sample of size two has the same probability of being selected.)

2.15 Seven applicants have applied for two jobs. How many ways can the jobs be filled if the following additional information is known?

a The first person chosen receives a higher salary than the second.

b There are no differences between the jobs.

2.16   A package of six light bulbs contains two defective bulbs. If three bulbs are selected for use, find the probability that none of the three is defective.

2.17   How many four-digit serial numbers can be formed if no digit is to be repeated within any one number? (The first digit may be a zero.)

2.18   A fleet of eight taxis is to be divided among three airports, A, B, and C, with two going to A, five to B, and one to C.

    a   In how many ways can this be done?

    b   What is the probability that the cab driven by Jones ends up at airport C?

2.19   Show that $\binom{n}{r} = \binom{n-1}{r-1} + \binom{n-1}{r}$, where $1 \le r \le n$.

2.20   Five employees of a firm are ranked from 1 to 5 in their abilities to program a computer. Three of these employees are selected to fill equivalent programming jobs. If all possible choices of three (out of the five) are equally likely, find the probabilities of the following events.

    a   The employee ranked number 1 is selected.

    b   The highest-ranked employee among those selected has rank 2 or lower.

    c   The employees ranked 4 and 5 are selected.

2.21   For a certain style of new automobile, the colors blue, white, black, and green are in equal demand. Three successive orders are placed for automobiles of this style. Find the probabilities of the following events.

    a   One blue, one white, and one green are ordered.

    b   Two blues are ordered.

    c   At least one black is ordered.

    d   Exactly two of the orders are for the same color.

2.22   A firm places three orders for supplies among five different distributors. Each order is randomly assigned to one of the distributors, and a distributor may receive multiple orders. Find the probabilities of the following events.

    a   All orders go to different distributors.

    b   All orders go to the same distributor.

    c   Exactly two of the three orders go to one particular distributor.

2.23   An assembly operation for a computer circuit board consists of four operations, which can be performed in any order.

    a   In how many ways can the assembly operation be performed?

    b   One of the operations involves soldering wire to a microchip. If all possible assembly orderings are equally likely, what is the probability that the soldering operation comes first or second?

2.24   Nine impact wrenches are to be divided evenly among three assembly lines.

    a   In how many ways can this be done?

    b   Two of the wrenches are used and seven are new. What is the probability that a particular line (line A) gets both used wrenches?

2.25   A quality improvement plan calls for daily inspection of ten items from a production process, with the items periodically sampled throughout the day. To see whether a "clumping" of the defects seems to be occurring, inspectors count the total number of runs of defectives and nondefectives. Would you suspect a nonrandom arrangement if, among three defectives and seven nondefectives, the following number of runs occurred?

    a   4

    b   3

    c   2

2.26 Among ten people traveling in a group, two have outdated passports. It is known that inspectors will check the passports of 20% of the people in any group passing their desks. The group can go as a whole (all ten) to one desk or can split into two groups of five and use two different desks. How should members of the group arrange themselves to maximize the probability of getting by the inspectors without having the outdated passports be detected?

2.5 Conditional Probability and Independence

2.5.1 Conditional Probability

Building on the connection between analysis of frequency data in tables and probability, let's consider the employment data in Table 2.2.

A common summary of these data is the "unemployment rate," which is the percentage of unemployed workers, given by

$$\frac{4,115,000}{101,735,000}(100) = 4.0$$

(This figure does not take into account persons no longer actively seeking work.) But, the overall unemployment rate does not tell us anything about the association between employment and education. To get at this question, we must calculate unemployment rates separately for each education category (each row of the table). Narrowing the focus to a single row is often referred to as **conditioning** on the row factor.

T A B L E **2.2**
Civilian Labor Force in the United
States, 1989 (Figures in
Thousands)

Education	Employed	Unemployed	Total
Elementary school	5299	406	5765
High school, 1–3 years	8144	705	8149
High school, 4 years	38171	1763	39,934
College, 1–3 years	19991	671	20,662
College, 4+ years	26015	570	26,585
Total	97,620	4115	101,735

Note: Figures are for civilians over 25 years of age.
Source: U.S. Bureau of Labor Statistics.

T A B L E **2.3**
1989 Unemployment Rates by
Education

Education	Employed	Unemployed
Elementary school	93%	7%
High school, 1–3 years	92%	8%
High school, 4 years	96%	4%
College, 1–3 years	97%	3%
College, 4 or more years	98%	2%

Source: Statistical Abstract of the United States.

The conditional relative frequencies for the data of Table 2.2 are given in Table 2.3.

Now it is apparent that unemployment is associated to some extent with educational level: categories of less education have higher unemployment rates. The conditional relative frequencies relate directly to **conditional probability.** If a national poll samples 1000 people from the national labor force, the expected percentage of unemployed workers it would find (in 1989) is about 4% of 1000—that is, 40 individuals. If, however, the 1000 people all have four or more years of college education, the expected percentage of unemployed workers drops to 2%, or 20 people.

EXAMPLE   **2.14**   Anticipated percentages of new workers in the labor force, for the years 1985 to 2000, are shown in Table 2.4. How do the relative frequencies for the three racial/immigrant categories compare between women and men?

Solution   Even though the data are expressed as percentages, rather than as frequencies, the relative frequencies can still be computed. Conditioning on women, the total number represents 65% of the population, while the number of whites represents 42% of the population. Therefore, (42/65) represents the proportion of whites among the women. Proceeding similarly across the other categories produces the **two** conditional distributions (one for women and one for men) shown in Table 2.5.

Notice that there is a fairly strong association between the two factors of gender and racial/immigrant status. Among the new members of the labor force, most of the women will be white, and men will include a high proportion of immigrants.   ∎

Conditioning can be represented in Venn diagrams as well. Of 100 students who completed an introductory statistics course, 20 were business majors. Further, 10 students received A's in the course, and 3 of these were business majors. These facts

T A B L E **2.4**
Net New Workers, 1985–2000

	Women	Men	Total
White	42%	15%	57%
Nonwhite	14%	7%	21%
Immigrant	9%	13%	22%
Total	65%	35%	100%

Note: Immigrants include both white and nonwhite.

T A B L E **2.5**
Proportions of Net New Workers, by Gender

	Women	Men
White	65%	43%
Nonwhite	21%	20%
Immigrant	14%	37%
Total	100%	100%

are easily displayed on a Venn diagram, such as Figure 2.10, where A represents students who received A's and B represents business majors.

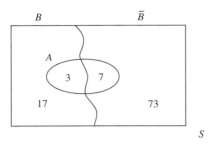

For a randomly selected student from this class, $P(A) = 0.1$ and $P(B) = 0.2$. But suppose that we already know that the randomly selected student is a business major. Then we may want to know the probability that the student received an A, *given* that she is a business major. Among the 20 business majors, 3 received A's. Thus, P (*A given B*)—written as $P(A|B)$—is 3/20.

From the Venn diagram, we see that the conditional (or given) information reduces the effective sample space to just the 20 business majors. Among them, 3 received A's.

Notice that

$$P(A|B) = \frac{3}{20} = \frac{P(AB)}{P(B)} = \frac{3/100}{20/100}.$$

This relationship motivates Definition 2.4.

D E F I N I T I O N   **2.4**

If A and B are any two events, then the **conditional probability** of A given B, denoted by $P(A|B)$, is

$$P(A|B) = \frac{P(AB)}{P(B)}$$

provided $P(B) > 0$. ∎

E X A M P L E   **2.15**   From among five motors, of which one is defective, two are to be selected at random for use on a particular day. Find the probability that the second motor selected is nondefective, given that the first was nondefective.

Solution   Let N_i denote that the ith motor selected is nondefective. We want to find $P(N_2|N_1)$. From Definition 2.5, we have

$$P(N_2|N_1) = \frac{P(N_1 N_2)}{P(N_1)}$$

Looking at the 20 possible outcomes given in Figure 2.8 (connection with Example 2.5), we can see that event N_1 contains 16 of these outcomes, and $N_1 N_2$ contains 12. Thus, since the 20 outcomes are equally likely,

$$P(N_2|N_1) = \frac{P(N_1 N_2)}{P(N_1)} = \frac{12/20}{16/20} = \frac{12}{16} = \frac{3}{4}$$

Does this answer seem intuitively reasonable?   ∎

Conditional probabilities satisfy the three axioms of probability (Definition 2.4), as can easily be shown. First, since $AB \subset B$, then $P(AB) \leq P(B)$. Also, $P(AB) \geq 0$ and $P(B) \geq 0$, so

$$0 \leq P(A|B) = \frac{P(AB)}{P(B)} \leq 1$$

Second,

$$P(S|B) = \frac{P(SB)}{P(B)} = \frac{P(B)}{P(B)} = 1$$

Third, if A_1, A_2, ... are mutually exclusive events, then so are $A_1 B$, $A_2 B$, ...; and

$$P\left(\bigcup_{i=1}^{\infty} A_i \Big| B\right) = \frac{P\left(\left(\bigcup_{i=1}^{\infty} A_i\right) B\right)}{P(B)}$$

$$= \frac{P\left(\bigcup_{i=1}^{\infty} (A_i B)\right)}{P(B)} = \frac{\sum_{i=1}^{\infty} P(A_i B)}{P(B)}$$

$$= \sum_{i=1}^{\infty} \frac{P(A_i B)}{P(B)} = \sum_{i=1}^{\infty} P(A_i|B).$$

Conditional probability plays a key role in many practical applications of probability. In these applications, important conditional probabilities are often drastically affected by seemingly small changes in the basic information from which the probabilities are derived. The following discussion of a medical application of probability illustrates the point.

A screening test indicates the presence or absence of a a particular disease; such tests are often used by physicians to detect diseases. Virtually all screening tests, however, have levels of error associated with their use. Two different kinds of errors are possible: the test could indicate that a person has the disease when he or she actually does not (false positive); or it could fail to show that a person has the disease when he or she actually does have it (false negative). Measures of these two types of errors are conditional probabilities called *sensitivity* and *specificity*.

The following diagram will help in defining and interpreting these measures, where the + indicates the presence of the disease under study and the − indicates absence of the disease. The true diagnosis may never be known, but often it can be determined by more intensive followup tests.

		True Diagnosis		
		+	−	Sum
Test	+	a	b	$a + b$
Result	−	c	d	$c + d$
Sum		$a + c$	$b + d$	$a + b + c + d = n$

In this scenario, n people are tested and the test results indicate that $a + b$ of them have the disease. Of these, a really have the disease and b do not (false positives). Of the $c + d$ who test negative, c actually do have the disease (false negatives). Using these labels,

$$\text{Sensitivity} = \frac{a}{a + c}$$

which represents the conditional probability of having a positive test, given that the person has the disease; and

$$\text{Specificity} = \frac{d}{b + d}$$

which represents the conditional probability of having a negative test, given that the person does not have the disease.

Obviously, a good test should have values for both sensitivity and specificity that are close to 1. If sensitivity is close to 1, then c (the number of false negatives) must be small. If specificity is close to 1, then b (the number of false positives) must be small. Even when sensitivity and specificity are both close to 1, a screening test can produce misleading results if it is not carefully applied. To illustrate this, let's look at one other important measure, the *predictive value* of a test, which is given by

$$\text{Predictive value} = \frac{a}{a + b}$$

The predictive value is the conditional probability of the person's actually having the disease, given that he or she tested positive. Clearly, a good test should have a high predictive value, but this is not always possible—even for highly sensitive and specific tests. The reason that all three measures cannot always be close to 1 simultaneously is because that predictive value is affected by the *prevalence rate* of the disease (that is, the proportion of the population under study that actually has the disease). We can show this with examples of three numerical situations (given next as diagrams I, II, and III).

			True Diagnosis		
			+	−	Sum
I.	Test	+	90	10	100
	Result	−	10	90	100
		Sum	100	100	200

			+	−	Sum
II.	Test	+	90	100	190
	Result	−	10	900	910
		Sum	100	1000	1100

			+	−	Sum
III.	Test	+	90	1000	1090
	Result	−	10	9000	9010
		Sum	100	10,000	10,100

Among the 200 people under study in diagram I, 100 have the disease (a prevalence rate of 50%). The sensitivity and the specificity of the test are each equal to 0.90, and the predictive value is $90/100 = 0.90$. This is a good situation; the test is a good one.

In diagram II, the prevalence rate changes to 100/1100, or 9%. Even though the sensitivity and specificity values are both still 0.90, the predictive value has dropped to $90/190 = 0.47$. In diagram III, the prevalence rate is 100/10,000, or about 1%, and the predictive value has dropped farther to 0.08. Thus, only 8% of those tested positive actually have the disease, even though the test has high sensitivity and high specificity. What does this imply about the use of screening tests on a large population in which the prevalence rate for the disease being studied is low? Assessing the answer to this question involves taking a careful look at conditional probabilities.

EXAMPLE **2.16**   The ELISA test for the presence of HIV antibodies was developed in the mid-1980s to screen blood samples. For test cases involving samples that were known to have been contaminated, ELISA was correct 98% of the time. In testing samples that were known to have been clean, ELISA declared 7% of them to be HIV-positive.

Suppose that a firm has 10,000 employees, all of whom are to be screened for HIV by ELISA. What can we say about the predictive value of such a screening procedure?

Solution   The result will depend on what we assume about the rate of actual HIV-positives among the 10,000 employees. Using the 15% at-risk figure from Table 2.1 and assuming that **all** of these were actually positive (an extremely rare case), we obtain the following diagram of expected outcomes.

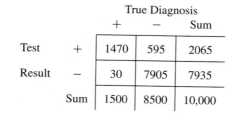

		+	−	Sum
Test	+	1470	595	2065
Result	−	30	7905	7935
	Sum	1500	8500	10,000

For this table, the predictive value is

$$\frac{1470}{2065} = 0.71$$

In other words, of those testing positive, 71% can be expected actually to be HIV-positive.

A more conservative view might be to select around 5% of the employees as potentially HIV-positive. The expected outcomes then become as shown in the following diagram.

		True Diagnosis		
		+	−	Sum
Test	+	490	665	1155
Result	−	10	8835	8845
	Sum	500	9500	10,000

For this arrangement, 0.42. What is the practical danger of running a procedure with such low predictive value?   ∎

2.5.2   Independence

Probabilities are usually very sensitive to the conditioning information. Sometimes, however, a probability does *not* change when conditioning information is supplied. If the extra information provided by knowing that an event B has occurred does not change the probability of A—that is, if $P(A|B) = P(A)$—then events A and B are said to be independent. Since

$$P(A|B) = \frac{P(AB)}{P(B)}$$

the condition $P(A|B) = P(A)$ is equivalent to

$$\frac{P(AB)}{P(B)} = P(A)$$
$$P(AB) = P(A)P(B)$$

DEFINITION   **2.5**

Two events A and B are said to be **independent** if

$$P(A|B) = P(A)$$

or

$$P(B|A) = P(B)$$

This is equivalent to stating that

$$P(AB) = P(A)P(B)$$

∎

EXAMPLE   **2.17**   Suppose that a foreman must select one worker from a pool of four available workers (numbered 1, 2, 3, and 4) for a special job. He selects the worker by mixing the four names and randomly selecting one. Let A denote the event that worker 1 or 2 is selected, let B denote the event that worker 1 or 3 is selected, and let C denote the event that worker 1 is selected. Are A and B independent? Are A and C independent?

Solution   Because the name is selected at random, a reasonable assumption for the probabilistic model is to assign a probability of 1/4 to each individual worker. Then $P(A) = 1/2$, $P(B) = 1/2$, and $P(C) = 1/4$. Because the intersection AB contains only worker 1, $P(AB) = 1/4$. Now $P(AB) = 1/4 = P(A)P(B)$, so A and B are independent. Since AC also contains only worker 1, $P(AC) = 1/4$. But $P(AC) = 1/4 \neq P(A)P(C)$, so A and C *are not* independent. A and C are said to be *dependent* because the fact that C occurs changes the probability that A occurs.   ∎

Most situations in which independence issues arise are not like the one portrayed in Example 2.17, where events were well defined and we merely calculated probabilities to check the definition. Often independence is *assumed* for two events, in order to calculate their joint probability. For example, let A denote that machine A does not break down today, and let B denote that machine B does not break down today. $P(A)$ and $P(B)$ can be approximated from the repair records of the machines. How do we find $P(AB)$, the probability that neither machine breaks down today? If we assume independence, $P(AB) = P(A)P(B)$—a straightforward calculation. If we do not assume independence, however, we cannot calculate $P(AB)$ unless we form a model for their dependence structure or collect data on their joint performance. Is independence a reasonable assumption? It may be, if the operation of one machine is not affected by the other; but it may not be, if the machines share the same room, the same power supply, or the same job foreman. Thus, independence is often used as a simplifying assumption and may not hold precisely in all cases where it is assumed. Remember, probabilistic models are simply models; they do not always precisely mirror reality. But all branches of science make simplifying assumptions when developing their models, whether these are probabilistic or deterministic.

2.6   Rules of Probability

Now we shall investigate how these definitions can be used to establish rules for computing probabilities of composite events. In each case, a similar rule for relative frequency data will be illustrated by reference to the **net** new worker data discussed in Example 2.14. The unconditional and conditional percentages from this example are repeated in Table 2.6 for convenience.

Categories like "women" and "men" are complementary in the sense that a worker must be in one or the other. The percentage of women among new workers is 65%, so the percentage of men must be 35%, since the two must add up to 100%.

When adding percentages (or relative frequencies), one must be careful to preserve the mutually exclusive character of the events. The percentage of whites—men or women—is clearly 42% + 15% = 57%. The percentage of new workers who are either white **or** female is **not** 65% + 57%, however. To find the fraction of workers who are white or female, we must subtract the 42% who are both white **and** female from the preceding total. Thus, the percentage of new workers who are either white or female can be written as

$$65\% + 57\% - 42\% = 80\%$$

which is the same as accumulating the percentages among the appropriate mutually exclusive categories:

$$9\% + 14\% + 42\% + 15\% = 80\%$$

What about the 15% that are both male **and** white? There is another way to view this situation, by making use of the conditional percentages. If we know that 35% of the

Unconditional

	Women	Men	Sum
White	42%	15%	57%
Nonwhite	14%	7%	21%
Immigrant	9%	13%	22%
Sum	65%	35%	100%

Conditional

	Women	Men
White	65%	43%
Nonwhite	21%	20%
Immigrant	14%	37%
Sum	100%	100%

workers are men and that 43% of the men are white, then 43% of the 35%, or 15% overall, must represent the percentage of white men among the new workers.

2.6.1   Complementary Events

In probabistic terms, the complement $\bar{A}$ of an event A is the set of all outcomes in a sample space S that are not in A. Thus, $\bar{A}$ and A are mutually exclusive and their union is S:

$$\bar{A} \cup A = S$$

It follows that

$$P(\bar{A} \cup A) = P(\bar{A}) + P(A) = P(S) = 1$$

or

$$P(\bar{A}) = 1 - P(A)$$

Thus, we have proved the following theorem.

THEOREM   2.5

If $\bar{A}$ is the complement of an event A in a sample space S, then

$$P(\bar{A}) = 1 - P(A)$$

∎

EXAMPLE   2.18   A quality control inspector has ten assembly lines from which to choose products for testing. Each morning of a five-day week, she randomly selects one of the lines to work on for the day. Find the probability that any of the ten lines is chosen more than once during the week.

Solution   It is easier, here, to think in terms of complements and to begin by finding the probability that no line is chosen more than once. If no line is repeated, five different lines must be chosen on successive days, which can be done in

$$P_5^{10} = \frac{10!}{5!} = 10(9)(8)(7)(6)$$

ways. The total number of possible outcomes for the selection of five lines without restriction is $(10)^5$, by an extension of the multiplication rule. Thus,

$$P(\text{No line is chosen more than once}) = \frac{P_5^{10}}{(10)^5}$$
$$= \frac{10(9)(8)(7)(6)}{(10)^5} = 0.30$$

and

$$P(\text{Some line is chosen more than once}) = 1.0 - 0.30 = 0.70$$

■

2.6.2 Additive Rule

Axiom 3 applies to $P(A \cup B)$ if A and B are disjoint. But, what happens when A and B are not disjoint? Theorem 2.6 gives the answer.

T H E O R E M **2.6**

If A and B are any two events, then

$$P(A \cup B) = P(A) + P(B) - P(AB).$$

If A and B are mutually exclusive, then

$$P(A \cup B) = P(A) + P(B).$$

Proof

From a Venn diagram for the union of A and B (refer to Figure 2.4), it is easy to see that

$$A \cup B = A\bar{B} \cup \bar{A}B \cup AB$$

and that the three events on the right-hand side of the equality are mutually exclusive. Hence,

$$P(A \cup B) = P(A\bar{B}) + P(\bar{A}B) + P(AB).$$

Now,

$$A = A\bar{B} \cup AB$$

and

$$B = \bar{A}B \cup AB$$

so

$$P(A) = P(A\bar{B}) + P(AB)$$

and

$$P(B) = P(\bar{A}B) + P(AB)$$

It follows that

$$P(A\bar{B}) = P(A) - P(AB)$$

and

$$P(\bar{A}B) = P(B) - P(AB)$$

Substituting into the first equation for $P(A \cup B)$, we have

$$P(A \cup B) = P(A) - P(AB) + P(B) - P(AB) + P(AB)$$
$$= P(A) + P(B) - P(AB),$$

and the proof is complete.   ▪

The formula for the probability of the union of k events $A_1, A_2, \ldots, A_k$ is derived in similar fashion. It is given by

$$P(A_1 \cup A_2 \cup \cdots \cup A_k) = \sum_{i=1}^{k} P(A_i) - \sum\sum_{i<j} P(A_i A_j)$$
$$+ \sum\sum\sum_{i<j<l} P(A_i A_j A_1) - \cdots + \cdots - (-1)^k P(A_1 A_2 \cdots A_k)$$

2.6.3   Multiplicative Rule

The next rule is actually just a rearrangement of the definition of conditional probability, for the case in which a conditional probability may be known and we want to find the probability of an intersection.

THEOREM   **2.7**

If A and B are any two events, then

$$P(AB) = P(A)P(B|A)$$
$$= P(B)P(A|B)$$

If A and B are independent, then

$$P(AB) = P(A)P(B)$$

▪

We illustrate the use of Theorems 2.5, 2.6, and 2.7 in the following three examples.

EXAMPLE   **2.19**   Records indicate that of all the parts produced by a hydraulic repair shop at an airplane rework facility, 20% have a shaft defect, 10% have a bushing defect, and 75% are defect-free. For an item chosen at random from this output, find the probabilities of the following events:

A: The item has at least one type of defect.

B: The item has only a shaft defect.

Solution   The percentages given imply that 5% of the items have both a shaft defect and a bushing defect. Let D_1 denote the event that an item has a shaft defect, and let D_2 denote the event that it has a bushing defect. Then,

$$A = D_1 \cup D_2$$

and

$$
\begin{aligned}
P(A) &= P(D_1 \cup D_2) \\
&= P(D_1) + P(D_2) - P(D_1 D_2) \\
&= 0.20 + 0.10 - 0.05 \\
&= 0.25
\end{aligned}
$$

Another possible solution is to observe that the complement of A is the event that an item has no defects. Thus,

$$
\begin{aligned}
P(A) &= 1 - P(\bar{A}) \\
&= 1 - 0.75 \\
&= 0.25
\end{aligned}
$$

To find $P(B)$, first notice that the event D_1 (that the item has a shaft defect) is the union of the event that it has *only* a shaft defect (B) and the event that it has *both* defects $(D_1 D_2)$; that is,

$$D_1 = B \cup D_1 D_2$$

where B and $D_1 D_2$ are mutually exclusive. Therefore,

$$P(D_1) = P(B) + P(D_1 D_2)$$

or

$$
\begin{aligned}
P(B) &= P(D_1) - P(D_1 D_2) \\
&= 0.20 - 0.05 \\
&= 0.15
\end{aligned}
$$

You should sketch these events on a Venn diagram and verify the results just derived.   ■

E X A M P L E   **2.20**   A section of an electrical circuit has two relays in parallel, as shown in Figure 2.11. The relays operate independently, and when a switch is thrown, each will close properly with probability only 0.8. If both relays are open, find the probability that current will flow from s to t when the switch is thrown.

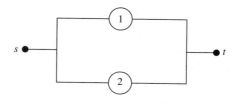

Solution   Let O denote an open relay, and let C denote a closed relay. The four outcomes for this experiment are shown in the following diagram.

		Relay 1	Relay 2
E_1	$=$	$\{(O,$	$O)\}$
E_2	$=$	$\{(O,$	$C)\}$
E_3	$=$	$\{(C,$	$O)\}$
E_4	$=$	$\{(C,$	$C)\}$

Since the relays operate independently, we can find the probabilities for each of these outcomes as follows:

$$P(E_1) = P(O)P(O) = (0.2)(0.2) = 0.04$$
$$P(E_2) = P(O)P(C) = (0.2)(0.8) = 0.16$$
$$P(E_3) = P(C)P(O) = (0.8)(0.2) = 0.16$$
$$P(E_4) = P(C)P(C) = (0.8)(0.8) = 0.64$$

If A denotes the event that current flows from s to t, then

$$A = E_2 \cup E_3 \cup E_4$$

or

$$\bar{A} = E_1$$

(At least one of the relays must close in order for current to flow.) Thus,

$$P(A) = 1 - P(\bar{A})$$
$$= 1 - P(E_1)$$
$$= 1 - 0.04$$
$$= 0.96$$

which is the same as $P(E_2) + P(E_3) + P(E_4)$.   ∎

EXAMPLE   2.21   Three different orders are to be mailed to three suppliers. However, an absent-minded secretary gets the orders mixed up, and sends them randomly to suppliers. If a match refers to the fact that a supplier receives the correct order, find the probability of the following events.

1   No matches occur.

2   Exactly one match occurs.

Solution This problem could be solved by listing outcomes, since only three orders and three suppliers are involved; but we will use a more general method of solution. Define the following events:

A_1: Match for supplier I.

A_2: Match for supplier II.

A_3: Match for supplier III.

There are $3! = 6$ equally likely ways of randomly sending the orders to suppliers, but there are only $2! = 2$ ways of sending the orders to suppliers if one particular supplier is required to have a match. Hence,

$$P(A_1) = P(A_2) = P(A_3) = 2/6 = 1/3$$

Similarly, it follows that

$$\begin{aligned}
P(A_1 A_2) &= P(A_1 A_3) \\
&= P(A_2 A_3) \\
&= P(A_1 A_2 A_3) \\
&= 1/6
\end{aligned}$$

1 Notice that

$$\begin{aligned}
P(\text{No matches}) &= 1 - P(\text{At least one match}) \\
&= 1 - P(A_1 \cup A_2 \cup A_3) \\
&= 1 - [P(A_1) + P(A_2) + P(A_3) \\
&\quad - P(A_1 A_2) - P(A_1 A_3) \\
&\quad - P(A_2 A_3) + P(A_1 A_2 A_3)] \\
&= 1 - [3(1/3) - 3(1/6) + (1/6)] \\
&= 1/3
\end{aligned}$$

2 It is left to the reader to show that

$$\begin{aligned}
P(\text{Exactly one match}) &= P(A_1) + P(A_2) + P(A_3) - 2[P(A_1 A_2) + P(A_1 A_3) \\
&\quad + P(A_2 A_3)] + 3P(A_1 A_2 A_3) \\
&= 3(1/3) - 2(3)(1/6) + 3(1/6) \\
&= 1/2
\end{aligned}$$

2.6.4   Bayes' Rule

The fourth rule we present in this section is based on the notion of a partition of a sample space. Events $B_1, B_2, \ldots, B_k$ are said to partition a sample space S if the following two conditions exist:

1 $B_i B_j = \emptyset$ for any pair i and j ($\emptyset$ denotes the null set.)

2 $B_1 \cup B_2 \cup \cdots \cup B_k = S$

For example, the set of tires in an auto assembly warehouse may be partitioned according to suppliers, or employees of a firm may be partitioned according to level of education. A partition for the case $k = 2$ is illustrated in Figure 2.12. The key idea with regard to a partition consists of observing that an event A can be written as the union of mutually exclusive events AB_1 and AB_2; that is,

$$A = AB_1 \cup AB_2$$

and thus

$$P(A) = P(AB_1) + P(AB_2)$$

If conditional probabilities $P(A|B_1)$ and $P(A|B_2)$ are known, then $P(A)$ can be found by writing

$$P(A) = P(B_1)P(A|B_1) + P(B_2)P(A|B_2)$$

In problems dealing with partitions, investigators frequently want to find probabilities of the form $P(B_1|A)$, which can be written as

$$P(B_1|A) = \frac{P(B_1 A)}{P(A)}$$

$$= \frac{P(B_1)P(A|B_1)}{P(B_1)P(A|B_1) + P(B_2)P(A|B_2)}$$

F I G U R E **2.12**
Partition of S into B_1 and B_2.

This result is a special case of *Bayes' Rule*, of which Theorem 2.8 is a general statement.

THEOREM **2.8**

If $B_1, B_2, \ldots, B_k$ form a partition of S, and A is any event in S, then

$$P(B_j|A) = \frac{P(B_j)P(A|B_j)}{\displaystyle\sum_{i=1}^{k} P(B_i)P(A|B_i)}$$

Proof

The proof, which will not be spelled out here, is an extension of the results shown earlier for $k = 2$. ∎

EXAMPLE **2.22**

A company buys tires from two suppliers—1 and 2. Supplier 1 has a record of delivering tires that contain 10% defectives, whereas supplier 2 has a defective rate of only 5%. Suppose that 40% of the current supply came from supplier 1. If a tire taken at random from this supply is observed to be defective, what is the probability that it came from supplier 1?

Solution

Let B_i denote the event that a tire comes from supplier i, where $i = 1, 2$. (Notice that B_1 and B_2 form a partition of the sample space for the experiment of selecting one tire.) Let A denote the event that the selected tire is defective. Then

$$\begin{aligned}
P(B_1|A) &= \frac{P(B_1)P(A|B_1)}{P(B_1)P(A|B_1) + P(B_2)P(A|B_2)} \\
&= \frac{0.40(0.10)}{0.40(0.10) + 0.60(0.05)} \\
&= \frac{0.04}{0.04 + 0.03} \\
&= \frac{4}{7}
\end{aligned}$$

Supplier 1 has a greater probability of being the party supplying the defective tire than does supplier 2. ∎

Let's return, now, to tables of real data. The data may not provide exact probabilities, but the approximations are frequently good enough to permit clearer insights into the problem at hand.

EXAMPLE **2.23**

The U.S. Bureau of Labor Statistics provided the data in Table 2.7 as a summary of employment in the United States for 1989.

TABLE 2.7
Civilian Employment in the
United States Among Persons
25 Years Old and Older, 1989

Classification	Total (in millions)
Civilian noninstitutional population	154
Civilian labor force	102
Employed	98
Unemployed	4
Not in the labor force	52

Source: U.S. Bureau of Labor Statistics.

Suppose that an arbitrarily selected U.S. resident was asked, in 1989, to fill out a questionnaire on employment. Find the probabilities of the following events.

1   The resident was in the labor force.

2   The resident was employed.

3   The resident was employed and in the labor force.

4   The resident was employed given that he or she was known to be in the labor force.

5   The resident was either not in the labor force or unemployed.

Solution   Let L denote the event that the resident was in the labor force, and let E denote the event that he or she was employed.

1   Since 154 million people constitute the population under study and since 102 million were in the labor force,

$$P(L) = \frac{102}{154}$$

2   Similarly, 98 million were employed; thus,

$$P(E) = \frac{98}{154}$$

3   Employed persons are a subset of all persons in the labor force; in other words, $EL = E$. Hence,

$$P(EL) = P(E) = \frac{98}{154}$$

4   Among the 102 million people in the labor force, 98 million were employed. Therefore,

$$P(E|L) = \frac{98}{102}$$

Notice that this result could also be found by using Definition 2.4, as follows:

$$P(E|L) = \frac{P(EL)}{P(L)} = \frac{98/154}{102/154} = \frac{98}{102}$$

5   The event that the resident was not in the labor force $\bar{L}$ is mutually exclusive from the event that he or she was unemployed. Therefore,

$$P(\bar{L} \cup \bar{E}) = P(\bar{L}) + P(\bar{E})$$
$$= \frac{52}{154} + \frac{4}{154}$$
$$= \frac{56}{154}$$

∎

The next example uses a data set that has a bit more complicated structure.

E X A M P L E   **2.24**   The National Fire Incident Reporting Sevice provided the information in Table 2.8 on fires reported in 1978. All figures in the table are percents, and the main body of the table shows percentages according to cause. Thus, the 22% in the upper left corner indicates that 22% of all fires in family homes were caused by heating. Suppose that a residential fire in 1978 was called into a fire station. Find the probabilities of the following events.

1   The fire was caused by heating.

2   The fire occurred in a family home.

3   The fire was caused by heating, given that it occurred in an apartment.

4   The fire occurred in an apartment and was caused by heating.

5   The fire in a family home, given that it was caused by heating.

Solution   **1**   Over all locations, 19% of fires were caused by heating. (See the right-hand column of Table 2.8.) Thus,

$$P(\text{Heating fire}) = 0.19$$

T A B L E   **2.8**
Causes of Fires in Residences, 1978

Cause of fire	Family homes	Apartments	Mobile homes	Hotels/ motels	Other	All locations
Heating	22%	6%	22%	8%	45%	19%
Cooking	15	24	13	7	0	16
Incendiary substance	10	15	7	16	8	11
Smoking	7	18	6	36	19	10
Electrical	8	5	15	7	28	8
Other	38	32	37	26	0	36
All causes	73	20	3	2	2	100

Source: National Fire Incident Reporting Service.

2 Over all causes, 73% of the fires occurred in family homes. (See the bottom row of Table 2.8.) Thus,

$$P(\text{Family home fire}) = 0.73$$

3 The third column in from the left deals exlusively with apartment fires. It indicates that 6% of all apartment fires were caused by heating. Notice that this is a conditional probability and can be written as

$$P(\text{Heating fire}|\text{Apartment fire}) = 0.06$$

4 In part 3, we know the fire was in an apartment, and the probability in question was conditional on that information. Here, we must find the probability of an intersection between "apartment fire" (say, event A) and "heating fire" (say, event H). We can identify $P(H|A)$ directly from the table, so

$$\begin{aligned} P(AH) &= P(A)P(H|A) \\ &= (0.20)(0.06) \\ &= 0.012 \end{aligned}$$

In other words, only 1.2% of all reported fires were caused by heating in apartments.

5 The figures recorded in the body of Table 2.8 give probabilities of causes, given the location of the fire. We are now asked to find the probability of a location, given the cause. This is exactly the situation to which Bayes' Rule (Theorem 2.8) applies. The locations form a partition of the set of all fires into five different groups. We have, then (with obvious shortcuts in wording),

$$P(\text{Family}|\text{Heating}) = \frac{P(\text{Family})P(\text{Heating}|\text{Family})}{P(\text{Heating})}$$

Now $P(\text{Heating}) = 0.19$ from Table 2.8. It might be informative, however, to examine how this figure can be derived from the other columns of the table. We have

$$\begin{aligned} P(\text{Heating}) =\ & P(\text{Family})P(\text{Heating}|\text{Family}) \\ & + P(\text{Apartment})P(\text{Heating}|\text{Apartment}) \\ & + P(\text{Mobile})P(\text{Heating}|\text{Mobile}) \\ & + P(\text{Hotel})P(\text{Heating}|\text{Hotel}) \\ & + P(\text{Other})P(\text{Heating}|\text{Other}) \\ =\ & (0.73)(0.22) + (0.20)(0.06) + (0.03)(0.22) \\ & + (0.02)(0.08) + (0.02)(0.45) \\ =\ & 0.19 \end{aligned}$$

Then,

$$P(\text{Family}|\text{Heating}) = \frac{(0.73)(0.22)}{0.19} = 0.85$$

In other words, 85% of all heating fires occurred in family homes.

∎

EXAMPLE **2.25**   Data from past studies are often used to anticipate results for future events, as we have seen. Suppose that three new employees are hired from a national pool of potential employees. Based on the information given in Example 2.3, find the probabilities of the following events.

1 Exactly one is at risk for HIV.

2 At least one is at risk for HIV.

3 Exactly two are at risk, given that at least one is known to be at risk.

Solution   As usual, some assumptions must be made in order to establish a random mechanism into the problem. Assuming that the three selected employees behave as a random sample from the general population with regard to HIV risk groups, we can assign a probability of 0.15 (refer to Table 2.1) to the event that any one new employee is at risk. If the three selections of new employees are unrelated to each other, they can be regarded as independent events.

1 Let R_i denote the event that person i is at risk ($i = 1, 2, 3$). A tree diagram can aid in showing the structure of outcomes from selecting three new employees. The probabilities along connected branches can be multiplied because of the independence assumption. By the rules of probability,

$$
\begin{aligned}
P(\text{exactly 1 is at risk}) &= P(\overline{R}_1\overline{R}_2 R_3 \cup \overline{R}_1 R_2\overline{R}_3 \cup R_1\overline{R}_2\overline{R}_3) \\
&= P(\overline{R}_1\overline{R}_2 R_3) + P(\overline{R}_1 R_2\overline{R}_3) + P(R_1\overline{R}_2\overline{R}_3) \\
&= P(\overline{R}_1)P(\overline{R}_2)P(R_3) + P(\overline{R}_1)P(R_2)P(\overline{R}_3) \\
&\quad + P(R_1)P(\overline{R}_2)P(\overline{R}_3) \\
&= (0.85)(0.85)(0.15) + (0.85)(0.15)(0.85) \\
&\quad + (0.15)(0.85)(0.85) \\
&= 3(0.15)(0.85)^2 \\
&= 0.325
\end{aligned}
$$

2 The probability that at least one new employee is at risk could be determined by repeating an argument similar to that in part 1 for "Exactly two are at risk" and "Exactly three are at risk," and adding the results. An easier method, however, is to use complements, so that

$$
\begin{aligned}
P(\text{At least 1 is at risk}) &= 1 - P(\text{None is at risk}) \\
&= 1 - P(\overline{R}_1\overline{R}_2\overline{R}_3) \\
&= 1 - (0.85)^3 \\
&= 0.386
\end{aligned}
$$

3   This part involves the conditional probability

$$P(\text{Exactly two are at risk}|\text{At least one is at risk}) = \frac{P(\text{Exactly two are at risk})}{P(\text{At least one is at risk})}$$
$$= \frac{3(0.15)^2(0.85)}{0.386}$$
$$= 0.148$$

Notice that the intersection of the two events in question is the same as the first event, "Exactly two are at risk." The probability in the numerator is found in a manner analogous to that used in part 1.

Suppose that the three new hires all came from high-risk cities, in which about 20% of the population is at risk for HIV. How would that fact change these probabilities?

■

Exercises

2.27   Vehicles coming into an intersection can turn left, turn right, or go straight ahead. Two vehicles enter an intersection in succession. Find the probability that at least one of the two turns left, given that at least one of the two vehicles turns. What assumptions have you made?

2.28   A purchasing office is to assign a contract for computer paper and another contract for microcomputer disks to any one of three firms bidding for these contracts. (Any one firm could receive both contracts.) Find the probabilities of the following events.

a   Firm I receives a contract, given that both contracts do not go to the same firm.

b   Firm I receives both contracts.

c   Firm I receives the contract for paper, given that it does not receive the contract for disks.

What assumptions have you made?

2.29   The data in the accompanying table identify the number of accidental deaths overall and for three specific causes, for the United States in 1984. You are told that a certain person recently died in an accident. Approximate the probabilities of the following events.

a   It was a motor vehicle accident.

b   It was a motor vehicle accident, given that the person was male.

c   It was a motor vehicle accident, given that the person was between 15 and 24 years of age.

d   It was a fall, given that the person was over age 75.

e   The person was male.

Accidental Deaths in the United States, 1984

		Cause of Accident		
Age Group	All Types	Motor Vehicle	Falls	Drowning
All ages	92,911	46,263	11,937	5388
Under 5	3652	1132	114	638
5–14	4198	2263	68	532
15–24	19,801	14,738	399	1353
25–44	25,498	15,036	963	1549
45–64	15,273	6954	1624	763
65–74	8424	3020	1702	281
75 and over	16,065	3114	7067	272
Male	64,053	32,949	6210	4420
Female	28,858	13,314	5727	968

Source: The World Almanac and Book of Facts, 1989 edition, copyright ©Newspaper Enterprise Association, Inc. 1988, New York, NY 10166. Used by permission.

2.30 The data in the accompanying table show the distribution of arrival times at work by mode of travel for workers in the central business district of a large city. The figures are percentages. (The columns should add to 100, but some do not because of rounding.) If a randomly selected worker is asked about his or her travel to work, find the probabilities of the following events.

a The worker arrives before 7:15, given that the worker drives alone.

b The worker arrives at or after 7:15, given that the worker drives alone.

c The worker arrives before 8:15, given that the worker rides in a car pool.

Can you find the probability that the worker drives alone, using only these data?

Arrival Times for Various Commuting Methods (by percentage)

Arrival Time	Transit	Drove Alone	Shared Ride with Family Member	Car Pool	All
Before 7:15	17%	16%	16%	19%	18%
7:15–7:45	35	30	30	42	34
7:45–8:15	32	31	43	35	33
8:15–8:45	10	14	8	2	10
After 8:45	5	11	3	2	6

Source: C. Hendrickson and E. Plank, "The Flexibility of Departure Times for Work Trips," *Transportation Research* 18A no. 1 (1984): 25–36. Used by permission.

2.31 Table 2.8 shows percentages of fires by cause for certain residential locations. If a fire is reported to be residential, find the probabilities of the following events.

a The fire was caused by smoking.

b The fire occurred in a mobile home.

c The fire was caused by smoking, given that it occurred in a mobile home.

d The fire occurred in a mobile home, given that it was caused by smoking.

2.32 An incoming lot of silicon wafers is to be inspected for defectives by an engineer in a microchip manufacturing plant. Suppose that, in a tray containing twenty wafers, four are defective. Two wafers are to be selected randomly for inspection. Find the probabilities of the following events.

a Neither is defective.

b At least one of the two is defective.

c Neither is defective, given that at least one is not defective.

2.33 In the setting of Exercise 2.32, answer the same three questions, assuming this time that only two among the twenty wafers are defective.

2.34 Show the following, for any events A and B.

a $P(AB) \geq P(A) + P(B) - 1$

b The probability that exactly one of the events occurs is $P(A) + P(B) - 2P(AB)$.

2.35 A certain firm produces resistors and markets them as 10-ohm resistors. However, the actual ohms of resistance produced by the resistors may vary. Research has established that 5% of the values are below 9.5 ohms and 10% are above 10.5 ohms. If two resistors, randomly selected, are used in a system, find the probabilies of the following events.

a Both resistors have actual values between 9.5 and 10.5 ohms.

b At least one resistor has an actual value in excess of 10.5 ohms.

2.36 Consider the following segment of an electric circuit equipped with three relays. Current will flow from a to b if at least one closed path exists when the relays are switched to "closed." However, the relays may malfunction. Suppose that they close properly only with probability 0.9 when the switch is thrown, and suppose that they operate independently of one another. Let A denote the event that current will flow from a to b when the relays are switched to "closed."

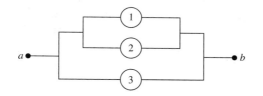

a Find $P(A)$.

b Find the probability that relay 1 is closed properly, given that current is known to be flowing from a to b.

2.37 With relays operating as in Exercise 2.36, compare the probability of current flowing from a to b in the following series system:

with the probability of current flowing in the following parallel system:

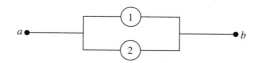

2.38 Electric motors coming off two assembly lines are pooled for storage in a common stockroom, which contains an equal number of motors from each line. Motors from that room are periodically sampled and tested. It is known that 10% of the motors from line I are defective and that 15% of the motors from line II are defective. If a motor is randomly selected from the stockroom and found to be defective, what is the probability that it came from line I?

2.39   Two methods, A and B, are available for teaching a certain industrial skill. The failure rate is 20% for method A and 10% for method B. Method B is more expensive, however, and hence is used only 30% of the time. (Method A is used the other 70% of the time.) A worker is taught the skill by one of the two methods, but he fails to learn it correctly. What is the probability that he was taught by method A?

2.40   A diagnostic test for a certain disease is said to be 90% accurate; that is, if a person has the disease, the test will detect it with probability 0.9. Moreover, if a person does not have the disease, the test will report that he or she doesn't have it with probability 0.9. Only 1% of the population has the disease in question. If the diagnostic test reports that a person chosen at random from the population has the disease, what is the conditional probability that the person does in fact, have the disease? Are you surprised by the size of the answer? Do you consider this diagnostic test reliable?

2.41   A proficiency examination for a certain skill was given to 100 employees of a firm. Forty of the employees were men. Sixty of the employees passed the examination (by scoring above a preset level for satisfactory performance). The breakdown of test results among men and women are shown in the accompanying diagram.

	Male (M)	Female (F)	Total
Pass (P)	24	36	60
Fail ($\bar{P}$)	16	24	40
Total	40	60	100

Suppose that an employee is selected at random from among the 100 who took the examination.

a   Find the probability that the employee passed, given that he was a man.

b   Find the probability that the employee was a man, given that a passing grade was received.

c   Are events P and M independent?

d   Are events P and F independent?

2.42   By using Venn diagrams or similar arguments, show that, for events A, B, and C,

$$P(A \cup B \cup C) = P(A) + P(B) + P(C) - P(AB) \\ -P(AC) - P(BC) + P(ABC)$$

2.43   By using the definition of conditional probability, show that

$$P(ABC) = P(A)P(B|A)P(C|AB)$$

2.44   There are 23 students in a classroom. What is the probability that at least 2 of them have the same birthday (day and month)? Assume that the year has 365 days. State your assumptions.

2.45   After packing k boxes (numbered 1, 2, ..., k) of m items each, workers discovered that one defective item had slipped in among the km items packed. In an attempt to find the defective item, they randomly sample n items from each box and examine these.

a   Find the probability that the defective item is in box i. What assumption is necessary for your answer to be valid?

b   Find the probability that the defective item is found in box 1, given that it was actually put in box 1.

c   Find the unconditional probability that the defective item is not found in box 1.

d   Find the conditional probability that the defective item is in box 1, given that it was not found in box 1.

e   Find the conditional probability that the defective item is in box 2, given that it was not found in box 1.

f   Comment on the behavior of these probabilities as $n \to m$. As $n \to 0$.

2.46   A single multiple-choice question has n choices, only one of which is correct. A student taking this test knows the answer with probability p. If the student does not know the answer, he or she guesses randomly. Find the conditional probability that the student knew the answer, given that the question was answered correctly.

2.47   A box contains M balls, of which W are white. A sample of n balls is drawn at random and without replacement. Let A_j, where $j = 1, \ldots, n$, denote the event that the ball drawn on the jth draw is white. Let B_k denote the event that the sample of n balls contains exactly k white balls.

a   Find the probability of A_j.

b   Show that $P(A_j | B_k) = \frac{k}{n}$.

c   Would the probability in part (b) change if the sampling were done with replacement?

2.7   Odds, Odds Ratios, and Relative Risk

"What are the odds that our team will win today?" This is a common way of talking about events whose unknown outcomes have probabilistic interpretations. The **odds in favor** of an event A is the ratio of the probability of A to the probability of $\overline{A}$; that is,

$$\text{Odds in favor of } A = \frac{P(A)}{P(\overline{A})}$$

The odds in favor of a balanced coin's coming up heads when flipped is $P(H)/P(T) = (\frac{1}{2})/(\frac{1}{2}) = 1$, often written as 1:1 (one to one). The odds in favor of a person's being at risk for HIV (if randomly selected from a high-risk city) is $(0.20/0.80) = 0.25$ or 1:4. The **odds against** that person's being at risk is $0.80/0.20 = 4$ or 4:1. The odds against an event B are the same as the odds in favor of $\overline{B}$.

Odds are not just a matter of betting and sports. They are a serious component of the analysis of frequency data, especially when researchers are comparing categorical variables on two-way frequency tables.

The Physicians' Health Study on the effects of aspirin on heart attacks randomly assigned over 22,000 physicians to either the "aspirin" or the "placebo" arm of the study. The data on myocardial infarctions (M.I.s) are summarized in Table 2.9.

For the aspirin group, the odds in favor of suffering an M.I. are

$$\frac{P(\text{M.I.})}{P(\overline{\text{M.I.}})} = \frac{139/11,037}{10,898/11,037} = \frac{139}{10,898} = 0.013$$

	M.I.	No M.I.	Total
Aspirin	139	10,898	11,037
Placebo	239	10,795	11,034
Total	378	21,693	22,071

Source: New England Journal of Medicine.

For the placebo group, the odds in favor of M.I. are

$$\frac{P(\text{M.I.})}{P(\overline{\text{M.I.}})} = \frac{239/11{,}034}{10{,}795/11{,}034} = \frac{239}{10{,}796} = 0.022$$

In such studies, odds are often interpreted as **risk**. Thus, the above results show that the risk of a heart attack with the placebo is considerably higher than the risk with aspirin. More specifically, the ratio of the two odds (risks) is called the *relative risk*:

$$\begin{aligned}
\text{Relative risk of M.I.} &= \frac{\text{Risk of M.I. with aspirin}}{\text{Risk of M.I. without aspirin}} \\
&= \frac{\text{Odds of M.I. with aspirin}}{\text{Odds of M.I. without aspirin}} \\
&= \frac{0.013}{0.022} \\
&= 0.59
\end{aligned}$$

Thus, the risk of suffering an M.I. for an individual in the aspirin group is 59% of the risk for an individual in the placebo group.

Odds ratios form a very useful single-number summary of the frequencies in a 2×2 (two-way) frequency table. In fact, the odds ratio (relative risk) has a simpler form for any 2×2 table, which can be written generically as

A	a	b
B	c	d

The odds in favor of A are a/b, and the odds in favor of B are c/d. Therefore, the odds ratio is simply

$$\frac{a/b}{c/d} = \frac{ad}{bc}$$

which is the ratio of the products of the diagonal elements.

E X A M P L E   **2.26**   Data on the civilian labor force can be recorded in a number of meaningful 2×2 tables. Two such tables track employment status by gender and employment status by education (elementary school only versus at least some high school) in Table 2.10. Interpret these tables by using odds ratios.

	Employed	Unemployed
Male	54,039	2207
Female	43,581	1908

	Employed	Unemployed
High School	46,315	2468
Elementary School	5299	406

Source: Statistical Abstract of the United States.

Solution   For the first table, the odds ratio is

$$\frac{(54,039)(1908)}{(43,581)(2207)} = 1.07$$

The odds in favor of being employed is about the same for males as for females. This result might be restated as indicating that the risk of being unemployed is close to the same for males and for females.

For the second table, the odds ratio is

$$\frac{(46,315)(406)}{(5299)(2468)} = 1.44$$

Conditioning on these two education groups, the odds in favor of employment for the high-school group is about 1.4 times greater than that for the elementary school group. Education does seem to have a significant impact. The risk of unemployment for those with some high-school education is $1/1.44 = 0.69$ or 69% of the risk of unemployment for those with only some elementary education. (Can you show that the odds ratio of 0.69 is correct?)

In short, odds and odds ratios are easy to compute, and they serve as a useful summary of the tabulated data.   ∎

Exercises

2.48   From the results of the Physicians' Health Study (discussed earlier in this section), an important factor is myocardial infarctions seems to be cholesterol level. The data in the accompanying table identify the number of M.I.s over the number in the cholesterol category for each arm of the study.

Cholesterol Level (mg per 100 ml)	Aspirin Group	Placebo Group
≤159	2/382	9/406
160–209	12/1587	37/1511
210–259	26/1435	43/1444
≥260	14/582	23/570

a Did the randomization in the study seem to do a good job of balancing the cholesterol levels between the two groups? Explain.

b Construct a 2 × 2 table of aspirin versus placebo M.I. response for each of the four cholesterol levels. Reduce the data in each table to the odds ratio.

c Compare the four odds ratios you found in part (b). Comment on the relationship between the effect of aspirin on heart attacks as it relates to different cholesterol levels. Do you see why odds ratios are handy tools for summarizing data in a 2 × 2 table?

2.49 Is a defendant's race associated with his or her chance of receiving the death penalty? This controversial issue has been studied by many researchers. One important data set was collected on 326 cases in which the defendant was convicted of homicide. The death penalty was imposed in 36 of these cases. The accompanying table shows the defendant's race, the homicide victim's race, and whether or not the death penalty was imposed.

Incidence of Death Penalty (D.P.)

	White Defendant			Black Defendant	
	D.P.	No D.P.		D.P.	No D.P.
White Victim	19	132	White Victim	11	52
Black Victim	0	9	Black Victim	6	97

Source: M. Radelet, "Racial Characteristics and Imposition of the Death Penalty," *American Sociological Review* 46 (1981) 918–27.

a Construct a single 2 × 2 table showing penalty versus defendant's race, across all homicide victims. Calculate the odds ratio and interpret it.

b Decompose the table in part (a) into two 2 × 2 tables of penalty versus defendant's race, one for white homicide victims and one for black homicide victims. Calculate the odds ratio for each table, and interpret each one.

c Do you see any inconsistency between the results of part (a) and the results of part (b)? Can you explain the apparent paradox?

2.50 Refer to parts (a) through (d) of Exercise 2.41. Construct and interpret a meaningful odds ratio for these data.

2.8 Activities for Students: Simulation

The long-run stability of relative frequencies for randomly generated events (Section 2.1) can be used to estimate probabilities through **simulation.** The techniques involved will be illustrated with a simple example before activities are suggested.

What is the probability that a three-child family has exactly two girls and one boy? Simulation of this event involves four steps:

1 Identify a random mechanism to generate the probability of a key component.

2   Define the number of key components in one trial. Call this number n.

3   Generate a large number of trials—say, t.

4   Estimate the probability in question by calculating the ratio of the number of successful trials s to the total number of trials t.

For the question posed above, a key component is the gender of a child, which has $P(G) = P(B) = 0.5$, approximately. Probabilities of 0.5 can be generated by considering whether a random digit is even. So, $P(G) = P(\text{Even digit}) = 0.5$. (Alternatively, coins could be used.) Since a three-child family was specified, there are $n = 3$ key components (children) in a trial. We must generate groups of three random digits, and observe the number of even digits. If the latter number is 2, the trial is a success. (That's what we are looking for.) For the illustration, we will take $t = 50$ trials from the random number table (Table 1) of the Appendix.

Beginning with the last three columns in the upper right-hand corner of the table (each person should choose a different random starting point), we see that the first set of three random digits is 700. Since 0 is considered even, this is a successful trial. The next few sets of random digits are

505
629
379
613

of which only 629 is a success. (It contains two evens and one odd). After 50 trials (50 rows of the table), we find that the number of successful trials is $s = 15$. Thus, the simulation estimate of the probability in question is $(s/t) = (15/50) = 0.30$. You may want to compare this number with the probability calculated according to the rules of this chapter.

2.8.1   Activity 1: What's the Chance of Driving?

A study of the need for parking spaces on campus begins with an estimate of the proportion of students who drive to school. It continues with the use of these data to anticipate what might happen next year.

1   Conduct a small survey among students to estimate what proportion drive to school. Carefully consider the questions to be asked. Use a randomization device to select the respondents.

2   Using your estimated probability that a student chosen at random will drive to school, design a simulation to approximate the probability that, out of five students sampled from next year's student body, at least two will drive to school. (The numbers here are kept small for the sake of simplicity. The idea generalizes to any sample size.)

2.8.2 Activity 2: Where Are the Defectives?

An assembly line is thought to produce defective items amounting to about 10% of its total yield. You want to examine a defective item to find the cause. If you randomly sample items coming off this line, what is the probability that the first defective item you see will be discovered only after four nondefective items have been inspected? What is the probability that you will have to inspect six or more items to find two defectives?

Construct a simulation to approximate these probabilities. Use the four steps identified above, and pay particular attention to how a trial is defined.

2.9 Summary

Data are the key to making sound and objective decisions, and **randomness** is the key to obtaining good data. The importance of randomness derives from the fact that the relative frequencies of random events tend to stabilize in the long run; this long-run relative frequency is called **probability.**

A more formal definition of probability allows for its use as a mathematical modeling tool to help explain and anticipate outcomes of events not yet seen. **Conditional probability, complements,** and rules for **addition** and **multiplication** of probabilities are essential parts of the modeling process.

The relative frequency notion of probability and the resulting rules governing probability calculations have direct parallels in the analysis of frequency data recorded in tables. Even with these rules, it is often more efficient to find an approximate answer through **simulation** than to labor over a theoretical probability calculation.

▪ ▪ ▪ ▪ ▪ ▪ ▪ ▪ ▪ ▪

Supplementary Exercises

2.51 A coin is tossed four times, and the outcome is recorded for each toss.

a List the outcomes for the experiment.

b Let A be the event that the experiment yields three heads. List the outcomes in A.

c Make a reasonable assignment of probabilities to the outcomes, and find $P(A)$.

2.52 A hydraulic rework shop in a factory turned out seven rebuilt pumps today. Suppose that three pumps are still defective. Two of the seven are selected for thorough testing and then classified as defective or not defective.

a List the outcomes for this experiment.

b Let A be the event that the selection includes no defectives. List the outcomes in A.

c Assign probabilities to the outcomes, and find $P(A)$.

2.53 The national maximum speed limit (NMSL) of 55 miles per hour was introduced in the United States in early 1974. The data in the accompanying table show the percent ages of vehicles found traveling at various speeds for three types of highways in 1973 (before the NMSL), in 1974 (the year the NMSL was introduced), and in 1975.

Vehicle Speeds on Various Highways, 1973–1975

Vehicle Speed (mph)	Rural Interstate						Rural Primary						Rural Secondary					
	Car			Truck			Car			Truck			Car			Truck		
	73	74	75	73	74	75	73	74	75	73	74	75	73	74	75	73	74	75
30–35	0	0	0	0	0	0	0	1	0	1	1	1	4	4	2	6	7	5
35–40	0	0	0	1	1	0	3	2	2	5	5	3	6	9	6	9	11	7
40–45	0	1	1	2	2	2	5	7	5	9	10	7	11	14	10	15	16	13
45–50	2	7	5	5	11	8	13	18	14	20	21	19	19	25	21	22	25	23
50–55	5	24	23	15	29	29	16	29	29	21	30	30	19	23	26	19	21	26
55–60	13	37	41	27	36	40	22	27	32	22	23	27	19	16	22	17	14	19
60–65	21	21	22	25	15	16	19	11	12	14	7	10	11	6	9	7	4	5
65–70	29	7	6	18	5	4	13	3	5	6	2	2	7	2	3	4	2	1
70–75	19	2	2	5	1	1	6	2	1	1	1	1	3	1	1	0	0	1
75–80	7	1	0	2	0	0	1	0	0	0	0	0	1	0	0	1	0	0
80–85	4	0	0	0	0	0	0	0	0	0	0	0	0	0	0	0	0	0

Source: D. B. Kamerud, "The 55 MPH Speed Limit: Costs, Benefits, and Implied Trade-offs," *Transportation Research* 17A, no. 1 (1983): 51–64.

    a   Find the probabilities that a randomly observed car on a rural interstate was traveling less than 55 mph in 1973, less than 55 mph in 1974, and less than 55 mph in 1975.

    b   Answer the questions in part (a) for a randomly observed truck.

    c   Answer the questions in part (a) for a randomly observed car on a rural secondary road.

2.54   An experiment consists of tossing a pair of dice.

    a   Use the combinatorial theorems to determine the number of outcomes in the sample space S.

    b   Find the probability that the sum of the numbers appearing on the dice is equal to 7.

2.55   Show that $\binom{3}{0} + \binom{3}{1} + \binom{3}{2} + \binom{3}{3} = 2^3$. Note that, in general, $\sum_{i=0}^{n} \binom{n}{i} = 2^n$.

2.56   Of the persons arriving at a small airport, 60% fly on major airlines, 30% fly on privately owned airplanes, and 10% fly on commercially owned airplanes not belonging to an airline. Of the persons arriving on major airlines, 50% are traveling for business reasons; the corresponding figures are 60% for persons arriving on private planes and 90% for persons arriving on other commercially owned planes. For a person selected at random from a group of arrivals, find the probabilities of the following events.

    a   The person is traveling on business.

    b   The person is traveling on business and on a private airplane.

    c   The person is traveling on business, given that he or she arrived on a commercial airliner.

    d   The person arrived on a private plane, given that he or she is traveling on business.

2.57   In how many ways can a committee of three be selected from among ten people?

2.58   How many different telephone numbers can be formed from a seven-digit number if the first digit cannot be zero?

2.59   A personnel director for a corporation has hired ten new engineers. If three (distinctly different) positions are open at a particular plant, in how many ways can the director fill the position?

2.60   An experimenter wishes to investigate the effects of three variables—pressure, temperature, and type of catalyst—on the yield in a refining process. If the experimenter intends to use three settings for temperature, three settings for pressure, and two types of catalysts, how many experimental runs must be conducted to run all possible combinations of pressure, temperature, and type of catalyst?

2.61   An inspector must perform eight tests on a randomly selected keyboard coming off an assembly line. The sequence in which the tests are conducted is important because the time lost between tests varies. If an efficiency expert wanted to study all possible sequences to find the one that required the minimum length of time, how many sequences would he include in his study?

2.62   **a**   Two cards are drawn from a fifty-two-card deck. What is the probability that the draw will yield an ace and a face card?

    b   Five cards are drawn from a fifty-two-card deck. What is the probability that all five cards will be spades? Will be of the same suit?

2.63   The quarterback on a certain football team completes 60% of his passes. He attempts three passes (assumed to be independent) in a given quarter.

    a   What is the probability that he will complete all three?

    b   What is the probability that he will complete at least one?

    c   What is the probability that he will complete at least two?

2.64   Two men toss one balanced coin each. They obtain a "match" if both coins come up heads or if both coins come up tails. Suppose that the process is repeated three times.

    a   What is the probability of three matches?

    b   What is the probability that all six tosses (three for each man) resulted in tails?

Coin tossing provides a model for many practical experiments. Suppose that the "coin tosses" represented the answers given by two students for three specific true/false questions on an exam. If the two students gave three matches for answers, would the low probability identified in part (a) suggest collusion?

2.65 Refer to Exercise 2.64. What is the probability that the pair of coins must be tossed four times before a match occurs (that is, a match occurs for the first time on the fourth toss)?

2.66 Suppose that the probability of exposure to the flu during an epidemic is 0.6. Experience has shown that a serum is 80% successful in preventing an inoculated person from acquiring the flu, if exposed to it. A person not inoculated faces a probability of 0.90 of acquiring the flu if exposed to it. Two persons—one inoculated and one not—can perform a highly specialized task in a business. Assume that they are not at the same location, are not in contact with the same people, and cannot give each other the flu. What is the probability that at least one will get the flu?

2.67 Two gamblers bet $1 each on successive tosses of a coin. Each has a bankroll of $6.

   a What is the probability that they break even after six tosses of the coin?

   b What is the probability that one particular player (say, Jones) wins all the money on the tenth toss of the coin?

2.68 Suppose that the streets of a city are laid out in a grid, with streets running north–south and east–west. Consider the following scheme for patrolling an area of sixteen blocks by sixteen blocks. A patrolman commences walking at the intersection in the center of the area. At the corner of each block, he randomly elects to go north, south, east, or west.

   a What is the probability that he will reach the boundary of his patrol area by the time he walks the first eight blocks?

   b What is the probability that he will return to the starting point after walking exactly four blocks?

2.69 Consider two mutually exclusive events A and B such that $P(A) > 0$ and $P(B) > 0$. Are A and B independent? Give a proof for your answer.

2.70 An accident victim will die unless, in the next 10 minutes, he receives a transfusion of type A Rh$^+$ blood, which can be supplied by a single donor. The medical team requires 2 minutes to type a prospective donor's blood and 2 minutes more to complete the transfer of blood. A large number of untyped donors are available, 40% of whom have type A Rh$^+$ blood. What is the probability that the accident victim will be saved, if only one blood-typing kit is available?

2.71 An assembler of electric fans uses motors from two sources. Company A supplies 90% of the motors, and company B supplies the other 10%. Suppose that 5% of the motors supplied by company A are defective and that 3% of the motors supplied by company B are defective. An assembled fan is found to have a defective motor. What is the probability that this motor was supplied by company B?

2.72 Show that, for three events A, B, and C,

$$P[(A \cup B)|C] = P(A|C) + P(B|C) - P[(A \cap B|C)]$$

2.73 If A and B are independent events, show that A and $\bar{B}$ are also independent.

2.74 Three events A, B, and C are said to be independent if the following equalities hold:

$$P(AB) = P(A)P(B)$$
$$P(AC) = P(A)P(C)$$
$$P(BC) = P(B)P(C)$$
$$P(ABC) = P(A)P(B)P(C)$$

Suppose that a balanced coin is independently tossed two times. Events A, B, and C are defined as follows:

   A: Heads comes up on the first toss.

   B: Heads comes up on the second toss.

   C: Both tosses yield the same outcome.

Are A, B, and C independent?

2.75   A line from a to b has midpoint c. A point is chosen at random on the line and marked x. (The fact that the point x was chosen at random implies that x is equally likely to fall in any subinterval of fixed length l.) Find the probability that the line segments ax, bx, and ac can be joined to form a triangle.

2.76   Eight tires of different brands are ranked from 1 to 8 (best to worst) according to mileage performance. If four of these tires are chosen at random by a customer, find the probability that the best tire among the four selected by the customer is actually ranked third among the original eight.

2.77   Suppose that n indistinguishable balls are to be arranged in N distinguishable boxes so that each distinguishable arrangement is equally likely. If $n \geq N$, show that the probability that no box will be empty is given by

$$\frac{\binom{n-1}{N-1}}{\binom{N+n-1}{N-1}}$$

2.78   Relays in a section of an electrical circuit operate independently, and each one has a probability of 0.9 of closing properly when a switch is thrown. The following two designs, each involving four relays, are presented for a section of a new circuit. Which design has the higher probability of permitting current to flow from a to b when the switch is thrown?

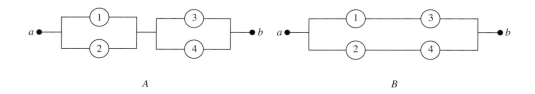

   A   B

2.79   A blood test for hepatitis yields the probabilities for true and false results listed in the accompanying table.

Test Accuracy for Hepatitis Blood Test

	Test Result	
Patient	$+$	$-$
With hepatitis	0.90	0.10
Without hepatitis	0.01	0.99

The disease rate in the general population is 1 in 10,000.

a   What is the probability that a person who receives a positive blood test result actually has hepatitis?

b   A patient is sent for a blood test because he has lost his appetite and has jaundice. The physician knows that this type of patient has a probability of 1/2 of having hepatitis. If this patient gets a positive result on his blood test, what is the probability that he has hepatitis?

3

Discrete Probability Distributions

3.1 Random Variables and Their Probability Distributions

Most of the experiments we encounter generate outcomes that can be interpreted in terms of real numbers, such as heights of children, numbers of voters favoring various candidates, tensile strength of wires, and numbers of accidents at specified intersections. These numerical outcomes, whose values can change from experiment to experiment, are called *random variables*. We will look at an illustrative example of a random variable before we attempt a more formal definition.

A section of an electrical circuit has two relays, numbered 1 and 2, operating in parallel. The current will flow when a switch is thrown if either one or both of the relays close. The probability that a relay will close properly is 0.8, and this probability is the same for each relay. The relays operate independently, we assume. Let E_i denote the event that relay i closes properly when the switch is thrown. Then, $P(E_i) = 0.8$.

When the switch is thrown, a numerical outcome of some interest to the operator of this system is X, the number of relays that close properly. Now, X can take on only three possible values, because the number of relays that close must be 0, 1, or 2. We can find the probabilities associated with these values of X by relating them to the underlying events E_i. Thus, we have

$$P(X = 0) = P(\overline{E}_1 \overline{E}_2)$$
$$= P(\overline{E}_1) P(\overline{E}_2)$$

$$= (0.2)(0.2)$$
$$= 0.04$$

because $X = 0$ means that neither relay closes and the relays operate independently. Similarly,

$$\begin{aligned}
P(X = 1) &= P(E_1 \overline{E}_2 \cup \overline{E}_1 E_2) \\
&= P(E_1 \overline{E}_2) + P(\overline{E}_1 E_2) \\
&= P(E_1)P(\overline{E}_2) + P(\overline{E}_1)P(E_2) \\
&= (0.8)(0.2) + (0.2)(0.8) \\
&= 0.32
\end{aligned}$$

and

$$\begin{aligned}
P(X = 2) &= P(E_1 E_2) \\
&= P(E_1)P(E_2) \\
&= (0.8)(0.8) \\
&= 0.64
\end{aligned}$$

The values of X, along with their probabilities, are more useful for keeping track of the operation of this system than are the underlying events E_i, because the *number* of properly closing relays is the key to whether the system will work. The current will flow if X is equal to at least 1, and this event has probability

$$\begin{aligned}
P(X \geq 1) &= P(X = 1 \text{ or } X = 2) \\
&= P(X = 1) + P(X = 2) \\
&= 0.32 + 0.64 \\
&= 0.96
\end{aligned}$$

Notice that we have mapped the outcomes of an experiment into a set of three meaningful real numbers and have attached a probability to each. Such situations provide the motivation for Definitions 3.1 and 3.2.

DEFINITION   **3.1**   A **random variable** is a real-valued function whose domain is a sample space.   ∎

Random variables will be denoted by upper-case letters such as X, Y, and Z. The actual values that random variables can assume will be denoted by lower-case letters such as x, y, and z. We can then talk about the "probability that X takes on the value x," $P(X = x)$, which is denoted by $p(x)$.

In the relay example, the random variable X has only three possible values, and it is a relatively simple matter to assign probabilities to these values. Such a random variable is called *discrete*.

DEFINITION   **3.2**

A random variable X is said to be **discrete** if it can take on only a finite number—or a countable infinity—of possible values x.
In this case,

1   $P(X = x) = p(x) \geq 0$.
2   $\sum_x P(X = x) = 1$, where the sum is over all possible values x.

The function $p(x)$ is called the *probability function* of X.   ∎

The probability function is sometimes called the *probability mass function* of X, to denote the idea that a mass of probability is piled up from values for discrete points.

It is often convenient to list the probabilities for a discrete random variable in a table. With X defined as the number of closed relays in the problem just discussed, the table is as follows:

x	$p(x)$
0	0.04
1	0.32
2	0.64
Total	1.00

This listing constitutes one way of representing the *probability distribution* of X.

Notice that the probability function $p(x)$ satisfies two properties:

1   $0 \leq p(x) \leq 1$ for any x.
2   $\sum_x p(x) = 1$, where the sum is over all possible values of x.

Functional forms for the probability function $p(x)$ of commonly occurring random variables will be given in later sections. We now illustrate another method for arriving at a tabular presentation of a discrete probability distribution.

EXAMPLE   **3.1**

Circuit boards from two assembly lines set up to produce identical boards are mixed in one storage tray. As inspectors examine the boards, they find that it is difficult to determine whether a board comes from line A or line B. A probabilistic assessment of this question is often helpful. Suppose that the storage tray contains ten circuit boards, of which six came from line A and four from line B. An inspector selects two of these identical-looking boards for inspection. He is interested in X, the number of inspected boards from line A. Find the probability for X.

Solution   The experiment consists of two selections, each of which can result in two outcomes. Let A_i denote the event that the ith board comes from line A, and let B_i denote the event that it comes from line B. The probability of selecting two boards from line A ($X = 2$) is

$$P(A_1 A_2) = P(A \text{ on } 1\text{st})P(A \text{ on } 2\text{nd} \mid A \text{ on } 1\text{st})$$

The multiplicative law of probability is used, and the probability for the second selection depends on what happened on the first selection. Other possibilities for outcomes will result in other values of X. These outcomes are conveniently listed on the tree in Figure 3.1. The probabilities for the various selections are given on the branches of the tree.

F I G U R E **3.1**
Outcomes for Example 3.1.

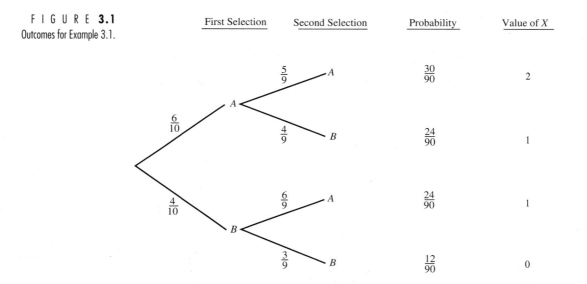

Clearly, X has three possible outcomes, with probabilities as follows:

x	$p(x)$
0	$\frac{12}{90}$
1	$\frac{48}{90}$
2	$\frac{30}{90}$
Total	1.00

Try to envision this concept extended to more selections from trays of various structure. ∎

We sometimes study the behavior of random variables by looking at their *cumulative* probabilities; that is, for any random variable X, we may look at $P(X \leq b)$ for any real number b. This is the cumulative probability for X evaluated at b. Thus, we can define a function $F(b)$ as

$$F(b) = P(X \leq b)$$

DEFINITION   **3.3**

The **distribution function** $F(b)$ for a random variable X is

$$F(b) = P(X \leq b)$$

If X is discrete,

$$F(b) = \sum_{x = -\infty}^{b} p(x)$$

where $p(x)$ is the probability function.

The distribution function is often called the *cumulative distribution function* (c.d.f.).   ∎

The random variable X, denoting the number of relays that close properly (as defined at the beginning of this section), has a probability distribution given by

$$P(X = 0) = 0.04$$
$$P(X = 1) = 0.32$$
$$P(X = 2) = 0.64$$

Notice that

$$P(X \leq 1.5) = P(X \leq 1.9) = P(X \leq 1) = 0.36$$

The distribution function for this random variable then has the form

$$F(b) = \begin{cases} 0 & b < 0 \\ 0.04 & 0 \leq b < 1 \\ 0.36 & 1 \leq b < 2 \\ 1.00 & 2 \leq b \end{cases}$$

This function is graphed in Figure 3.2.

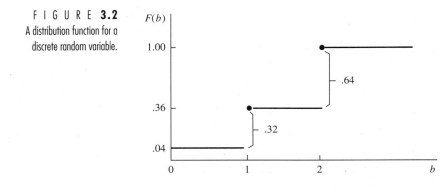

F I G U R E **3.2**
A distribution function for a
discrete random variable.

Exercises

3.1   Among ten applicants for an open position, six are women and four are men. Suppose that three applicants are randomly selected from the applicant pool for final interviews. Find the probability distribution for X, the number of female applicants among the final three.

3.2   The median annual income for heads of household in a certain city is $18,000. Four such heads of household are randomly selected for inclusion in an opinion poll.

a   Find the probability distribution of X, the number (out of the four) who have annual incomes below $18,000.

b   Is it unusual to see all four below $18,000 in this type of poll? (What is the probability of this event?)

3.3   Wade Boggs (then of the Boston Red Sox) hit .363 in 1987, meaning that he got a hit on 36.3% of his official times at bat. In a typical game, he had three official at bats. Find the probability distribution for X, the number of hits Boggs got in a typical game. What assumptions are involved in the answer? Are the assumptions reasonable? Is it unusual for a good hitter to go 0 for 3 in one game?

3.4   A commercial building has two entrances, numbered I and II. Three people enter the building at 9:00 A.M. Let X denote the number who select entrance I. Assuming that the people choose entrances independently and at random, find the probability distribution for X. Were any additional assumptions necessary for your answer?

3.5   Table 2.8 on page 50 gives information on causes of residential fires. Suppose that four independent residential fires are reported in one day, and let X denote the number, out of the four, that are in family homes.

a   Find the probability distribution for X, in tabular form.

b   Find the probability that at least one of the four fires is in a family home.

3.6   Observers noted that 40% of the vehicles crossing a certain toll bridge are commercial trucks. Four vehicles will cross the bridge in the next minute. Find the probability distribution for X, the number of commercial trucks among the four, if the vehicle types are independent of one another.

3.7   Of the people who enter a blood bank to donate blood, 1 in 3 have type O^+ blood, and 1 in 15 have type O^- blood. For the next three people entering the blood bank, let X denote the number with O^+ blood, and let Y denote the number with O^- blood. Assuming independence among the people with respect to blood type, find the probability distributions for X and Y. Then find the probability distribution for $X + Y$, the number of people with type O blood.

3.8   Daily sales records for a computer manufacturing firm show that it will sell 0, 1, or 2 mainframe computer systems, with probabilities as listed:

Number of Sales	0	1	2
Probability	0.7	0.2	0.1

a   Find the probability distribution for X, the number of sales in a two-day period, assuming that sales are independent from day to day.

b   Find the probability that at least one sale is made in the two-day period.

3.9   Four microchips are to be placed in a computer. Two of the four chips are randomly selected for inspection before the computer is assembled. Let X denote the number of defective chips found among the two inspected. Find the probability distribution for X for the following events.

a   Two of the microchips were defective.

b   One of the four microchips was defective.

c   None of the microchips was defective.

3.10   When turned on, each of the three switches in the accompanying diagram works properly with probability 0.9. If a switch is working properly, current can flow through it when it is turned on. Find the probability distribution for Y, the number of closed paths from a to b, when all three switches are turned on.

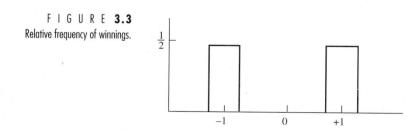

3.2 Expected Values of Random Variables

Because a probability can be thought of as the long-run relative frequency of occurrence for an event, a probability distribution can be interpreted as showing the long-run relative frequency of occurrence for numerical outcomes associated with an experiment. Suppose, for example, that you and a friend are matching balanced coins. Each of you tosses a coin. If the upper faces match, you win $1.00; if they do not match, you lose $1.00 (your friend wins $1.00). The probability of a match is 0.5 and, in the long run, you should win about half of the time. Thus, a relative frequency distribution of your winnings should look like the one shown in Figure 3.3. The -1 under the leftmost bar indicates a loss by you.

F I G U R E **3.3**
Relative frequency of winnings.

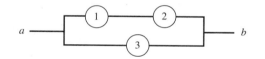

On the average, how much will you win per game over the long run? If Figure 3.3 presents a correct display of your winnings, you win -1 half of the time and $+1$ half of the time, for an average of

$$(-1)\left(\frac{1}{2}\right) + (1)\left(-\frac{1}{2}\right) = 0$$

This average is sometimes called your expected winnings per game, or the *expected value* of your winnings. (An expected value of 0 indicates that this is a fair game.) The general definition of expected value is given in Definition 3.4.

DEFINITION   3.4

The **expected value*** of a discrete random variable X with probability distribution $p(x)$ is given by

$$E(X) = \sum_x xp(x)$$

(The sum is over all values of x for which $p(x) > 0$.)
   We sometimes use the notation

$$E(X) = \mu$$

for this equivalence.   ∎

Now payday has arrived, and you and your friend up the stakes to $10 per game of matching coins. You now win -10 or $+10$ with equal probability. Your expected winnings per game is

$$(-10)\left(\frac{1}{2}\right) + (10)\left(\frac{1}{2}\right) = 0$$

and the game is still fair. The new stakes can be thought of as a function of the old in the sense that, if X represents your winnings per game when you were playing for $1.00, then $10X$ represents your winnings per game when you play for $10.00. Such functions of random variables arise often. The extension of the definition of expected value to cover these cases is given in Theorem 3.1.

THEOREM   3.1

If X is a discrete random variable with probability distribution $p(x)$ and if $g(x)$ is any real-valued function of X, then

$$E[g(X)] = \sum_x g(x)p(x)$$

(The proof of this theorem will not be given.)   ∎

---

*We assume absolute convergence when the range of X is countable; we talk about an *expectation* only when it is assumed to exist.

You and your friend decide to complicate the payoff picture in the coin-matching game by agreeing to let you win $1 if the match is tails and $2 if the match is heads. You still lose $1 if the coins do not match. Quickly you see that this is not a fair game, because your expected winnings are

$$(-1)\left(\frac{1}{2}\right) + (1)\left(\frac{1}{4}\right) + (2)\left(\frac{1}{4}\right) = 0.25$$

You compensate for this by agreeing to pay your friend $1.50 if the coins do not match. Then, your expected winnings per game are

$$(-1.5)\left(\frac{1}{2}\right) + (1)\left(\frac{1}{4}\right) + (2)\left(\frac{1}{4}\right) = 0$$

and the game is again fair. What is the difference between this game and the original one, in which all payoffs were $1? The difference certainly cannot be explained by the expected value, since both games are fair. You can win more but also lose more with the new payoffs, and the difference between the two games can be explained to some extent by the increased variability of your winnings across many games. This increased variability can be seen in Figure 3.4, which displays the relative frequency for your winnings in the new game; the winnings are more spread out than they were in Figure 3.3. Formally, variation is often measured by the *variance* and by a related quantity called the *standard deviation*.

FIGURE 3.4
Relative frequency of winnings.

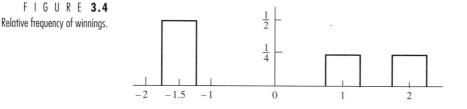

DEFINITION 3.5

The **variance** of a random variable X with expected value μ is given by

$$V(X) = E(X - \mu)^2$$

We sometimes use the notation

$$E(X - \mu)^2 = \sigma^2$$

for this equivalence.   ∎

The smallest value that σ^2 can assume is zero; it occurs when all the probability is concentrated at a single point (that is, when X takes on a constant value with probability 1). The variance becomes larger as the points with positive probability spread out more.

The variance squares the units in which we are measuring. A measure of variation that maintains the original units is the standard deviation.

D E F I N I T I O N   **3.6**

The **standard deviation** of a random variable X is the square root of the variance and is given by

$$\sigma = \sqrt{\sigma^2} = \sqrt{E(X - \mu)^2} \quad \blacksquare$$

For the game represented in Figure 3.3, the variance of your winnings (with $\mu = 0$) is

$$\sigma^2 = E(X - \mu)^2$$
$$= (-1)^2 \left(\frac{1}{2}\right) + (1)^2 \left(\frac{1}{2}\right) = 1$$

It follows that $\sigma = 1$, as well. For the game represented in Figure 3.4, the variance of your winnings is

$$\sigma^2 = (-1.5)^2 \left(\frac{1}{2}\right) + (1)^2 \left(\frac{1}{4}\right) + (2)^2 \left(\frac{1}{4}\right)$$
$$= 2.375$$

and the standard deviation is

$$\sigma = 1.54$$

Which game would you rather play?

The standard deviation can be thought of as the size of a "typical" deviation between an observed outcome and the expected value. For the situation described in Figure 3.3, each outcome (-1 or $+1$) deviates by precisely one standard deviation from the expected value. For the situation described in Figure 3.4, the positive values average 1.5 units from the expected value of 0 (as do the negative values), and so 1.5 units is approximately one standard deviation here.

The mean and the standard deviation often yield a useful summary of the probability distribution for a random variable that can assume many values. An illustration is provided by the age distribution of the U.S. population for 1990 and 2050 (projected, as shown in Table 3.1).

Age is actually a continuous measurement, but since it is reported in categories, we can treat it as a discrete random variable for purposes of approximating its key functions. To move from continuous age intervals to discrete age classes, we assign each interval the value of its midpoint (rounded). Thus, the data in Table 3.1 are interpreted as showing that 7.6% of the 1990 population were around 3 years of age and that 22.5% of the 2050 population is anticipated to be around 55 years of age. (The open intervals at the upper end were stopped at 100 for convenience.)

Age Interval	Age Midpoint	1990	2050
Under 5	3	7.6%	6.4%
5–13	9	12.8	11.6
14–17	16	5.3	5.2
18–24	21	10.8	9.0
25–34	30	17.3	12.5
35–44	40	15.1	12.2
45–64	55	18.6	22.5
65–84	75	11.3	16.0
85 and over	92	1.2	4.6

Source: U.S. Bureau of the Census.

Interpreting the percentages as probabilities, we see that the mean age for 1990 is approximated by

$$\mu = \sum xp(x)$$
$$= 3(0.076) + 9(0.128) + \cdots + 92(0.012)$$
$$= 35.5$$

(How does this compare with the median age for 1990, as approximated from Table 3.1?) For 2050, the mean age is approximated by

$$\mu = \sum xp(x)$$
$$= 3(0.064) + 9(0.116) + \cdots + 92(0.046)$$
$$= 41.2$$

Over the projected period, the mean age increases rather markedly (as does the median age).

The variations in the two age distributions can be approximated by the standard deviations. For 1990, this is

$$\sigma = \sqrt{\sum (x - \mu)^2 p(x)}$$
$$= \sqrt{(3 - 35.5)^2(0.76) + \cdots + (92 - 35.5)^2(0.012)}$$
$$= 22.5$$

A similar calculation for the 2050 data yields $\sigma = 25.4$. These results are summarized in Table 3.2.

Statistic	1990	2050
Mean	35.5	41.2
Standard deviation	22.5	25.4

Not only is the population getting older, on the average, but its variability is increasing. What are some of the implications of these trends?

We now provide other examples and extensions of these basic results.

E X A M P L E **3.2** The manager of a stockroom in a factory knows from his study of records that the daily demand (number of times used) for a certain tool has the following probability distribution:

Demand	0	1	2
Probability	0.1	0.5	0.4

(In other words, 50% of the daily records show that the tool was used one time.) Letting X denote the daily demand, find $E(X)$ and $V(X)$.

Solution From Definition 3.4, we see that

$$E(X) = \sum_x xp(x)$$
$$= 0(0.1) + 1(0.5) + 2(0.4) = 1.3$$

The tool is used an average of 1.3 times per day.

From Definition 3.5, we see that

$$V(X) = E(X - f\mu)^2$$
$$= \sum_x (x - \mu)^2 p(x)$$
$$= (0 - 1.3)^2(0.1) + (1 - 1.3)^2(0.5) + (2 - 1.3)^2(0.4)$$
$$= (1.69)(0.1) + (0.09)(0.5) + (0.49)(0.4)$$
$$= 0.410 \quad \blacksquare$$

Our work in manipulating expected values can be greatly facilitated by making use of the two results of Theorem 3.2. Often, $g(X)$ is a linear function; and when this is the case, the calculations of expected value and variance are especially simple.

T H E O R E M **3.2** For any random variable X and constants a and b,

1 $E(aX + b) = aE(X) + b$

2 $V(aX + b) = a^2 V(X)$

Proof We sketch a proof of this theorem for a discrete random variable X having a probability distribution given by $p(x)$. By Theorem 3.1,

$$E(aX + b) = \sum_x (ax + b)p(x)$$

$$= \sum_x [(ax)p(x) + bp(x)]$$

$$= \sum_x axp(x) + \sum_x bp(x)$$

$$= a \sum_x xp(x) + b \sum_x p(x)$$

$$= aE(X) + b$$

Notice that $\sum_x p(x)$ must equal unity. Also, by Definition 3.5,

$$V(aX + b) = E[(aX + b) - E(aX + b)]^2$$
$$= E[aX + b - (aE(X) + b)]^2$$
$$= E[aX - aE(X)]^2$$
$$= E[a^2(X - E(X))^2]$$
$$= a^2 E[X - E(X)]^2$$
$$= a^2 V(X) \quad \blacksquare$$

An important special case of Theorem 3.2 involves establishing a "standardized" variable. If X has mean μ and standard deviation σ, then the "standardized" form of X is given by

$$Y = \frac{X - \mu}{\sigma}$$

Employing Theorem 3.2, one can easily show that $E(Y) = 0$ and $V(Y) = 1$. This idea will be used often in later chapters.

We illustrate the use of these results in the following example.

E X A M P L E   **3.3**   In Example 3.2, suppose that it costs the factory $10 each time the tool is used. Find the mean and the variance of the daily costs of using this tool.

Solution   Recall that the X of Example 3.2 is the daily demand. The daily cost of using this tool is the $10X$. By Theorem 3.2, we have

$$E(10X) = 10E(X)$$
$$= 10(1.3)$$
$$= 13$$

Thus, the factory should budget $13 per day to cover the cost of using the tool. Also, by Theorem 3.2,

$$V(10X) = (10)^2 V(X)$$
$$= 100(0.410)$$
$$= 41$$

We will make use of this value in a later example.   ∎

Theorem 3.2 leads us to a more efficient computational formula for variance, as given in Theorem 3.3.

THEOREM   **3.3**   If X is a random variable with mean μ, then

$$V(X) = E(X^2) - \mu^2$$

Proof   Starting with the definition of variance, we have

$$
\begin{aligned}
V(X) &= E(X - \mu)^2 \\
&= E(X^2 - 2X\mu + \mu^2) \\
&= E(X^2) - E(2X\mu) + E(\mu^2) \\
&= E(X^2) - 2\mu E(X) + \mu^2 \\
&= E(X^2) - 2\mu^2 + \mu^2 \\
&= E(X^2) - \mu^2 \quad \blacksquare
\end{aligned}
$$

EXAMPLE   **3.4**   Use the result of Theorem 3.3 to compute the variance of X as given in Example 3.2.

Solution   In Example 3.2, X had a probability distribution given by

x	0	1	2
$p(x)$	0.1	0.5	0.4

and we saw that $E(X) = 1.3$. Now,

$$
\begin{aligned}
E(X^2) &= \sum_x x^2 p(x) \\
&= (0)^2(0.1) + (1)^2(0.5) + (2)^2(0.4) \\
&= 0 + 0.5 + 1.6 \\
&= 2.1
\end{aligned}
$$

By Theorem 3.3,

$$
\begin{aligned}
V(X) &= E(X^2) - \mu^2 \\
&= 2.1 - (1.3)^2 = 0.41 \quad \blacksquare
\end{aligned}
$$

We have computed means and variances for a number of probability distributions and noted that these two quantities give us some useful information on the center and spread of the probability mass. Now suppose that we know only the mean and the variance for a probability distribution. Can we say anything specific about probabilities for certain intervals about the mean? The answer is "yes," and a useful result of the relationship among mean, standard deviation, and relative frequency will now be discussed.

THEOREM 3.4

Tchebysheff's Theorem   Let X be a random variable with mean μ and a variance σ^2. Then for any positive k,

$$P(|X - \mu| < k\sigma) \geq 1 - \frac{1}{k^2}$$

Proof

We begin with the definition of $V(X)$ and then make substitutions in the sum defining this quantity. Now,

$$V(X) = \sigma^2$$

$$= \sum_{-\infty}^{\infty} (x - \mu)^2 p(x)$$

$$= \sum_{-\infty}^{\mu - k\sigma} (x - \mu)^2 p(x) + \sum_{\mu - k\sigma}^{\mu + k\sigma} (x - \mu)^2 p(x) + \sum_{\mu + k\sigma}^{\infty} (x - \mu)^2 p(x)$$

(The first sum stops at the largest value of x smaller than $\mu - k\sigma$, and the third sum begins at the smallest value of x larger than $\mu - k\sigma$; the middle sum collects the remaining terms.) Observe that the middle sum is never negative; and for both of the outside sums,

$$(x - \mu)^2 \geq k^2 \sigma^2$$

Eliminating the middle sum and substituting for $(x - \mu)^2$ in the other two, we get

$$\sigma^2 \geq \sum_{-\infty}^{\mu - k\sigma} k^2 \sigma^2 p(x) + \sum_{\mu + k\sigma}^{\infty} k^2 \sigma^2 p(x)$$

or

$$\sigma^2 \geq k^2 \sigma^2 \left[\sum_{-\infty}^{\mu - k\sigma} p(x) + \sum_{\mu + k\sigma}^{\infty} p(x) \right]$$

or

$$\sigma^2 \geq k^2 \sigma^2 P(|X - \mu| \geq k\sigma)$$

It follows that

$$P(|X - \mu| \geq k\sigma) \leq \frac{1}{k^2}$$

or

$$P(|X - \mu| < k\sigma) \geq 1 - \frac{1}{k^2} \quad \blacksquare$$

The inequality in the statement of the theorem is equivalent to

$$P(\mu - k\sigma < X < \mu + k\sigma) \geq 1 - \frac{1}{k^2}$$

To interpret this result, let $k = 2$, for example. Then the interval from $\mu - 2\sigma$ to $\mu + 2\sigma$ must contain at least $1 - 1/k^2 = 1 - 1/4 = 3/4$ of the probability mass for the random variable. We consider more specific illustrations in the following two examples.

E X A M P L E   **3.5**   The daily production of electric motors at a certain factory averaged 120, with a standard deviation of 10.

1   What can be said about the fraction of days on which the production level falls between 100 and 140?

2   Find the shortest interval certain to contain at least 90% of the daily production levels.

Solution   1   The interval from 100 to 140 represents $\mu - 2\sigma$ to $\mu + 2\sigma$, with $\mu = 120$ and $\sigma = 10$. Thus, $k = 2$ and

$$1 - \frac{1}{k^2} = 1 - \frac{1}{4} = \frac{3}{4}$$

At least 75% of all days, therefore, will have a total production value that falls in this interval. (This percentage could be closer to 95% if the daily production figures show a mound-shaped, symmetric relative frequency distribution.)

2   To find k, we must set $(1 - 1/k^2)$ equal to 0.9 and solve for k; that is,

$$1 - \frac{1}{k^2} = 0.9$$
$$\frac{1}{k^2} = 0.1$$
$$k^2 = 10$$
$$k = \sqrt{10}$$
$$= 3.16$$

The interval

$$\mu - 3.16\sigma \text{ to } \mu + 3.16\sigma$$

or

$$120 - 3.16(10) \text{ to } 120 + 3.16(10)$$

or

$$88.4 \text{ to } 151.6$$

will then contain at least 90% of the daily production levels.   ▪

E X A M P L E   **3.6**   The daily cost for use of a certain tool has a mean value of $13 and a variance of 41 (see Example 3.3). How often will this cost exceed $30?

Solution   First, we must find the distance between the mean and 30, in terms of the standard deviation of the distribution of costs. We have

$$\frac{30 - \mu}{\sqrt{\sigma^2}} = \frac{30 - 13}{\sqrt{41}} = \frac{17}{6.4} = 2.66$$

Thus, 30 is 2.66 standard deviations above the mean. Letting $k = 2.66$ in Theorem 3.4, we can find that the interval

$$\mu - 2.66\sigma \text{ to } \mu + 2.66\sigma$$

or

$$13 - 2.66(6.4) \text{ to } 13 + 2.66(6.4)$$

or

$$-4 \text{ to } 30$$

must contain at least

$$1 - \frac{1}{k^2} = 1 - \frac{1}{(2.66)^2} = 1 - 0.14 = 0.86$$

of the probability. Since the daily cost cannot be negative, at most 0.14 of the probability mass can exceed 30. Thus the cost cannot exceed $30 more than 14% of the time.   ∎

Exercises

3.11   You are to pay $1.00 to play a game that consists of drawing one ticket at random from a box of unnumbered tickets. You win the amount (in dollars) of the number on the ticket you draw. The following two boxes of numbered tickets are available:

I | 0, 1, 2 | II | 0, 0, 0, 1, 4 |

a   Find the expected value and variance of your *net* gain per play with box I.

b   Repeat part (a) for box II.

c   Given that you have decided to play, which box should you choose, and why?

Size Distribution of U.S. Families

Number of Persons	Percentage
1	24.5%
2	32.3
3	17.3
4	15.5
6	2.3
7 or more	1.4

Source: U.S. Bureau of the Census.

3.12　The size distribution of U.S. families is shown in the accompanying table.

　　a　Calculate the mean and the standard deviation of family size. Are these exact values or approximations?

　　b　How does the mean family size compare to the median family size?

3.13　The accompanying graph shows the age distribution for AIDS deaths in the United States, through 1989. Approximate the mean and the standard deviation of this age distribution. How does the mean age compare to the approximate median age?

Distribution of AIDS Deaths (by age), 1982–1989

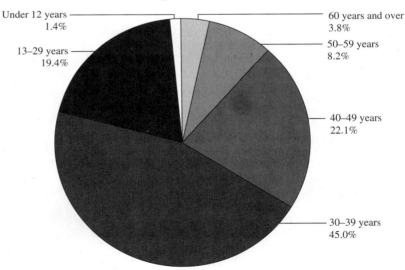

77,350 Deaths

Under 12 years — 1.4%

13–29 years — 19.4%

60 years and over — 3.8%

50–59 years — 8.2%

40–49 years — 22.1%

30–39 years — 45.0%

Source: Chart prepared by U.S. Bureau of the Census.

3.14   How old are our drivers? The accompanying table gives the age distribution of licensed drivers in the United States. Describe this age distribution in terms of median, mean, and standard deviation.

Licensed U.S. Drivers in 1989

Age	Number (in millions)
19 and under	9.7
20–24	16.9
25–29	20.6
30–34	20.5
35–39	18.6
40–44	16.1
45–49	12.6
50–54	10.2
55–59	9.5
60–64	9.3
65–69	8.3
70 and over	13.3
Total	165.6

Source: Statistical Abstract of the United States.

3.15   Daily sales records for a computer manufacturing firm show that it will sell zero, one, or two mainframe computer systems with the following probabilities:

Number of Sales	0	1	2
Probability	0.7	0.2	0.1

Find the expected value, the variance, and the standard deviation for daily sales.

3.16   Approximately 10% of the glass bottles coming off a production line have serious defects in the glass. Two bottles are randomly selected for inspection. Find the expected value and the variance of the number of inspected bottles with serious defects.

3.17   Two construction contracts are to be randomly assigned to one or more of three firms—I, II, and III. A firm may receive more than one contract. If each contract has a potential profit of $90,000, find the expected potential profit for firm I. Then find the expected potential profit for firms I and II together.

3.18   Two balanced coins are tossed. What are the expected value and the variance of the number of heads observed?

3.19   The number of breakdowns by a university computer system is closely monitored by the director of the computing center, since it is critical to the efficient operation of the center. The number averages 4 per week, with a standard deviation of 0.8 per week.

    a   Find an interval that includes at least 90% of the weekly figures for number of breakdowns.

    b   The center director promises that the number of breakdowns will rarely exceed 8 in a one-week period. Is the director safe in making this claim? Why?

3.20   Keeping an adequate supply of spare parts on hand is an important function of the parts department of a large electronics firm. The monthly demand for microcomputer printer boards was studied for some months and found to average 28 with a standard deviation of 4. How many printer boards should be stocked at the beginning of each month to ensure that the demand will exceed the supply with a probability of less than 0.10?

3.21   An important feature of golf cart batteries is the number of minutes they will perform before needing to be recharged. A certain manufacturer advertises batteries that will run, under a 75-amp discharge test, for an average of 100 minutes, with a standard deviation of 5 minutes.

   a   Find an interval that contains at least 90% of the performance periods for batteries of this type.

   b   Would you expect many batteries to die out in less than 80 minutes? Why?

3.22   Costs of equipment maintenance are an important part of a firm's budget. Each visit by a field representative to check out a malfunction in a word-processing system costs $40. The word-processing system is expected to malfunction approximately 5 times per month, and the standard deviation of the number of malfunctions is 2.

   a   Find the expected value and standard deviation of the monthly cost of visits by the field representative.

   b   How much should the firm budget per month to ensure that the costs of these visits are covered at least 75% of the time?

At this point, it may seem that every problem has its own unique probability distribution, and that we must start from basics to construct such a distribution each time a new problem comes up. Fortunately, this is not the case. Certain basic probability distributions can be developed as models for a large number of practical problems. In the remainder of this chapter, we shall consider some fundamental discrete distributions, looking at the theoretical assumptions that underlie these distributions as well as at the means, variances, and applications of the distributions.

3.3   The Bernoulli Distribution

Consider the inspection of a single item taken from an assembly line. Suppose that a 0 is recorded if the item is not defective, and a 1 is recorded if the item is defective. If X is the random variable denoting the condition of the inspected item, then $X = 0$ with probability $(1 - p)$ and $X = 1$ with probability p, where p denotes the probability of observing a defective item. The probability distribution of X, then, is given by

$$p(x) = p^x(1 - p)^{1-x}, \qquad x = 0, 1$$

where $p(x)$ denotes the probability that $X = x$. Such a random variable is said to have a *Bernoulli distribution* or to represent the outcome of a single Bernoulli trial. Any random variable that denotes the presence or absence of a certain condition in an observed phenomenon possesses a distribution of this type. Frequently, one outcome is termed *success* and the other is termed *failure*.

Suppose that we repeatedly observe items of this type, recording a value of X for each item observed. What average of X should we expect to see? By Definition 3.4, the expected value of X is given by

$$E(X) = \sum_x xp(x)$$
$$= 0p(0) + 1p(1)$$
$$= 0(1 - p) + 1(p) = p$$

Thus, if 10% of the items are defective, we should expect to observe an *average* of 0.1 defective items per item inspected. (In other words, we should expect to see one defective item for every ten items inspected.)

For the Bernoulli random variable X, the variance (see Theorem 3.3) is

$$V(X) = E(X^2) - [E(X)]^2$$
$$= \sum_x x^2 p(x) - p^2$$
$$= 0(1 - p) + 1(p) - p^2$$
$$= p - p^2 = p(1 - p)$$

The Bernoulli random variable will be used as a building block to form other probability distributions, such as the binomial distribution of Section 3.4. The properties of the Bernoulli distribution are summarized next.

The Bernoulli Distribution

$$p(x) = p^x(1 - p)^{1-x}, \qquad x = 0, 1 \quad \text{for} \quad 0 \le p \le 1$$
$$E(X) = p \qquad\qquad\qquad V(X) = p(1 - p)$$

3.4   The Binomial Distribution

3.4.1   Probability Function

Instead of inspecting a single item, as we do with the Bernoulli random variable, suppose that we now independently inspect n items and record values for $X_1, X_2, \ldots, X_n$, where $X_i = 1$ if the ith inspected item is defective and $X_i = 0$ otherwise. We, in fact, have observed a sequence of n independent events, each producing Bernoulli random variable. One especially interesting function of $X_1, \ldots, X_n$ is the sum

$$Y = \sum_{i=1}^{n} X_i$$

which denotes the number of defectives among the n sampled items.

We can easily find the probability distribution for Y under the assumption that $P(X_i = 1) = p$, where p remains constant over all trials. For the sake of simplicity, let's look at the specific case of $n = 3$. The random variable Y can then take on four possible values: 0, 1, 2, and 3. For Y to be 0, all three X_i values must be 0. Thus,

$$P(Y = 0) = P(X_1 = 0, \ X_2 = 0, \ X_3 = 0)$$
$$= P(X_1 = 0) P(X_2 = 0) P(X_3 = 0)$$
$$= (1 - p)^3$$

Now if $Y = 1$, then exactly one value of X_i is 1 and the other two are 0. Thus,

$$P(Y = 1) = P[(X_1 = 1, \ X_2 = 0, \ X_3 = 0) \quad \cup \quad (X_1 = 0, \ X_2 = 1, \ X_3 = 0)$$

$$\cup (X_1 = 0, \ X_2 = 0, \ X_3 = 1)]$$
$$= P(X_1 = 1, \ X_2 = 0, \ X_3 = 0) + P(X_1 = 0, \ X_2 = 1, \ X_3 = 0)$$
$$+ P(X_1 = 0, \ X_2 = 0, \ X_3 = 1)$$

(because the three possibilities are mutually exclusive)

$$= P(X_1 = 1) P(X_2 = 0) P(X_3 = 0) + P(X_1 = 0) P(X_2 = 1) P(X_3 = 0)$$
$$+ P(X_1 = 0) P(X_2 = 0) P(X_3 = 1)$$
$$= p(1 - p)^2 + p(1 - p)^2 + p(1 - p)^2$$
$$= 3p(1 - p)^2$$

For $Y = 2$, two values of X_i must be 1 and one must be 0, which can also occur in three mutually exclusive ways. Hence,

$$P(Y = 2) = P(X_1 = 1, \ X_2 = 1, \ X_3 = 0) + P(X_1 = 1, \ X_2 = 0, \ X_3 = 1)$$
$$+ P(X_1 = 0, \ X_2 = 1, \ X_3 = 1)$$
$$= 3p^2(1 - p)$$

The event $Y = 3$ can occur only if all values of X_i are 1, so

$$P(Y = 3) = P(X_1 = 1, \ X_2 = 1, \ X_3 = 1)$$
$$= P(X_1 = 1) P(X_2 = 1) P(X_3 = 1)$$
$$= p^3$$

Notice that the coefficient in each of the expressions for $P(Y = y)$ is the number of ways of selecting y positions, in the sequence, in which to place 1's. Because there are three positions in the sequence, this number amounts to $\binom{3}{y}$. Thus, we can write

$$P(Y = y) = \binom{3}{y} p^y (1 - p)^{3 - y} \qquad y = 0, 1, 2, 3, \ \text{when } n = 3$$

For general values of n, the probability that Y will take on a specific value—say, y—is given by the term $p^y(1 - p)^{n-y}$ multiplied by the number of possible outcomes that result in exactly y defectives being observed. This number, which represents the number of possible ways of selecting y positions for defectives in the n possible positions of the sequence, is given by

$$\binom{n}{y} = \frac{n!}{y!(n - y)!}$$

where $n! = n(n - 1) \cdots 1$ and $0! = 1$. Thus, in general,

$$P(Y = y) = p(y) = \binom{n}{y} p^y (1 - p)^{n-y} \qquad y = 0, 1, \ldots, n$$

The probability distribution just given is referred to as the *binomial distribution*.

> To summarize, a random variable Y possesses a binomial distribution if the following conditions are satisfied:
>
> **1** The experiment consists of a fixed number n of identical trials.
>
> **2** Each trial can result in one of only two possible outcomes, called *success* or *failure*.
>
> **3** The probability of success p is constant from trial to trial.
>
> **4** The trials are independent.
>
> **5** Y is defined to be the number of successes among the n trials.

Many experimental situations involve random variables that can be adequately modeled by the binomial distribution. In addition to the number of defectives in a sample of n items, examples include the number of employees who favor a certain retirement policy out of n employees interviewed, the number of pistons in an eight-cylinder engine that are misfiring, and the number of electronic systems sold this week out of the n that were manufactured. A general formula for $p(x)$ identifies a family of distributions indexed by certain constants called *parameters*. Thus n and p are the parameters of the binomial distribution.

E X A M P L E **3.7** Suppose that a large lot of fuses contains 10% defectives. If four fuses are randomly sampled from the lot, find the probability that exactly one fuse is defective. Find the probability that at least one fuse in the sample of four is defective.

Solution We assume that the four trials are independent and that the probability of observing a defective is the same (0.1) for each trial. This would be approximately true if the lot indeed were *large*. (If the lot contained only a few fuses, removing one fuse would substantially change the probability of observing a defective on the second draw.) Thus the binomial distribution provides a reasonable model for this experiment, and we have (with Y denoting the number of defectives)

$$p(1) = \binom{4}{1}(0.1)^1(0.9)^3 = 0.2916$$

To find $P(Y \geq 1)$, we observe that

$$
\begin{aligned}
P(Y \geq 1) &= 1 - P(Y = 0) = 1 - p(0) \\
&= 1 - \binom{4}{0}(0.1)^0(0.9)^4 \\
&= 1 - (0.9)^4 \\
&= 0.3439 \quad \blacksquare
\end{aligned}
$$

Discrete distributions, like the binomial, can arise in situations where the underlying problem involves a continuous (that is, nondiscrete) random variable. The following example provides an illustration.

E X A M P L E 3.8   In a study of life lengths for a certain type of battery, researchers found that the probability that a battery life X will exceed 4 hours is 0.135. If three such batteries are in use in independently operating systems, find the probability that only one of the batteries will last 4 hours or more.

Solution   Letting Y denote the number of batteries lasting 4 hours or more, we can reasonably assume Y to have a binomial distribution, with $p = 0.135$. Hence,

$$P(Y = 1) = p(1) = \binom{3}{1}(0.135)^1(0.865)^2$$
$$= 0.303 \quad \blacksquare$$

3.4.2 Mean and Variance

There are numerous ways of finding $E(Y)$ and $V(Y)$ for a binomially distributed random variable Y. We might use the basic definition and compute

$$E(Y) = \sum_x yp(y)$$
$$= \sum_{y=0}^{n} y\binom{n}{y}p^y(1-p)^{n-y}$$

but direct evaluation of this expression is a bit tricky. Another approach is to make use of the results on linear functions of random variables, which will be presented in Chapter 5. We shall see in Chapter 5 that, because the binomial Y arose as a sum of independent Bernoulli random variables $X_1, X_2, \ldots, X_n$,

$$E(Y) = E\left[\sum_{i=1}^{n} X_i\right] = \sum_{i=1}^{n} E(X_i)$$
$$= \sum_{i=1}^{n} p$$
$$= np$$

and

$$V(Y) = \sum_{i=1}^{n} V(X_i) = \sum_{i=1}^{n} p(1-p) = np(1-p)$$

E X A M P L E 3.9   Referring to Example 3.7, suppose that the four fuses sampled from the lot were shipped to a customer, before being tested, on a guarantee basis. Assume that the cost of making the shipment good is given by $C = 3Y^2$, where Y denotes the number of defectives in the shipment of four. Find the expected repair cost.

Solution   We know that

$$E(C) = E(3Y^2) = 3E(Y^2)$$

and it now remains for us to find $E(Y^2)$. From Theorem 3.3,

$$V(Y) = E(Y - \mu)^2 = E(Y^2) - \mu^2$$

Since $V(Y) = np(1 - p)$ and $\mu = E(Y) = np$, we see that

$$E(Y^2) = V(Y) + \mu^2$$
$$= np(1 - p) + (np)^2$$

For Example 3.7, $p = 0.1$ and $n = 4$; hence,

$$E(C) = 3E(Y^2) = 3[np(1 - p) + (np)^2]$$
$$= 3[(4)(0.1)(0.9) + (4)^2(0.1)^2]$$
$$= 1.56$$

If the costs were originally expressed in dollars, we could expect to pay an average of $1.56 in repair costs for each shipment of four fuses.   ∎

3.4.3   Applications

Table 2 in the Appendix gives cumulative binomial probabilities for selected values of n and p. The entries in the table are values of

$$\sum_{y=0}^{a} p(y) = \sum_{y=0}^{a} \binom{n}{y} p^y (1 - p)^{n-y}, \qquad a = 0, 1, \ldots, n - 1$$

The following example illustrates the use of Table 2.

EXAMPLE   **3.10**   An industrial firm supplies ten manufacturing plants with a certain chemical. The probability that any one firm will call in an order on a given day is 0.2, and this probability is the same for all ten plants. Find the probability that, on the given day, the number of plants calling in orders is as follows.

1   at most 3

2   at least 3

3   exactly 3

Solution   Let Y denote the number of plants that call in orders on the day in question. If the plants order independently, then Y can be modeled to have a binomial distribution with $p = 0.2$.

1   We then have

$$
\begin{aligned}
P(Y \le 3) &= \sum_{y=0}^{3} p(y) \\
&= \sum_{y=0}^{3} \binom{10}{y} (0.2)^y (0.8)^{10-y} \\
&= 0.879
\end{aligned}
$$

from Appendix Table 2(b), at column $p = 0.2$ and row $a = 3$.

2   Notice that

$$
\begin{aligned}
P(Y \ge 3) &= 1 - P(Y \le 2) \\
&= 1 - \sum_{y=0}^{2} \binom{10}{y} (0.2)^y (0.8)^{10-y} \\
&= 1 - 0.678 = 0.322
\end{aligned}
$$

3   Observe that

$$
\begin{aligned}
P(Y = 3) &= P(Y \le 3) - P(Y \le 2) \\
&= 0.879 - 0.678 = 0.201
\end{aligned}
$$

from the results just established.   ∎

The examples used to this point have specified n and p in order to calculate probabilities or expected values. Sometimes, however, it is necessary to choose n so as to achieve a specified probability. Example 3.11 illustrates the point.

E X A M P L E   **3.11**   The guidance system for a rocket operates correctly with probability p when called upon. Independent but identical backup systems are installed in the rocket so that the probability that at least one system will operate correctly when called upon is no less than 0.99. Let n denote the number of guidance systems in the rocket. How large must n be to achieve the specified probability of at least one guidance system's operating, if

1 $p = 0.9$?

2 $p = 0.8$?

Solution   Let Y denote the number of correctly operating systems. If the systems are identical and independent, Y has a binomial distribution. Thus,

$$
\begin{aligned}
P(Y \ge 1) &= 1 - P(Y = 0) \\
&= 1 - \binom{n}{0} p^0 (1 - p)^n
\end{aligned}
$$

$$= 1 - (1 - p)^n$$

The conditions specify that n must be such that $P(Y \geq 1) = 0.99$ or more.

1 When $p = 0.9$,

$$P(Y \geq 1) = 1 - (1 - 0.9)^n \geq 0.99$$

results in

$$1 - (0.1)^n \geq 0.99$$

or

$$(0.1)^n \leq 1 - 0.99 = 0.01$$

so $n = 2$; that is, installing two guidance systems will satisfy the specifications.

2 When $p = 0.8$,

$$P(Y \geq 1) = 1 - (1 - 0.8)^n \geq 0.99$$

results in

$$(0.2)^n \leq 0.01$$

Now $(0.2)^2 = 0.04$, and $(0.2)^3 = 0.008$, so we must go to $n = 3$ systems to ensure that

$$P(Y \geq 1) = 1 - (0.2)^3 = 0.992 > 0.99$$

Note: We cannot achieve the 0.99 probability exactly, because Y can assume only integer values. ∎

E X A M P L E   **3.12**   Virtually any process can be improved by the use of statistics, including the law. A much-publicized case that involved a debate about probability was the Collins case, which began in 1964. An incident of purse snatching in the Los Angeles area led to the arrest of Michael and Janet Collins. At their trial, an "expert" presented the following probabilities on characteristics possessed by the couple seen running from the crime. The chance that a couple had all of these characteristics together is 1 in 12 million. Since the Collinses had all of the specified characteristics, they must be guilty. What, if anything, is wrong with this line of reasoning?

Man with beard	$\frac{1}{10}$
Blond woman	$\frac{1}{4}$
Yellow car	$\frac{1}{10}$
Woman with ponytail	$\frac{1}{10}$
Man with mustache	$\frac{1}{3}$
Interracial couple	$\frac{1}{1000}$

Solution   First, no background data are offered to support the probabilities used. Second, the six events are not independent of one another and, therefore, the probabilities cannot be multiplied. Third, and most interesting, the wrong question is being addressed. The question of interest is not "What is the probability of finding a couple with these characteristics?" Since one such couple has been found (the Collinses), the proper question is: "What is the probability that *another* such couple exists, given that we have found one?" Here is where the binomial distribution comes into play. In the binomial model, let

n = Number of couples who could have committed the crime

p = Probability that any one couple possesses the six listed characteristics

x = Number of couples who possess the six characteristics

Form the binomial distribution, we know that

$$P(X = 0) = (1 - p)^n$$
$$P(X = 1) = np(1 - p)^{n-1}$$
$$P(X \geq 1) = 1 - (1 - p)^n$$

Then, the answer to the conditional question posed above is

$$P(X > 1 | X \geq 1) = \frac{1 - (1 - p)^n - np(1 - p)^{n-1}}{1 - (1 - p)^n}$$

Substituting $p = 1/12$ million and $n = 12$ million, which are plausible but not well-justified guesses, we get

$$P(X > 1 | X \geq 1) = 0.42$$

So the probability of seeing another such couple, given that we've already seen one, is much larger than the probability of seeing such a couple in the first place. This holds true even if the numbers are dramatically changed. For instance, if n is reduced to 1 million, the conditional probability becomes 0.05, which is still much larger than 1/12 million.

The important lessons illustrated here are that the correct probability question is sometimes difficult to determine and that conditional probabilities are *very* sensitive to the conditions. ■

We shall soon move on to a discussion of other discrete random variables; but the binomial distribution, summarized next, will be used frequently throughout the remainder of the text.

The Binomial Distribution

$$p(y) = \binom{n}{y} p^y (1-p)^{n-y}, \qquad y = 0, 1, \ldots, n \quad \text{for} \quad 0 \le p \le 1$$

$$E(Y) = np \qquad V(Y) = np(1-p)$$

Exercises

3.23   Let X denote a random variable that has a binomial distribution with $p = 0.2$ and $n = 4$. Find the following values.

a $P(X = 2)$

b $p(X \ge 2)$

c $P(X \le 2)$

d $E(X)$

e $V(X)$

3.24   Let X denote a random variable that has a binomial distribution with $p = 0.4$ and $n = 20$. Use Table 2 of the Appendix to evaluate the following probabilities.

a $P(X \le 6)$

b $P(X \ge 12)$

c $P(X = 8)$

3.25   A machine that fills boxes of cereal underfills a certain proportion p. If 25 boxes are randomly selected from the output of this machine, find the probability that no more than 2 boxes are underfilled when

a $p = 0.1$

b $p = 0.2$

3.26   In testing the lethal concentration of a chemical found in polluted water, researchers have determined that a certain concentration will kill 20% of the fish that are subjected to it for 24 hours. If 20 fish are placed for 24 hours in a tank containing this concentration of the chemical, find the probabilities of the following events.

a   Exactly 14 survive.

b   At least 10 survive.

c   At most 16 survive.

3.27   Refer to Exercise 3.26.

a   Find the number expected to survive, out of 20.

b   Find the variance of the number of survivors, out of 20.

3.28   Among persons donating blood to a clinic, 80% have Rh$^+$ blood (that is, the Rhesus factor is present in their blood). Five people donate blood at the clinic on a particular day.

a   Find the probability that at least one of the five does not have the Rh factor.

b   Find the probability that at most four of the five have Rh$^+$ blood.

3.29   The clinic in Exercise 3.28 needs five Rh$^+$ donors on a certain day. How many people must donate blood to have the probability of obtaining blood from at least five Rh$^+$ donors over 0.90?

3.30   The *U.S. Statistical Abstract* reports that the median family income in the United States for 1989 was \$34,200. Among four randomly selected families, find the probabilities of the following events.

a   All four had incomes above \$34,200 in 1989.

b One of the four had an income below $34,200 in 1989.

3.31 According to B. E. Sullivan (1984, p. 119), 55% of U.S. corporations say that one of the most important factors to consider in locating a corporate headquarters is the "quality of life" the locality offers employees. If five firms are contacted by the governor of Florida regarding possible relocation to the state, find the probability that at least three say that quality of life is an important factor in their decision. What assumption have you made in finding this answer?

3.32 Goranson and Hall (1980, pp. 279–80) explain that the probability of detecting a crack in an airplane wing is the product of p_1, the probability of inspecting a plane with a wing crack; p_2, the probability of inspecting the detail in which the crack is located; and p_3, the probability of detecting the damage.

a What assumptions justify the multiplication of these probabilities?

b Suppose that $p_1 = 0.9$, $p_2 = 0.8$, and $p_3 = 0.5$ for a certain fleet of planes. If three planes are inspected from this fleet, find the probability that a wing crack will be detected in at least one of them.

3.33 A missile protection system consists of n radar sets operating independently, each with a probability of 0.9 of detecting an aircraft entering a specified zone. (All radar sets cover the same zone.) If an airplane enters the zone, find the probability that it will be detected under the following conditions.

a $n = 2$ **b** $n = 4$

3.34 Refer to Exercise 3.33. How large must n be to have a 0.99 probability of detecting an aircraft entering the zone?

3.35 A complex electronic system is built with a certain number of backup components in its subsystems. One subsystem has four identical components, each with a probability of 0.2 of failing in less than 1000 hours. The subsystem will operate if any two or more of the four components are operating. Assuming that the components operate independently, find the probabilities of the following events.

a Exactly two of the four components last longer than 1000 hours.

b The subsystem operates for longer than 1000 hours.

3.36 An oil exploration firm is to drill ten wells, each of which has a probability of 0.1 of successfully striking recoverable oil. It costs the firm $10,000 to drill each well. A successful well will bring in oil worth $500,000.

a Find the firm's expected gain from the ten wells.

b Find the standard deviation of the firm's gain.

3.37 A firm sells four items randomly selected from a large lot that is known to contain 10% defectives. Let Y denote the number of defectives among the four sold. The purchaser of the items will return the defectives for repair, and the repair cost is given by

$$C = 3Y^2 + Y + 2$$

Find the expected repair cost.

3.38 From a large lot of new tires, n are to be sampled by a potential buyer, and the number of defectives X is to be observed. If at least one defective is observed in the sample of n, the entire lot is to be rejected by the potential buyer. Find n so that the probability of detecting at least one defective is approximately 0.90 if the following percentages are correct.

a 10% of the lot is defective.

b 5% of the lot is defective.

3.39 Ten motors are packaged for sale in a certain warehouse. The motors sell for $100 each, but a double-your-money-back guarantee is in effect for any defectives the purchaser might receive. Find the expected net gain for the seller if the probability of any one motor's being defective is 0.08. (Assume that the quality of any one motor is independent of the quality of the others.)

3.5 The Geometric Distribution

3.5.1 Probability Function

Suppose that a series of test firings of a rocket engine can be represented by a sequence of independent Bernoulli random variables, with $X_i = 1$ if the ith trial results in a successful firing and with $X_i = 0$ otherwise. Assume that the probability of a successful firing is constant for the trials, and let this probability be denoted by p. For this problem we might be interested in the number of the trial on which the first successful firing occurs. If Y denotes the number of the trial on which the first success occurs, then

$$\begin{aligned} P(Y = y) = p(y) &= P(X_1 = 0,\ X_2 = 0,\ \ldots,\ X_{y-1} = 0,\ X_y = 1) \\ &= P(X_1 = 0)P(X_2 = 0) \cdots P(X_{y-1} = 0)P(X_y = 1) \\ &= (1 - p)^{y-1}p, \qquad y = 1, 2, \ldots \end{aligned}$$

because of the independence of the trials. This formula is referred to as the *geometric probability distribution*. Notice that this random variable can take on a countably infinite number of possible values.

In addition to the rocket-firing example just given, other situations may result in a random variable whose probability can be modeled by a geometric distribution: the number of customers contacted before the first sale is made; the number of years a dam is in service before it overflows; and the number of automobiles going through a radar check before the first speeder is detected. The following example illustrates the use of the geometric distribution.

EXAMPLE **3.13**

A recruiting firm finds that 30% of the applicants for a certain industrial job have received advanced training in computer programming. Applicants are interviewed sequentially and selected at random from the pool. Find the probability that the first applicant who has received advanced training in programming is found on the fifth interview.

Solution

The probability of finding a suitably trained applicant will remain relatively constant from trial to trial if the pool of applicants is reasonably large. It then makes sense to define Y as the number of the trial on which the first applicant having advanced training in programming is found, and to model Y as having a geometric distribution. Thus,

$$P(Y = 5) = p(5) = (0.7)^4(0.3)$$
$$= 0.072 \quad \blacksquare$$

3.5.2 Mean and Variance

From the basic definition,

$$E(Y) = \sum_y yp(y) = \sum_{y=1}^{\infty} yp(1-p)^{y-1}$$

$$= p \sum_{y=1}^{\infty} y(1-p)^{y-1}$$

$$= p[1 + 2(1-p) + 3(1-p)^2 + \cdots]$$

The infinite series can be split up into a triangular array of series as follows:

$$E(Y) = p[1 + (1-p) + (1-p)^2 + \cdots$$
$$+(1-p) + (1-p)^2 + \cdots$$
$$+(1-p)^2 + \cdots$$
$$+\cdots]$$

Each line on the right side is an infinite, decreasing geometric progression with common ratio $(1-p)$. Thus, the first line inside the bracket sums to $1/p$; the second, to $(1-p)/p$; the third, to $(1-p)^2/p$; and so on.* On accumulating these totals, we then have

$$E(Y) = p\left[\frac{1}{p} + \frac{1-p}{p} + \frac{(1-p)^2}{p} + \cdots\right]$$

$$= 1 + (1-p) + (1-p)^2 + \cdots$$

$$= \frac{1}{1-(1-p)}$$

$$= \frac{1}{p}$$

This answer for $E(Y)$ should seem intuitively realistic. For example, if 10% of a certain lot of items are defective, and if an inspector looks at randomly selected items one at a time, she should expect to wait until the tenth trial to see the first defective item.

The variance of the geometric distribution will be derived in Section 3.9 and in Chapter 5. The result, however, is

$$V(Y) = \frac{1-p}{p^2}$$

E X A M P L E **3.14** Referring to Example 3.13, let Y denote the number of the interview on which the first applicant having advanced training in computer programming is found. Suppose that the first applicant with advanced training is offered the position, and the applicant

---

*Recall that $a + ax + ax^2 + ax^3 + \cdots = a/(1-x)$, if $|x| < 1$.

accepts. If each interview costs $30.00, find the expected value and the variance of the total cost of interviewing until the job is filled. Within what interval should this cost be expected to fall?

Solution   Because Y is the number of the trial on which the interviewing process ends, the total cost of interviewing is $C = 30Y$. Now,

$$
\begin{aligned}
E(C) &= 30E(Y) \\
&= 30\left(\frac{1}{p}\right) \\
&= 30\left(\frac{1}{0.3}\right) \\
&= 100
\end{aligned}
$$

and

$$
\begin{aligned}
V(C) &= (30)^2 V(Y) \\
&= \frac{900(1-p)}{p^2} \\
&= \frac{900(0.7)}{(0.3)^2} \\
&= 7000
\end{aligned}
$$

The standard deviation of C is then $\sqrt{V(C)} = \sqrt{7000} = 83.67$. Tchebysheff's Theorem (see Section 3.2) says that C will lie within two standard deviations of its mean at least 75% of the time. Thus, it is quite likely that C will be between

$$100 - 2(83.67) \quad \text{and} \quad 100 + 2(83.67)$$

or

$$-67.34 \quad \text{and} \quad 267.34$$

Since the lower bound is negative, that end of the interval is meaningless. Nonetheless, we can say that the total cost of such an interviewing process is quite likely to be less than $267.34.   ▪

The Geometric Distribution

$$
p(y) = p(1-p)^{y-1}, \qquad y = 1, 2, \ldots \quad \text{for} \quad 0 < p < 1
$$
$$
E(Y) = \frac{1}{p} \qquad\qquad\qquad V(Y) = \frac{1-p}{p^2}
$$

3.6   The Negative Binomial Distribution

3.6.1   Probability Function

In Section 3.5, we saw that the geometric distribution models the probabilistic behavior of the number of the trial on which the *first success* occurs in a sequence of independent Bernoulli trials. But what if we were interested in the number of the trial for the second success, or the third success, or (in general) the rth success? The distribution governing probabilistic behavior in these cases is called the *negative binomial distribution*.

Let Y denote the number of the trial on which the rth success occurs in a sequence of independent Bernoulli trials, with p denoting the common probability of success. We can derive the distribution of Y from known facts. Now,

$$P(Y = y) = P[(1\text{st }(y-1)\text{ trials contain }(r-1)\text{ successes and }y\text{th trial is a success}]$$
$$= P[(1\text{st }(y-1)\text{ trials contain }(r-1)\text{ successes}] \times P[y\text{th trial is a success}]$$

Because the trials are independent, the joint probability can be written as a product of probabilities. The first probability statement is identical to the one that results in a binomial model; and hence,

$$P(Y = y) = p(y)$$
$$= \binom{y-1}{r-1} p^{r-1}(1-p)^{y-r} \times p$$
$$= \binom{y-1}{r-1} p^{r}(1-p)^{y-r} \qquad y = r, r+1, \ldots$$

E X A M P L E   **3.15**   As in Example 3.13, 30% of the applicants for a certain position have received advanced training in computer programming. Suppose that three jobs that require advanced programming training are open. Find the probability that the third qualified applicant is found on the fifth interview, if the applicants are interviewed sequentially and at random.

Solution   Again we assume independent trials, with 0.3 being the probability of finding a qualified candidate on any one trial. Let Y denote the number of the trial on which the third qualified candidate is found. Y can reasonably be assumed to have a negative binomial distribution, so

$$P(Y = 5) = p(5) = \binom{4}{2}(0.3)^3(0.7)^2$$
$$= 6(0.3)^3(0.7)^2$$
$$= 0.079 \quad \blacksquare$$

3.6.2   Mean and Variance

The expected value, or mean, and the variance for the negative binomially distributed Y can easily be found by analogy with the geometric distribution. Recall that Y denotes the number of the trial on which the rth success occurs. Let W_1 denote the number of the trial on which the first success occurs; let W_2 denote the number of trials between the first success and the second success, including the trial of the second success; let W_3 denote the number of trials between the second success and the third success; and so forth. The results of the trials can then be represented as follows (where F stands for failure and S stands for success):

$$\underbrace{FF \cdots FS}_{W_1}, \quad \underbrace{F \cdots FS}_{W_2}, \quad \underbrace{F \cdots FS}_{W_3}$$

Clearly, $Y = \sum_{i=1}^{r} W_i$, where the W_i values are independent and each has a geometric distribution. Thus, by results to be derived in Chapter 5,

$$E(Y) = \sum_{i=1}^{r} E(W_i) = \sum_{i=1}^{r} \left(\frac{1}{p} \right) = \frac{r}{p}$$

and

$$V(Y) = \sum_{i=1}^{r} V(W_i) = \sum_{i=1}^{r} \frac{1-p}{p^2} = \frac{r(1-p)}{p^2}$$

EXAMPLE   **3.16**   A large stockpile of used pumps contains 20% that are currently unusable and need to be repaired. A repairman is sent to the stockpile with three repair kits. He selects pumps at random and tests them one at a time. If a pump works, he goes on to the next one. If a pump doesn't work, he uses one of his repair kits on it. Suppose that it takes 10 minutes to test whether a pump works, and 20 minutes to repair a pump that does not work. Find the expected value and variance of the total time it takes the repairman to use up his three kits.

Solution   Letting Y denote the number of the trial on which the third defective pump is found, we see that Y has a negative binomial distribution, with $p = 0.2$. The total time T taken to use up the three repair kits is therefore

$$T = 10(Y - 3) + 3(30) = 10Y + 3(20)$$

(Each test takes 10 minutes, but the repairs take 20 extra minutes.)
    It follows that

$$\begin{aligned}
E(T) &= 10E(Y) + 3(20) \\
&= 10 \left(\frac{3}{0.2} \right) + 3(20) \\
&= 150 + 60 \\
&= 210
\end{aligned}$$

and

$$V(T) = (10)^2 V(Y)$$
$$= (10)^2 \left[\frac{3(0.8)}{(0.2)^2} \right]$$
$$= 100[60]$$
$$= 6000$$

Thus, the total time needed to use up the kits has an expected value of 210 minutes, with a standard deviation of $\sqrt{6000} = 77.46$ minutes.   ∎

The negative binomial distribution is used to model a wide variety of phenomena, from the number of defects per square yard in fabrics to the number of individuals in an insect population after many generations.

y = trials
r = success

The Negative Binomial Distribution

$$p(y) = \binom{y-1}{r-1} p^r (1-p)^{y-r}, \qquad y = r, r+1, \ldots \quad \text{for} \quad 0 < p < 1$$

$$E(Y) = \frac{r}{p} \qquad\qquad\qquad V(Y) = \frac{r(1-p)}{p^2}$$

Exercises

3.40   Let Y denote a random variable that has a geometric distribution, with a probability of success on any trial denoted by p.

a   Find $P(Y \geq 2)$ if $p = 0.1$.

b   Find $P(Y > 4 | Y > 2)$ for general p. Compare this result with the unconditional probability $P(Y > 2)$. [This property is referred to as "lack of memory."]

3.41   Let Y denote a negative binomial random variable, with $p = 0.4$. Find $P(Y \geq 4)$ if the following values for r exist.

a $r = 2$ **b** $r = 4$

3.42   Suppose that 10% of the engines manufactured on a certain assembly line are defective. If engines are randomly selected one at a time and tested, find the probability that the first nondefective engine will be found on the second trial.

3.43   Referring to Exercise 3.42, find the probability that the third nondefective engine will be found as follows.

a   On the fifth trial.   **b**   On or before the fifth trial.

3.44   Referring to Exercise 3.42, given that the first two engines are defective, find the probability that at least two more engines must be tested before the first nondefective engine is found.

3.45   Referring to Exercise 3.42, find the mean and the variance of the number of the trial on which the following events occur.

    a   The first nondefective engine is found.

    b   The third nondefective engine is found.

3.46   The employees of a firm that manufactures insulation are being tested for indications of asbestos in their lungs. The firm is asked to send three employees who have positive indications of asbestos on to a medical center for further testing. If 40% of the employees have positive indications of asbestos in their lungs, find the probability that ten employees must be tested to find three positives.

3.47   Referring to Exercise 3.46, if each test costs $20, find the expected value and the variance of the total cost of conducting the tests to locate three positives. Is it highly likely that the cost of completing these tests will exceed $350?

3.48   If one-third of the persons donating blood at a clinic have O^+ blood, find the probabilities of the following events.

    a   The first O^+ donor is the fourth donor of the day.

    b   The second O^+ donor is the fourth donor of the day.

3.49   A geological study indicates that an exploratory oil well drilled in a certain region should strike oil with probability 0.2. Find the probabilities of the following events.

    a   The first strike of oil comes on the third well drilled.

    b   The third strike of oil comes on the fifth well drilled.

    What assumptions must be true for your answers to be correct?

3.50   In the setting of Exercise 3.49, suppose that a company wants to set up three producing wells. Find the expected value and the variance of the number of wells that must be drilled to find three successful ones.

3.51   A large lot of tires contains 10% defectives. Four tires are to be chosen from the lot and placed on a car.

    a   Find the probability that six tires must be selected from the lot to get four good ones.

    b   Find the expected value and the variance of the number of selections that must be made to get four good tires.

3.52   The telephone lines coming into an airline reservation office are all occupied about 60% of the time.

    a   If you are calling this office, what is the probability that you will complete your call on the first try? the second try? the third try?

    b   If both you and a friend must complete separate calls to this reservation office, what is the probability that it will take a total of four tries for the two of you?

3.53   An appliance comes in two colors, white and brown, which are in equal demand. A certain dealer in these appliances has three of each color in stock, although this is not known to the customers. Customers arrive and independently order these appliances. Find the probabilities of the following events.

    a   The third white appliance is ordered by the fifth customer.

    b   The third brown appliance is ordered by the fifth customer.

    c   All of the white appliances are ordered before any of the brown appliances.

    d   All of the white appliances are ordered before all of the brown appliances.

3.7 The Poisson Distribution

3.7.1 Probability Function

A number of probability distributions come about through limiting arguments applied to other distributions. One very useful distribution of this type is called the *Poisson distribution*.

Consider the development of a probabilistic model for the number of accidents that occur at a particular highway intersection in a period of one week. We can think of the time interval as being split up into n subintervals such that

$$P(\text{One accident in a subinterval}) = p$$
$$P(\text{No accidents in a subinterval}) = 1 - p$$

Here we are assuming that the same value of p holds for all subintervals, and that the probability of more than one accident's occurring in any one subinterval is zero. If the occurrence of accidents can be regarded as independent from subinterval to subinterval, the total number of accidents in the time period (which equals the total number of subintervals that contain one accident) will have a binomial distribution.

Although there is no unique way to choose the subintervals—and we therefore know neither n nor p—it seems reasonable to assume that, as n increases, p should decrease. Thus, we want to look at the limit of the binomial probability distribution as $n \to \infty$ and $p \to 0$. To get something interesting, we take the limit under the restriction that the mean (np in the binomial case) remains constant at a value we will call λ.

Now, with $np = \lambda$ or $p = \lambda/n$, we have

$$\lim_{n \to \infty} \binom{n}{y}\left(\frac{\lambda}{n}\right)^y\left(1 - \frac{\lambda}{n}\right)^{n-y}$$

$$= \lim_{n \to \infty} \frac{\lambda^y}{y!}\left(1 - \frac{\lambda}{n}\right)^n \frac{n(n-1)\cdots(n-y+1)}{n^y}\left(1 - \frac{\lambda}{n}\right)^{-y}$$

$$= \frac{\lambda^y}{y!} \lim_{n \to \infty}\left(1 - \frac{\lambda}{n}\right)^n\left(1 - \frac{\lambda}{n}\right)^{-y}\left(1 - \frac{1}{n}\right)\left(1 - \frac{2}{n}\right)\cdot\left(1 - \frac{y-1}{n}\right)$$

Noting that

$$\lim_{n \to \infty}\left(1 - \frac{\lambda}{n}\right)^n = e^{-\lambda}$$

and that all other terms involving n tend toward unity, we have the limiting distribution

$$p(y) = \frac{\lambda^y}{y!}e^{-\lambda}, \qquad y = 0, 1, 2, \ldots$$

Recall that λ denotes the mean number of occurrences in one time period (a week, for the example under consideration); hence if t nonoverlapping time periods were considered, the mean would be λt.

This distribution, called the *Poisson distribution with parameter* λ, can be used to model counts in areas or volumes, as well as in time. For example, we may use this distribution to model the number of flaws in a square yard of textile, the number of bacteria colonies in a cubic centimeter of water, or the number of times a machine fails in the course of a workday. We illustrate the use of the Poisson distribution in the following example.

E X A M P L E   **3.17**   For a certain manufacturing industry, the number of industrial accidents averages three per week. Find the probability that no accidents will occur in a given week.

Solution   If accidents tend to occur independently of one another, and if they occur at a constant rate over time, the Poisson model provides an adequate representation of the probabilities. Thus,

$$p(0) = \frac{3^0}{0!}e^{-3} = e^{-3} = 0.05 \quad \blacksquare$$

Table 3 of the Appendix gives values for cumulative Poisson probabilities of the form

$$\sum_{y=0}^{a} e^{-\lambda}\frac{\lambda^y}{y!}, \qquad a = 0, 1, 2, \ldots$$

for selected values of λ.

The following example illustrates the use of Table 3.

E X A M P L E   **3.18**   Refer to Example 3.17, and let Y denote the number of accidents in the given week. Find $P(Y \leq 4)$, $P(Y \geq 4)$, and $P(Y = 4)$.

Solution   From Table 3,

$$P(Y \leq 4) = \sum_{y=0}^{4} \frac{(3)^y}{y!}e^{-3} = 0.815$$

Also,

$$P(Y \geq 4) = 1 - P(Y \leq 3) = 1 - 0.647 = 0.353$$

and

$$P(Y = 4) = P(Y \leq 4) - P(Y \leq 3) = 0.815 - 0.647 = 0.168 \quad \blacksquare$$

3.7.2   Mean and Variance

We can intuitively determine what the mean and the variance of a Poisson distribution should be by recalling the mean and the variance of a binomial distribution and the relationship between the two distributions. A binomial distribution has mean np and variance $np(1 - p) = np - (np)p$. Now, if n gets large and p remains at $np = \lambda$, the variance $np - (np)p = \lambda - \lambda p$ should tend toward λ. In fact, the Poisson distribution does have both its mean and its variance equal to λ.

The mean of the Poisson distribution can easily be derived formally if one remembers a simple Taylor series expansion of e^x—namely,

$$e^x = 1 + x + \frac{x^2}{2!} + \frac{x^3}{3!} + \cdots$$

Then,

$$
\begin{aligned}
E(Y) &= \sum_y yp(y) \\
&= \sum_{y=0}^{\infty} y \frac{\lambda^y}{y!} e^{-\lambda} \\
&= \sum_{y=1}^{\infty} y \frac{\lambda^y}{y!} e^{-\lambda} \\
&= \lambda e^{-\lambda} \sum_{y=1}^{\infty} \frac{\lambda^{y-1}}{(y-1)!} \\
&= \lambda e^{-\lambda} \left(1 + \lambda + \frac{\lambda^2}{2!} + \frac{\lambda^3}{3!} + \cdots \right) \\
&= \lambda e^{-\lambda} e^{\lambda} \\
&= \lambda
\end{aligned}
$$

The formal derivation of the fact that

$$V(Y) = \lambda$$

is left as a challenge for the interested reader. [*Hint:* First, find $E[Y(Y-1)]$.]

E X A M P L E   **3.19**   The manager of an industrial plant is planning to buy a new machine of either type A or type B. For each day's operation, the number of repairs X that machine A requires is a Poisson random variable with mean $0.10t$, where t denotes the time (in hours) of daily operation. The number of daily repairs Y for machine B is a Poisson random variable with mean $0.12t$. The daily cost of operating A is $C_A(t) = 10t + 30X^2$; for B, the cost is $C_B(t) = 8t + 30Y^2$. Assume that the repairs take negligible time and that each night the machines are to be cleaned, so that they operate like new machines at the start of each day. Which machine minimizes the expected daily cost if a day consists of the following time spans?

1 10 hours

2 20 hours

Solution The expected cost for machine A is

$$E[C_A(t)] = 10t + 30E(X^2)$$
$$= 10t + 30[V(X) + (E(X))^2]$$
$$= 10t + 30[0.10t + 0.01t^2]$$
$$= 13t + 0.3t^2$$

Similarly,

$$E[C_B(t)] = 8t + 30E(Y^2)$$
$$= 8t + 30[0.12t + 0.0144t^2]$$
$$= 11.6t + 0.432t^2$$

1 Here,

$$E[C_A(10)] = 13(10) + 0.3(10)^2 = 160$$

and

$$E[C_B(10)] = 11.6(10) + 0.432(10)^2 = 159.2$$

which results in the choice of machine B.

2 Here,

$$E[C_A(20)] = 380$$

and

$$E[C_B(20)] = 404.8$$

which results in the choice of machine A.

In conclusion, B is more economical for short time periods because of its smaller hourly operating cost. For long time periods, however, A is more economical because it tends to be repaired less frequently. ∎

The Poisson Distribution

$$p(y) = \frac{\lambda^y}{y!}e^{-\lambda}, \qquad y = 0, 1, 2, \ldots$$
$$E(Y) = \lambda \qquad V(Y) = \lambda$$

Exercises

3.54   Let Y denote a random variable that has a Poisson distribution with mean $\lambda = 2$. Find the following probabilities.

    a $P(Y = 4)$ **b** $P(Y \geq 4)$

    c $P(Y < 4)$ **d** $P(Y \geq 4 | Y \geq 2)$

3.55   The number of telephone calls coming into the central switchboard of an office building averages four per minute.

    a   Find the probability that no calls will arrive in a given one-minute period.

    b   Find the probability that at least two calls will arrive in a given one-minute period.

    c   Find the probability that at least two calls will arrive in a given two-minute period.

3.56   The quality of computer disks is measured by sending the disks through a certifier, which counts the number of missing pulses. A certain brand of computer disks has averaged 0.1 missing pulse per disk.

    a   Find the probability that the next inspected disk will have no missing pulse.

    b   Find the probability that the next inspected disk will have more than one missing pulse.

    c   Find the probability that neither of the next two inspected disks will contain any missing pulse.

3.57   The national maximum speed limit (NMSL) of 55 miles per hour was imposed in the United States in early 1974. The benefits of this law have been studied by D. B. Kamerud (1983), who reported that the fatality rate for interstate highways with the NMSL in 1975 was approximately 16 per 10^9 vehicle miles.

    a   Find the probability that at most 15 fatalities occurred in a given block of 10^9 vehicle miles.

    b   Find the probability that at least 20 fatalities occurred in a given block of 10^9 vehicle miles.

    (Assume that fatalities per vehicle mile follows a Poisson distribution.)

3.58   In the article cited in Exercise 3.57, the projected fatality rate for 1975 if the NMSL had not been in effect was 25 per 10^9 vehicle miles. Assume that these conditions had prevailed.

    a   Find the probability that at most 15 fatalities occurred in a given block of 10^9 vehicle miles.

    b   Find the probability that at least 20 fatalities occurred in a given block of 10^9 vehicle miles.

    c   Compare your answers in parts (a) and (b) to the corresponding answers in Exercise 3.55.

3.59   In a timesharing computer system, the number of teleport inquiries averages 0.2 per millisecond and follows a Poisson distribution.

    a   Find the probability that no inquiries are made during the next millisecond.

    b   Find the probability that no inquiries are made during the next 3 milliseconds.

3.60   Rebuilt ignition systems leave an aircraft rework facility at a rate of three per hour, on average. The assembly line needs four ignition systems in the next hour. What is the probability that they will be available?

3.61   Customer arrivals at a checkout counter in a department store have a Poisson distribution with an average of eight per hour. For a given hour, find the probabilities of the following events.

    a   Exactly eight customers arrive.

    b   No more than three customers arrive.

    c   At least two customers arrive.

3.62   Referring to Exercise 3.61, if it takes approximately 10 minutes to service each customer, find the mean and the variance of the total service time connected to the customer arrivals for one hour. (Assume that an unlimited number of servers are available, so that no customer has to wait for service.) Is total service time highly likely to exceed 200 minutes?

3.63 Referring to Exercise 3.61, find the probabilities that exactly two customers will arrive in the following two-hour periods of time.

a Between 2:00 P.M. and 4:00 P.M. (one continuous two-hour period).

b Between 1:00 P.M. and 2:00 P.M. and between 3:00 P.M. and 4:00 P.M. (two separate one-hour periods for a total of two hours).

3.64 The number of imperfections in the weave of a certain textile has a Poisson distribution with a mean of four per square yard.

a Find the probability that a 1-square-yard sample will contain at least one imperfection.

b Find the probability that a 3-square-yard sample will contain at least one imperfection.

3.65 Referring to Exercise 3.64, the cost of repairing the imperfections in the weave is $10 per imperfection. Find the mean and the standard deviation of the repair costs for an 8-square-yard bolt of the textile in question.

3.66 The number of bacteria colonies of a certain type in samples of polluted water has a Poisson distribution with a mean of two per cubic centimeter.

a If four 1-cubic-centimeter samples of this water are independently selected, find the probability that at least one sample will contain one or more bacteria colonies.

b How many 1-cubic-centimeter samples should be selected to establish a probability of approximately 0.95 of containing at least one bacteria colony?

3.67 Let Y have a Poisson distribution with mean λ. Find $E[Y(Y-1)]$, and use the result to show that $V(Y) = \lambda$.

3.68 A food manufacturer uses an extruder (a machine that produces bite-size foods like cookies and many snack foods) that has a revenue-producing value to the firm of $200 per hour when it is in operation. However, the extruder breaks down an average of twice every 10 hours of operation. If Y denotes the number of breakdowns during the time of operation, the revenue generated by the machine is given by

$$R = 200t - 50Y^2$$

where t denotes hours of operation. The extruder is shut down for routine maintenance on a regular schedule, and it operates like a new machine after this maintenance. Find the optimal maintenance interval t_0 to maximize the expected revenue between shutdowns.

3.69 The number of cars entering a parking lot is a random variable that has a Poisson distribution, with a mean of four per hour. The lot holds only twelve cars.

a Find the probability that the lot will fill up in the first hour. (Assume that all cars stay in the lot longer than one hour.)

b Find the probability that fewer than twelve cars will arrive during an eight-hour day. ■

3.8 The Hypergeometric Distribution

3.8.1 The Probability Function

The distributions already discussed in this chapter have as their basic building block a series of *independent* Bernoulli trials. The examples, such as sampling from large lots, depict situations in which the trials of the experiment generate, for all practical purposes, independent outcomes. But suppose that we have a relatively small lot consisting of N items, of which k are defective. If two items are sampled sequentially, the outcome for the second draw is very significantly influenced by what happened on the first draw, provided that the first item drawn remains out of the lot. A new distribution must be developed to handle this situation involving *dependent* trials.

In general, suppose that a lot consists of N items, of which k are of one type (called *successes*) and $N - k$ are of another type (called *failures*). Suppose that n items are sampled randomly and sequentially from the lot, and suppose that none of the sampled items is replaced. (This is called *sampling without replacement*.) Let $X_i = 1$ if the ith draw results in a success, and let $X_i = 0$ otherwise, where $i = 1, \ldots, n$. Let Y denote the total number of successes among the n sampled items. To develop the probability distribution for Y, let us start by looking at a special case for $Y = y$. One way for y successes to occur is to have

$$X_1 = 1, \quad X_2 = 1, \quad \ldots, \quad X_y = 1, \quad X_{y+1} = 0, \quad \ldots, \quad X_n = 0$$

We know that

$$P(X_1 = 1, \quad X_2 = 1) = P(X_1 = 1)P(X_2 = 1 | X_1 = 1)$$

and this result can be extended to give

$$P(X_1 = 1, \quad X_2 = 1, \quad \ldots, \quad X_y = 1, \quad X_{y+1} = 0, \quad \ldots, \quad X_n = 0)$$
$$= P(X_1 = 1)P(X_2 = 1 | X_1 = 1)P(X_3 = 1 | X_2 = 1, \quad X_1 = 1) \cdots$$
$$P(X_n = 0 | X_{n-1} = 0, \quad \ldots, \quad X_{y+1} = 0, \quad X_y = 1, \quad \ldots, \quad X_1 = 1)$$

Now,

$$P(X_1 = 1) = \frac{k}{N}$$

if the item is randomly selected; and similarly,

$$P(X_2 = 1 | X_1 = 1) = \frac{k - 1}{N - 1}$$

because, at this point, one of the k successes has been removed. Using this idea repeatedly, we see that

$$P(X_1 = 1, \ldots, X_y = 1, X_{y+1} = 0, \ldots, X_n = 0)$$
$$= \left(\frac{k}{N} \right) \left(\frac{k - 1}{N - 1} \right) \cdots \left(\frac{k - y + 1}{N - y + 1} \right) \times \left(\frac{N - k}{N - y} \right) \cdots \left(\frac{N - k - n + y + 1}{N - n + 1} \right)$$

provided that $y \le k$. A more compact way to write the preceding expression is to employ factorials, arriving at the formula

$$\frac{\frac{k!}{(k-y)!} \times \frac{(N-k)!}{(N-k-n+y)!}}{\frac{N!}{(N-n)!}}$$

(The reader can check the equivalence of the two expressions.)

Any specified arrangement of y successes and $(n - y)$ failures will have the same probability as the one just derived for all successes followed by all failures; the terms will merely be rearranged. Thus, to find $P(Y = y)$, we need only count how many of these arrangements are possible. Just as in the binomial case, the number of such arrangements is $\binom{n}{y}$. Hence, we have

$$P(Y = y) = \binom{n}{y} \frac{\frac{k!}{(k-y)!} \times \frac{(N-k)!}{(N-k-n+y)!}}{\frac{N!}{(N-n)!}}$$

$$= \frac{\binom{k}{y}\binom{N-k}{n-y}}{\binom{N}{n}}$$

Of course, $0 \le y \le k \le N$ and $0 \le y \le n \le N$. This formula is referred to as the *hypergeometric probability distribution*. Notice that it arises from a situation quite similar to the binomial, except that the trials here are *dependent*.

Experiments that result in a random variable's possessing a hypergeometric distribution usually involve counting the number of successes in a sample taken from a small lot. An example might be counting the number of men who show up on a committee of five randomly selected from among twenty employees or counting the number of Brand A alarm systems sold in three sales from a warehouse containing two Brand A and four Brand B systems.

EXAMPLE **3.20**   A personnel director selects two employees for a certain job from a group of six employees, of which one is female and five are male. Find the probability that the woman is selected for one of the jobs.

Solution   If the selections are made at random, and if Y denotes the number of women selected, then the hypergeometric distribution would provide a good model for the behavior of Y. Hence,

$$P(Y = 1) = p(1) = \frac{\binom{1}{1}\binom{5}{1}}{\binom{6}{2}} = \frac{1 \times 5}{15} = \frac{1}{3}$$

Here, $N = 6, k = 1, n = 2$, and $y = 1$.

It might be instructive to see this calculation from basic principles, letting $X_i = 1$ if the ith draw results in selection of the woman, and letting $X_i = 0$ otherwise. Then,

$$P(Y = 1) = P(X_1 = 1, \quad X_2 = 0) + P(X_1 = 0, \quad X_2 = 1)$$
$$= P(X_1 = 1)P(X_2 = 0|X_1 = 1)$$
$$\quad + P(X_1 = 0)P(X_2 = 1|X_1 = 0)$$
$$= \left(\frac{1}{6}\right)\left(\frac{5}{5}\right) + \left(\frac{5}{6}\right)\left(\frac{1}{5}\right) = \frac{1}{3} \quad \blacksquare$$

3.8.2   Mean and Variance

The techniques needed to derive the mean and the variance of the hypergeometric distribution will be given in Chapter 5. The results are

$$E(Y) = n\left(\frac{k}{N}\right)$$

$$V(Y) = n\left(\frac{k}{N}\right)\left(1 - \frac{k}{N}\right)\left(\frac{N-n}{N-1}\right) .$$

Because the probability of selecting a success on one draw is k/N, the mean of the hypergeometric distribution has the same form as the mean of the binomial distribution. Likewise, the variance of the hypergeometric matches the variance of the binomial, multiplied by $(N-n)/(N-1)$, a correction factor for dependent samples.

E X A M P L E   **3.21**   In an assembly-line production of industrial robots, gear-box assemblies can be installed in 1 minute each if the holes have been properly drilled in the boxes, and in 10 minutes each if the holes must be redrilled. Twenty gear boxes are in stock, and it is assumed that two will possess improperly drilled holes. Five gear boxes must be selected from the twenty available for installation in the next five robots in line.

1   Find the probability that all five gear boxes will fit properly.

2   Find the expected value, the variance, and the standard deviation of the time it will take to install these five gear boxes.

Solution   1   In this problem, $N = 20$; and the number of nonconforming boxes is assumed to be $k = 2$, according to the manufacturer's usual standards. Let Y denote the number of nonconforming boxes (that is, the number with improperly drilled holes) in the sample of five. Then,

$$P(Y = 0) = \frac{\binom{2}{0}\binom{18}{5}}{\binom{20}{5}}$$

$$= \frac{(1)(8568)}{15,504}$$

$$= 0.55$$

2   The total time T taken to install the boxes (in minutes) is

$$T = 10Y + (5 - Y)$$
$$= 9Y + 5$$

since each of Y nonconforming boxes takes 10 minutes to install, and the others take only 1 minute. To find $E(T)$ and $V(T)$, we first need to calculate $E(Y)$ and $V(Y)$:

$$E(Y) = n\left(\frac{k}{N}\right) = 5\left(\frac{2}{20}\right) = 0.5$$

and

$$V(Y) = n\left(\frac{k}{N}\right)\left(1 - \frac{k}{N}\right)\left(\frac{N-n}{N-1}\right)$$
$$= 5(.1)(1 - 0.1)\left(\frac{20-5}{20-1}\right)$$
$$= 0.355$$

It follows that

$$E(T) = 9E(Y) + 5$$
$$= 9(0.5) + 5$$
$$= 9.5$$

and

$$V(T) = (9)^2 V(Y)$$
$$= 81(0.355)$$
$$= 28.755$$

Thus, installation time should average 9.5 minutes, with a standard deviation of $\sqrt{28.755} = 5.4$ minutes.   ∎

The Hypergeometric Distribution

$$p(y) = \frac{\binom{k}{y}\binom{N-k}{n-y}}{\binom{N}{n}}, \qquad y = 0, 1, \ldots, k \text{ with } \binom{b}{a} = 0 \text{ if } a > b$$

$$E(Y) = n\left(\frac{k}{N}\right) \qquad V(Y) = n\left(\frac{k}{N}\right)\left(1 - \frac{k}{N}\right)\left(\frac{N-n}{N-1}\right)$$

Exercises

3.70   From a box containing four white and three red balls, two balls are selected at random, without replacement. Find the probabilities of the following events.

a   Exactly one white ball is selected.

b   At least one white ball is selected.

c   Two white balls are selected, given that at least one white ball is selected.

d   The second ball drawn is white.

3.71 A warehouse contains ten printing machines, four of which are defective. A company randomly selects five of the machines for purchase. What is the probability that all five of the machines are not defective?

3.72 Referring to Exercise 3.71, the company purchasing the machines returns the defective ones for repair. If it costs $50 to repair each machine, find the mean and the variance of the total repair cost. By Tchebysheff's Theorem, in what interval should you expect the repair costs on these five machines to lie?

3.73 A corporation has a pool of six firms (four of which are local) from which they can purchase certain supplies. If three firms are randomly selected without replacement, find the probabilities of the following events.

a At least one selected firm is not local.

b All three selected firms are local.

3.74 A foreman has ten employees from among whom he must select four to perform a certain undesirable task. Three of the ten employees belong to a minority ethnic group. The foreman selected all three minority employees (plus one other) to perform the undesirable task. The minority employees then protested to the union steward that they were discriminated against by the foreman. The foreman claimed that the selection had been completely random. What do you think?

3.75 Specifications call for a type of thermistor to test out at between 9000 and 10,000 ohms at 25°C. Ten thermistors are available, and three of these are to be selected for use. Let Y denote the number among the three that do not conform to specifications. Find the probability distribution for Y (in tabular form) if the following conditions prevail.

a The ten contain two thermistors not conforming to specifications.

b The ten contain four thermistors not conforming to specifications.

3.76 Used photocopying machines are returned to the supplier, cleaned, and then sent back out on lease agreements. Major repairs are not made, however; and as a result, some customers receive malfunctioning machines. Among eight used photocopiers in supply today, three are malfunctioning. A customer wants to lease four machines immediately. Hence, four of the eight machines are quickly selected and sent out, with no further checking. Find the probabilities of the following events.

a The customer receives no malfunctioning machines.

b The customer receives at least one malfunctioning machine.

c The customer receives three malfunctioning machines.

3.77 An eight-cylinder automobile engine has two misfiring spark plugs. If all four plugs are removed from one side of the engine, what is the probability that the two misfiring plugs are among them?

3.78 The "worst-case" requirements are defined in the design objectives for a brand of computer terminal. A quick preliminary test indicates that four out of a lot of ten such terminals failed to meet the worst-case requirements. Five of the ten are randomly selected for further testing. Let Y denote the number, among the five, that failed the preliminary test. Find the following probabilities.

a $P(Y \geq 1)$ **b** $P(Y \geq 3)$

c $P(Y \geq 4)$ **d** $P(Y \geq 5)$

3.79 An auditor checking the accounting practices of a firm samples three accounts from an accounts receivable list of eight. Find the probability that the auditor sees at least one past-due account under the following conditions.

a There are two such accounts among the eight.

b There are four such accounts among the eight.

c There are seven such accounts among the eight.

3.80 Six software packages available to solve a linear programming problem have been ranked from 1 to 6 (best to worst). An engineering firm selects two of these packages for purchase, without looking at the ratings. Let Y denote the number of packages purchased by the firm that are ranked from 3 to 6. Show the probability distribution for Y in tabular form.

3.81 Lot acceptance sampling procedures for an electronics manufacturing firm call for sampling n items from a lot of N items and accepting the lot if $Y \leq c$, where Y is the number of non-conforming items in the sample. For an incoming lot of 20 printer covers, 5 are to be sampled. Find the probability of accepting the lot if $c = 1$ and the actual number of nonconforming covers in the lot is as follows.

a 0   **b** 1   **c** 2   **d** 3   **e** 4

3.82 In the setting and terminology of Exercise 3.81, answer the same questions if $c = 2$.

3.83 Two assembly lines (I and II) have the same rate of defectives in their production of voltage regulators. Five regulators are sampled from each line and tested. Among the total of ten tested regulators, four are defective. Find the probability that exactly two of the defectives came from line I.

3.9 The Moment-generating Function

We saw in earlier sections that, if $g(Y)$ is a function of a random variable Y, with probability distribution given by $p(y)$, then

$$E[g(Y)] = \sum_y g(y)p(y)$$

A special function with many theoretical uses in probability theory is the expected value of e^{tY}, for a random variable Y, and this expected value is called the *moment-generating function* (mgf). We denote an mgf by $M(t)$. Thus,

$$M(t) = E(e^{tY})$$

The expected values of powers of a random variable are often called *moments*. Thus, $E(Y)$ is the first moment of Y, and $E(Y^2)$ is the second moment of Y. One use for the moment-generating function is that, in fact, it does generate moments of Y. When $M(t)$ exists, it is differentiable in a neighborhood of the origin $t = 0$, and the derivatives may be taken inside the expectation. Thus,

$$
\begin{aligned}
M^{(1)}(t) &= \frac{dM(t)}{dt} \\
&= \frac{d}{dt} E[e^{tY}] \\
&= E\left[\frac{d}{dt} e^{tY}\right] \\
&= E[Ye^{tY}]
\end{aligned}
$$

Now, if we set $t = 0$, we have

$$M^{(1)}(0) = E(Y)$$

Going on to the second derivative,

$$M^{(2)}(t) = E(Y^2 e^{tY})$$

and

$$M^{(2)}(0) = E(Y^2)$$

In general,

$$M^{(k)}(0) = E(Y^k)$$

It often is easier to evaluate $M(t)$ and its derivatives than to find the moments of the random variable directly. Other theoretical uses of the mgf will be discussed in later chapters.

E X A M P L E   **3.22**   Evaluate the moment-generating function for the geometric distribution, and use it to find the mean and the variance of this distribution.

Solution   For the geometric variable Y, we have

$$M(t) = E(e^{tY})$$

$$= \sum_{y=1}^{\infty} e^{ty} p(1-p)^{y-1}$$

$$= pe^t \sum_{y=1}^{\infty} (1-p)^{y-1} (e^t)^{y-1}$$

$$= pe^t \sum_{y=1}^{\infty} [(1-p)e^t]^{y-1}$$

$$= pe^t \{1 + [(1-p)e^t] + [(1-p)e^t]^2 + \cdots$$

$$= pe^t \left[\frac{1}{1-(1-p)e^t} \right]$$

because the series is geometric with a common ratio of $(1-p)e^t$.
  To evaluate the mean, we have

$$M^{(1)}(t) = \frac{[1-(1-p)e^t]pe^t - pe^t[-(1-p)e^t]}{[1-(1-p)e^t]^2}$$

$$= \frac{pe^t}{[1-(1-p)e^t]^2}$$

and

$$M^{(1)}(0) = \frac{p}{[1-(1-p)]^2} = \frac{1}{p}$$

To evaluate the variance, we first need

$$E(Y^2) = M^{(2)}(0)$$

Now,

$$M^{(2)}(t) = \frac{[1 - (1-p)e^t]^2 pe^t - pe^t\{2[1 - (1-p)e^t](-1)(1-p)e^t\}}{[1 - (1-p)e^t]^4}$$

and

$$M^{(2)}(0) = \frac{p^3 + 2p^2(1-p)}{p^4} = \frac{p + 2(1-p)}{p^2}$$

Hence,

$$\begin{aligned}
V(Y) &= E(Y^2) - [E(Y)]^2 \\
&= \frac{p + 2(1-p)}{p^2} - \frac{1}{p^2} \\
&= \frac{p + 2(1-p) - 1}{p^2} \\
&= \frac{1-p}{p^2} \quad \blacksquare
\end{aligned}$$

Moment-generating functions have important properties that make them extremely useful in finding expected values and in determining the probability distributions of random variables. These properties will be discussed in detail in Chapters 4 and 5, but one such property is given in Exercise 3.85.

3.10 The Probability-generating Function

In an important class of discrete random variables, Y takes integral values ($Y = 0, 1, 2, 3, \ldots$) and consequently represents a count. The binomial, geometric, hypergeometric, and Poisson random variables all fall in this class. The following examples present practical situations involving integral-valued random variables. One, tied to the theory of queues (waiting lines), is concerned with the number of persons (or objects) awaiting service at a particular point in time. Understanding the behavior of this random variable is important in designing manufacturing plants where production consists of a sequence of operations, each of which requires a different length of time to complete. An insufficient number of service stations for a particular production operation can result in a bottleneck—the forming of a queue of products waiting to be serviced—which slows down the entire manufacturing operation. Queuing theory is also important in determining the number of checkout counters needed for a supermarket and in designing hospitals and clinics.

Integer-valued random variables are extremely important in studies of population growth, too. For example, epidemiologists are interested in the growth of bacterial populations and also in the growth of the number of persons afflicted by a particular disease. The number of elements in each of these populations is an integral-valued random variable.

A mathematical device that is very useful in finding the probability distributions and other properties of integral-valued random variables is the probability-generating function $P(t)$, which is defined by

$$P(t) = E(t^Y)$$

If Y is an integer-valued random variable, with

$$P(Y = i) = p_i, \qquad i = 0, 1, 2, \ldots$$

then

$$P(t) = E(t^Y) = p_0 + p_1 t + p_2 t^2 + \cdots$$

The reason for calling $P(t)$ a probability-generating function is clear when we compare $P(t)$ with the moment-generating function $M(t)$. Particularly, the coefficient of t^i in $P(t)$ is the probability p_i. If we know $P(t)$ and can expand it into a series, we can determine $p(y)$ as the coefficient of t^y. Repeated differentiation of $P(t)$ yields *factorial moments* for the random variable Y.

DEFINITION   **3.7**

The kth **factorial moment** for a random variable Y is defined to be

$$\mu_{|k|} = E[Y(Y-1)Y-2)\cdots(Y-k+1)]$$

where k is a positive integer.   ■

When a probability-generating function exists, it can be differentiated in a neighborhood of $t = 1$. Thus, with

$$P(t) = E(t^Y)$$

we have

$$
\begin{aligned}
P^{(1)}(t) &= \frac{dP(t)}{dt} \\
&= E\left[\frac{dt^Y}{dt}\right] \\
&= E(Yt^{Y-1})
\end{aligned}
$$

and

$$P^{(1)}(1) = E(Y)$$

Similarly,

$$P^{(2)}(t) = E[Y(Y-1)t^{Y-2}]$$

and

$$P^{(2)}(1) = E[Y(Y-1)] = \mu_{|2|}$$

In general,

$$P^{(k)}(t) = E[Y(Y-1)\cdots(Y-k+1)t^{Y-k}]$$

and

$$P^{(k)}(1) = E[Y(Y-1)\cdots(Y-k+1)]$$
$$= \mu_{|k|}$$

E X A M P L E   3.23   Find the probability-generating function for the geometric random variable, and use this function to find the mean.

Solution   Notice that $p_0 = 0$ because Y cannot assume this value. Then,

$$P(t) = E(t^Y) = \sum_{y=1}^{\infty} t^y q^{y-1} p$$
$$= \sum_{y=1}^{\infty} \frac{p}{q}(qt)^y$$
$$= \frac{p}{q}[qt + (qt)^2 + (qt)^3 + \cdots]$$

The terms of the series are those of an infinite geometric progression. We can let $t \leq 1$, so that $qt < 1$. Then,

$$P(t) = \frac{p}{q}\left\{\frac{qt}{1-qt}\right\} = \frac{pt}{1-qt}$$

Now,

$$P^{(1)}(t) = \frac{d}{dt}\left\{\frac{pt}{1-qt}\right\} = \frac{(1-qt)p - (pt)(-q)}{(1-qt)^2}$$

Setting $t = 1$,

$$P^{(1)}(1) = \frac{p^2 + pq}{p^2} = \frac{p(p+q)}{p^2} = \frac{1}{p}$$

which is the mean of a geometric random variable.   ∎

Since we already have the moment-generating function to assist us in finding the moment of a random variable, we might ask how knowing $P(t)$ can help us. The answer is that in some instances it may be exceedingly difficult to find $M(t)$ but easy to find $P(t)$. Alternatively, $P(t)$ may be easier to work with in a particular setting. Thus, $P(t)$ simply provides an additional tool for finding the moments of a random variable. It may or may not be useful in a given situation.

Finding the moments of a random variable is not the major use of the probability-generating function. Its primary application is in deriving the probability function (and hence the probability distribution) for related integral-valued random variables. For these applications, see Feller (1968), Parzen (1964), and Section 6.7.

Exercises

3.84   Find the moment-generating function for the Bernoulli random variable.

3.85   Show that the moment-generating function for the binomial random variable is given by

$$M(t) = [pe^t + (1 - p)]^n$$

Use this result to derive the mean and the variance for the binomial distribution.

3.86   Show that the moment-generating function for the Poisson random variable with mean λ is given by

$$M(t) = e^{\lambda(e^t - 1)}$$

Use this result to derive the mean and the variance for the Poisson distribution.

3.87   If X is a random variable with moment-generating function $M(t)$, and Y is a function of X given by $Y = aX + b$, show that the moment-generating function for Y is $e^{tb} M(at)$.

3.88   Use the result of Exercise 3.85 to show that

$$E(Y) = aE(X) + b$$

and

$$V(Y) = a^2 V(X)$$

3.89   Find the probability-generating function for a binomial random variable of n trials, with probability of success p. Use this function to find the mean and the variance of the binomial.

3.11   Markov Chains

Consider a system that can be in any of a finite number of states, and assume that it moves from state to state according to some prescribed probability law. The system, for example, could record weather conditions from day to day, with the possible states being clear, partly cloudy, and cloudy. Observing conditions over a long period of time would allow one to find the probability of its being clear tomorrow given that it is partly cloudy today.

Let X_i denote the state of the system at time point i, and let the possible states be denoted by $S_1, \ldots, S_m$, for a finite integer m. We are interested not in the elapsed time between transitions from one state to another, but only in the states and the probabilities of going from one state to another—that is, in the *transition probabilities*. We assume that

$$P(X_i = S_k | X_{i-1} = S_j) = p_{jk}$$

where p_{jk} is the transition probability from S_j to S_k; and this probability is independent of i. Thus, the transition probabilities depend not on the time points, but only on the states. The event $(X_i = S_k | X_{i-1} = S_j)$ is assumed to be independent of the past history of the process. Such a process is called a *Markov chain* with stationary transition probabilities. The transition probabilities can conveniently be displayed in a matrix:

$$\mathbf{P} = \begin{bmatrix} p_{11} & p_{12} & \cdots & p_{1m} \\ p_{21} & p_{22} & \cdots & p_{2m} \\ \vdots & & & \vdots \\ p_{m1} & p_{m2} & \cdots & p_{mm} \end{bmatrix}$$

Let X_0 denote the starting state of the system, with probabilities given by

$$p_k^{(0)} = P(X_0 = S_k)$$

and let the probability of being in state S_k after n steps be given by $p_k^{(n)}$. These probabilities are conveniently displayed by vectors:

$$\mathbf{p}^{(0)} = [p_1^{(0)}, \ p_2^{(0)}, \ \ldots, \ p_m^{(0)}]$$

and

$$\mathbf{p}^{(n)} = [p_1^{(n)}, \ p_2^{(n)}, \ \ldots, \ p_m^{(n)}]$$

To see how $\mathbf{p}^{(0)}$ and $\mathbf{p}^{(1)}$ are related, consider a Markov chain with only two states, so that

$$\mathbf{P} = \begin{bmatrix} p_{11} & p_{12} \\ p_{21} & p_{22} \end{bmatrix}$$

There are two ways to get to state 1 after one step: either the chain starts in state 1 and stays there, or the chain starts in state 2 and then moves to state 1 in one step. Thus,

$$\mathbf{p}_1^{(1)} = p_1^{(0)} p_{11} + p_2^{(0)} p_{21}$$

Similarly,

$$\mathbf{p}_2^{(1)} = p_1^{(0)} p_{12} + p_2^{(0)} p_{22}$$

In terms of matrix multiplication,

$$\mathbf{p}^{(1)} = \mathbf{p}^{(0)} \mathbf{P}$$

and, in general,

$$\mathbf{p}^{(n)} = \mathbf{p}^{(n-1)} \mathbf{P}$$

$\mathbf{P}$ is said to be *regular* if some power of $\mathbf{P}$ ($\mathbf{P}^n$ for some n) has all positive entries. Thus, one can get from state S_j to state S_k, eventually, for any pair $(j, \ k)$. (Notice that the condition of regularity rules out certain chains that periodically return to certain states.) If $\mathbf{P}$ is regular, the chain has a stationary (or equilibrium) distribution that gives the probabilities of its being in the respective states after many transitions have evolved. In other words, $p_j^{(n)}$ must have a limit π_j, as $n \to \infty$. Suppose that such limits exist; then $\pi = (\pi_1, \ \ldots, \ \pi_m)$ must satisfy

$$\pi = \pi \mathbf{P}$$

because $\mathbf{p}^{(n)} = \mathbf{p}^{(n-1)} \mathbf{P}$ will have the same limit as $\mathbf{p}^{(n-1)}$.

E X A M P L E   **3.24**   A supermarket stocks three brands of coffee—A, B, and C—and customers switch from brand to brand according to the transition matrix

$$\mathbf{P} = \begin{bmatrix} 3/4 & 1/4 & 0 \\ 0 & 2/3 & 1/3 \\ 1/4 & 1/4 & 1/2 \end{bmatrix}$$

where S_1 corresponds to a purchase of brand A, S_2 to brand B, and S_3 to brand C; that is, 3/4 of the customers buying brand A also buy brand A the next time they purchase coffee, whereas 1/4 of these customers switch to brand B.

1   Find the probability that a customer who buys brand A today will again purchase brand A two weeks from today, assuming that he or she purchases coffee once a week.

2   In the long run, what fractions of customers purchase the respective brands?

Solution   1   Assuming that the customer is chosen at random, his or her transition probabilities are given by **P**. The given information indicates that $\mathbf{p}^{(0)} = (1, 0, 0)$; that is, the customer starts with a purchase of brand A. Then

$$\mathbf{p}^{(1)} = \mathbf{p}^{(0)}\mathbf{P} = \left(\frac{3}{4}, \frac{1}{4}, 0 \right)$$

gives the probabilities for the next week's purchase. The probabilities for two weeks from now are given by

$$\mathbf{p}^{(2)} = \mathbf{p}^{(1)}\mathbf{P} = \left(\frac{9}{16}, \frac{17}{48}, \frac{1}{12} \right)$$

that is, the chance of the customer's purchasing A two weeks from now is only 9/16.

2   The answer to the long-run frequency ratio is given by π, the stationary distribution. The equation

$$\pi = \pi \mathbf{P}$$

yields the system

$$\pi_1 = \left(\frac{3}{4} \right) \pi_1 + \left(\frac{1}{4} \right) \pi_3$$

$$\pi_2 = \left(\frac{1}{4} \right) \pi_1 + \left(\frac{2}{3} \right) \pi_2 + \left(\frac{1}{4} \right) \pi_3$$

$$\pi_3 = \left(\frac{1}{3} \right) \pi_2 + \left(\frac{1}{2} \right) \pi_3$$

Combining these equations with the fact that $\pi_1 + \pi_2 + \pi_3 = 1$ yields

$$\pi = \left(\frac{2}{7}, \frac{3}{7}, \frac{2}{7} \right)$$

Thus, the store should stock more brand B coffee than either brand A or brand C.   ∎

E X A M P L E   **3.25**   Markov chains are used in the study of probabilities connected to genetic models. Genes come in pairs; and for any trait governed by a pair of genes, an individual may have genes of the geno type GG (dominant), Gg (hybrid), or gg (recessive). Each offspring inherits one gene of a pair from each parent, at random and independently.

1   Suppose that an individual of unknown genetic makeup is mated with a hybrid. Set up a transition matrix to describe the possible states of a resulting offspring and their probabilities.

2   What will happen to the genetic makeup of the offspring after many generations of mating with a hybrid?

Solution   **1**   If the unknown is dominant (GG) and is mated with a hybrid (Gg), the offspring has a probability of $\frac{1}{2}$ of being dominant and a probability of $\frac{1}{2}$ of being hybrid. If two hybrids are mated, the offspring may be dominant, hybrid, or recessive with probabilities $\frac{1}{4}$, $\frac{1}{2}$, and $\frac{1}{4}$, respectively. If the unknown is recessive (gg), the offspring of it and a hybrid has a probability of $\frac{1}{2}$ of being recessive and a probability of $\frac{1}{2}$ of being hybrid. Following along these lines, a transition matrix from the unknown parent to an offspring is given by

$$\mathbf{P} = \begin{array}{c} \\ d \\ h \\ r \end{array} \begin{array}{ccc} d & h & r \\ \left[\begin{array}{ccc} \frac{1}{2} & \frac{1}{2} & 0 \\ \frac{1}{4} & \frac{1}{2} & \frac{1}{4} \\ 0 & \frac{1}{2} & \frac{1}{2} \end{array} \right] \end{array}$$

2   The matrix $\mathbf{P}^2$ has all positive entries ($\mathbf{P}$ is regular); and hence, a stationary distribution exists. From the matrix equation

$$\pi = \pi \mathbf{P}$$

we obtain

$$\pi_1 = \tfrac{1}{2}\pi_1 + \tfrac{1}{4}\pi_2$$
$$\pi_2 = \tfrac{1}{2}\pi_1 + \tfrac{1}{2}\pi_2 + \tfrac{1}{2}\pi_3$$
$$\pi_3 = \qquad \tfrac{1}{4}\pi_2 + \tfrac{1}{2}\pi_3$$

Since $\pi_1 + \pi_2 + \pi_3 = 1$, the second equation yields $\pi_2 = \frac{1}{2}$. It is then easy to establish that $\pi_1 = \pi_3 = \frac{1}{4}$. Thus,

$$\pi = \left(\frac{1}{4}, \frac{1}{2}, \frac{1}{4} \right)$$

No matter what the genetic makeup of the unknown parent happened to be, the ratio of dominant to hybrid to recessive offspring among its descendants, after many generations of mating with hybrids, should be 1:2:1.   ∎

An interesting example of a transition matrix that is not regular is formed by a Markov chain with *absorbing states*. A state S_i is said to be *absorbing* if $p_{ii} = 1$ and $p_{ij} = 0$ for $j \neq 1$. That is, once the system is in state S_i, it cannot leave it. The transition matrix for such a chain can always be arranged in a standard form, with the absorbing states listed first. For example, suppose that a chain has five states, of which two are absorbing. Then **P** can be written as

$$\mathbf{P} = \begin{bmatrix} 1 & 0 & 0 & 0 & 0 \\ 0 & 1 & 0 & 0 & 0 \\ p_{31} & p_{32} & p_{33} & p_{34} & p_{35} \\ p_{41} & p_{42} & p_{43} & p_{44} & p_{45} \\ p_{51} & p_{52} & p_{53} & p_{54} & p_{55} \end{bmatrix}$$

$$= \begin{bmatrix} \mathbf{I} & \mathbf{0} \\ \mathbf{R} & \mathbf{Q} \end{bmatrix}$$

where **I** is a 2×2 identity matrix and **0** is a matrix of zeros. Such a transition matrix is **not** regular. Many interesting properties of these chains can be expressed in terms of **R** and **Q** (see Kemeny et al. 1962).

The following discussion will be restricted to the case in which **R** and **Q** are such that it is possible to get to an absorbing state from every other state, eventually. In that case, the Markov chain eventually will end up in an absorbing state. Questions of interest then involve the expected number of steps to absorption and the probability of absorption in the various absorbing states.

Let m_{ij} denote the expected (or mean) number of times the system is in state S_j, given that it started in S_i, for nonabsorbing states S_i and S_j. From S_i, the system could go to an absorbing state in one step, or it could go to a nonabsorbing state—say, S_k—and eventually be absorbed from there. Thus, m_{ij} must satisfy

$$m_{ij} = \partial_{ij} + \sum_k p_{ik} m_{kj}$$

where the summation is over all nonabsorbing states and

$$\partial_{ij} = \begin{cases} 1 & \text{if } i = j \\ 0 & \text{otherwise} \end{cases}$$

The term ∂_{ij} accounts for the fact that, if the chain goes to an absorbing state in one step, it was in state S_i one time.

If we denote the matrix of m_{ij} terms by **M**, the preceding equation can then be generalized to

$$\mathbf{M} = \mathbf{I} + \mathbf{QM}$$

or

$$\mathbf{M} = (\mathbf{I} - \mathbf{Q})^{-1}$$

(Matrix operations, such as inversion, will not be discussed here. The equations can be solved directly if matrix operations are unfamiliar to the reader.)

The expected number of steps to absorption, from the nonabsorbing starting state S_i, will be denoted by $\mathbf{m}_i$ and given simply by

$$m_i = \sum_k m_{ik}$$

again summing over nonabsorbing states.

Turning now to the probability of absorption into the various absorbing states, we let a_{ij} denote the probability of the system's being absorbed in state S_j, given that it started in state S_i, for nonabsorbing S_i and absorbing S_j. Repeating the preceding argument, the system could move to S_j in one step, or it could move in a nonabsorbing state S_k and be absorbed from there. Thus, a_{ij} satisfies

$$a_{ij} = p_{ij} + \sum_k p_{ik} a_{kj}$$

where the summation occurs over the nonabsorbing states. If we denote the matrix of a_{ij} terms by $\mathbf{A}$, the preceding equation then generalizes to

$$\mathbf{A} = \mathbf{R} + \mathbf{QA}$$

or

$$\mathbf{A} = (\mathbf{I} - \mathbf{Q})^{-1}\mathbf{R}$$
$$= \mathbf{MR}$$

The following example illustrates the computations.

E X A M P L E **3.26**   A manager of one section of a plant has different employees working at level I and at level II. New employees may enter his section at either level. At the end of each year, the performance of each employee is evaluated; employees can be reassigned to their level I or II jobs, terminated, or promoted to level III, in which case they never go back to I or II. The manager can keep track of employee movement as a Markov chain. The absorbing states are termination (S_1) and employment at level III (S_2); the nonabsorbing states are employment at level I (S_3) and employment at level II (S_4). Records over a long period of time indicate that the following is a reasonable assignment of probabilities:

$$\mathbf{P} = \begin{bmatrix} 1 & 0 & 0 & 0 \\ 0 & 1 & 0 & 0 \\ 0.2 & 0.1 & 0.2 & 0.5 \\ 0.1 & 0.3 & 0.1 & 0.5 \end{bmatrix}$$

Thus, if an employee enters at a level I job, the probability is 0.5 that she will jump to level II work at the end of the year, but the probability is 0.2 that she will be terminated.

1 Find the expected number of evaluations an employee must go through in this section.

2 Find the probabilities of being terminated or promoted to level III eventually.

Solution   **1**   For the **P** matrix,

$$R = \begin{bmatrix} 0.2 & 0.1 \\ 0.1 & 0.3 \end{bmatrix}$$

and

$$Q = \begin{bmatrix} 0.2 & 0.5 \\ 0.1 & 0.5 \end{bmatrix}$$

Thus,

$$I - Q = \begin{bmatrix} 0.8 & -0.5 \\ -0.1 & 0.5 \end{bmatrix}$$

and

$$M = (I - Q)^{-1} = \begin{bmatrix} 10/7 & 10/7 \\ 2/7 & 16/7 \end{bmatrix} = \begin{bmatrix} m_{33} & m_{34} \\ m_{43} & m_{44} \end{bmatrix}$$

It follows that

$$m_3 = \frac{20}{7} \quad \text{and} \quad m_4 = \frac{18}{7}$$

In other words, a new employee in this section can expect to remain there through 20/7 evaluations periods if she enters at level I, whereas she can expect to remain there through 18/7 evaluations if she enters at level II.

2   The fact that

$$A = MR = \begin{bmatrix} 3/7 & 4/7 \\ 2/7 & 5/7 \end{bmatrix} = \begin{bmatrix} a_{31} & a_{32} \\ a_{41} & a_{42} \end{bmatrix}$$

implies that an employee entering at level I has a probability of 4/7 of reaching level III, whereas an employee entering at level II has a probability of 5/7 of reaching level III. The probabilities of termination at levels I and II are therefore 3/7 and 2/7, respectively. ∎

EXAMPLE **3.27**   Continuing the genetic example of Example 3.25, suppose that an individual of unknown genetic makeup is mated with a known dominant (GG) individual. The matrix of transition probabilities for the first-generation offspring then becomes

$$P = \begin{matrix} & \begin{matrix} d & h & r \end{matrix} \\ \begin{matrix} d \\ h \\ r \end{matrix} & \begin{bmatrix} 1 & 0 & 0 \\ \frac{1}{2} & \frac{1}{2} & 0 \\ 0 & 1 & 0 \end{bmatrix} \end{matrix}$$

which has one absorbing state. Find the mean number of generations until all offspring become dominant.

Solution   In the notation used earlier,

$$Q = \begin{bmatrix} \frac{1}{2} & 0 \\ 1 & 0 \end{bmatrix}$$

$$I - Q = \begin{bmatrix} \frac{1}{2} & 0 \\ -1 & 1 \end{bmatrix}$$

and

$$(I - Q)^{-1} = \begin{bmatrix} 2 & 0 \\ 2 & 1 \end{bmatrix} = M$$

Thus, if the unknown is hybrid, we should expect all offspring to be dominant after two generations. If the unknown was recessive, we should expect all offspring to be dominant after three generations. Notice here that

$$A = MR = \begin{bmatrix} 2 & 0 \\ 2 & 1 \end{bmatrix} \begin{bmatrix} \frac{1}{2} \\ 0 \end{bmatrix} = \begin{bmatrix} 1 \\ 1 \end{bmatrix}$$

which simply indicates that we are guaranteed to reach the fully dominant state eventually, no matter what the genetic makeup of the unknown parent may be.   ∎

Exercises

3.90   For a Markov chain, show that $P^{(n)} = P^{(n-1)}P$.

3.91   A certain city prides itself on having sunny days. If it rains one day, there is a 90% chance that it will be sunny the next day. If it is sunny one day, there is a 30% chance that it will rain the following day. (Assume that there are only sunny or rainy days.) Does the city have sunny days most of the time? In the long run, what fraction of all days are sunny?

3.92   Suppose that a particle moves in unit steps along a straight line. At each step, the particle either remains where it is, moves one step to the right, or moves one step to the left. The line along which the particle moves has barriers at 0 and at b, a positive integer, and the particle only moves between these barriers; it is absorbed if it lands on either barrier. Now, suppose that the particle moves to the right with probability p and to the left with probability $1 - p = q$.

a   Set up the general form of the transition matrix for a particle in this system.

b   For the case $b = 3$ and $p \neq q$, show that the absorption probabilities are as follows:

$$a_{10} = \frac{\left(\frac{q}{p}\right) - \left(\frac{q}{p}\right)^3}{1 - \left(\frac{q}{p}\right)^3}$$

$$a_{20} = \frac{\left(\frac{q}{p}\right)^2 - \left(\frac{q}{p}\right)^3}{1 - \left(\frac{q}{p}\right)^3}$$

c   In general, it can be shown that

$$a_{10} = \frac{\left(\frac{q}{p}\right)^j - \left(\frac{q}{p}\right)^b}{1 - \left(\frac{q}{p}\right)^b}, \qquad p \neq q$$

By taking the limit of a_{j0} as $p \to \frac{1}{2}$, show that

$$a_{j0} = \frac{b - j}{b}, \qquad p = q$$

d   For the case $b = 3$, find an expression for the mean time to absorption from state j, with $p \neq q$. Can you generalize this result?

3.93   Suppose that n white balls and n black balls are placed in two urns so that each urn contains n balls. A ball is randomly selected from each urn and placed in the opposite urn. [This is one possible model for the diffusion of gases.]

a   Number the urns 1 and 2. The state of the system is the number of black balls in urn 1. Show that the transition probabilities are given by the following quantities:

$$P_{jj-1} = \left(\frac{j}{n}\right)^2, \qquad j > 0$$

$$P_{jj} = \frac{2j(n - j)}{n^2}$$

$$P_{jj+1} = \left(\frac{n - j}{n}\right)^2, \qquad j < n$$

$$P_{jk} = 0, \qquad \text{otherwise}$$

b   After many transitions, show that the stationary distribution is satisfied by

$$P_j = \frac{\binom{n}{j}^2}{\binom{2n}{n}}$$

Give an intuitive argument as to why this looks like a reasonable answer.

3.94   Suppose that two friends, A and B, toss a balanced coin. If the coin comes up heads, A wins \$1 from B. If it comes up tails, B wins \$1 from A. The game ends only when one player has all the other's money. If A starts with \$1 and B with \$3, find the expected duration of the game and the probability that A will win.

3.12   Activities for Students: Simulation

Computers lend themselves nicely to use in the area of probability. Not only can computers be used to calculate probabilities, but they can also be used to simulate random variables from specified probability distributions. A simulation, performed on the computer, permits the user to analyze both theoretical and applied problems. A simulated model attempts to copy the behavior of a situation under consideration; practical applications include models of inventory control problems, queuing systems, production lines, medical systems, and flight patterns of major jets. Simulation can also be used to determine the behavior of a complicated random variable whose precise probability distribution function is difficult to evaluate mathematically.

Generating observations from a probability distribution is based on random numbers on [0, 1]. A random number R_i on the interval [0, 1] is chosen, on the condition that each number between 0 and 1 has the same probability of being selected. A sequence of numbers that appears to follow a certain pattern or trend should not be considered random. Most computer languages have built-in random generators that give a number on [0, 1]. If a built-in generator is not available, algorithms are available for setting up one. One basic technique used is a congruential method, such as a linear congruential generator (LCG). The generator most commonly used is

$$x_i = (ax_{i-1} + c) \ (\text{mod } m), \qquad\qquad i = 1, 2, \ldots$$

where x_i, a, c, and m are integers and $0 \le x_i < m$. If $c = 0$, this method is called a multiplicative-congruential sequence. The values of y_i generated by the LCG are contained in the interval $[0, m)$. To obtain a value in $[0, 1)$, we evaluate x_i/m.

A random number generator should have certain characteristics, such as the following:

1   The numbers produced (x_i/m) should appear to be distributed uniformly on [0, 1]; that is, the probability function for the numbers should be constant over the [0, 1]. The numbers should also be independent of each other. Many statistical tests can be applied to check for uniformity and independence, such as comparing the number of occurrences of a digit in a sequence with the expected number of occurrences for that digit.

2   The random number generator should have a long, full period; that is, the sequence of numbers should not begin repeating or "cycling" too quickly. A sequence has a full period (p) if $p = m$. It has been shown that an LCG has a full period if and only if the following are true (Kennedy and Gentle 1980, p. 137):

   a   The only positive integer that (exactly) divides both m and c is 1.

   b   If q is a prime number that divides m, then q divides $a - 1$.

   c   If 4 divides m, then 4 divides $a - 1$.

   This implies that, if c is odd and $a - 1$ is divisible by 4, when a full-period generator is used, x_0 can be any integer between 0 and $m - 1$ without affecting the generator's period. A good choice for m is 2^b, where b is the number of bits. Based on the just-mentioned relationships of a, c, and m, the following values have been found to be satisfactory for use on a microcomputer: $a = 25,173$; $c = 13,849$; $m = 2^{16} = 65,536$ (Yang and Robinson 1986, p. 6).

3   It is beneficial to be able to reuse the same random numbers in a different simulation run. The preceding LCG has this ability, because all values of x_i are determined by the initial value (seed) x_0. Because of this, the computed random numbers are referred to as *pseudo-random numbers*. Even though these numbers are not truly random, careful selection of a, c, m, and x_0 will yield values for x_i that behave as random numbers and pass the appropriate statistical tests, as described in characteristic 1.

4   The generator should be efficient—that is, fast and in need of little storage.

Given that a random number R_i on [0, 1] can be generated, we will now consider a brief description of generating discrete random variables for the distributions discussed in this chapter.

3.12.1   Bernoulli Distribution

Let p represent the probability of success. If $R_i \le p$, then $X_i = 1$; otherwise, $X_i = 0$.

3.12.2 Binomial Distribution

A binomial random variable X_i can be expressed as the sum of independent Bernoulli random variables Y_j; that is, $X_i = \sum Y_j$, where $j = 1, \ldots, n$. Thus, to simulate X_i with parameters n and p, we simulate n Bernoulli random variables, as stated previously. X_i is equal to the sum of the n Bernoulli variables.

3.12.3 Geometric Distribution

Let X_i represent the number of trials necessary for the first success, with p being probability of success. $X_i = m$, where m is the number of R_i's generated until $R_i \leq p$.

3.12.4 Negative Binomial Distribution

Let X_i represent the number of trials necessary until the rth success, with p being the probability of success. A negative binomial random variable X_i can be expressed as the sum of r independent geometric random variables Y_j; that is, $X_i = \sum Y_j$, where $j = 1, \ldots, r$. Thus, to simulate X_i with parameter p, we simulate r geometric random variables, as stated previously. X_i is equal to the sum of the r geometric variables.

3.12.5 Poisson Distribution

Generating Poisson random variables will be discussed at the end of Chapter 4.

Let us consider some simple examples of possible uses for simulating discrete random variables. Suppose that n_1 items are to be inspected from one production line and that n_2 items are to be inspected from another. Let p_1 represent the probability of a defective from line 1, and let p_2 represent the probability of a defective from line 2. Let X be a binomial random variable with parameters n_1 and p_1. Let Y be a binomial random variable with parameters n_2 and p_2. A variable of interest is W, which represents the total number of defective items observed in both production lines. Let $W = X + Y$. Unless $p_1 = p_2$, the distribution of W will not be binomial. To see how the distribution of W will behave, we can perform a simulation. Useful information could be obtained from the simulation by looking at a histogram of the values of W_i generated and considering the values of the sample mean and sample variance. Let's consider the following random variables X and Y: X is binomial with $n_1 = 7$, $p_1 = 0.2$; and Y is binomial with $n_2 = 8$, $p_2 = 0.6$. Defining $W = X + Y$, a simulation produced the histogram shown in Simulation 1.

The sample mean was 6.2, with a sample standard deviation of 1.76. In Chapter 5, we will be able to show that these values are very close to the expected values of $\mu_w = 6.2$ and $\sigma_w = 1.74$. (This calculation will make sense after we discuss the linear functions of random variables.) From the histogram, we see that the probability that the total number of defective items is at least 9 is given by 0.09.

Another example of interest might be the coupon-collector problem, which incorporates the geometric distribution. Suppose that there are n distinct colors of coupons. We assume that, each time someone obtains a coupon, it is equally likely to be any one of the n colors and the selection of the coupon is independent of any previously

Simulation 1

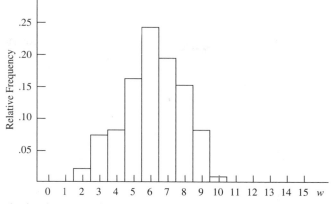

obtained coupon. Suppose that an individual can redeem a set of coupons for a prize if each possible color coupon is represented in the set. We define the random variable X as representing the total number of coupons that must be selected to complete a set of each color coupon at random. Questions of interest might include the following:

1   What is the expected number of coupons needed in order to obtain this complete set; that is, what is $E(X)$?

2   What is the standard deviation of X?

3   What is the probability that one must select at most x coupons to obtain this complete set?

Instead of answering these questions by deriving the distribution function of X, one might try simulation. Two simulations (Simulations 2 and 3) follow. The first histogram represents a simulation where n, the number of different color coupons, is equal to 5.

Simulation 2

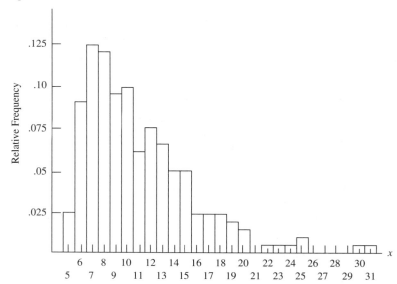

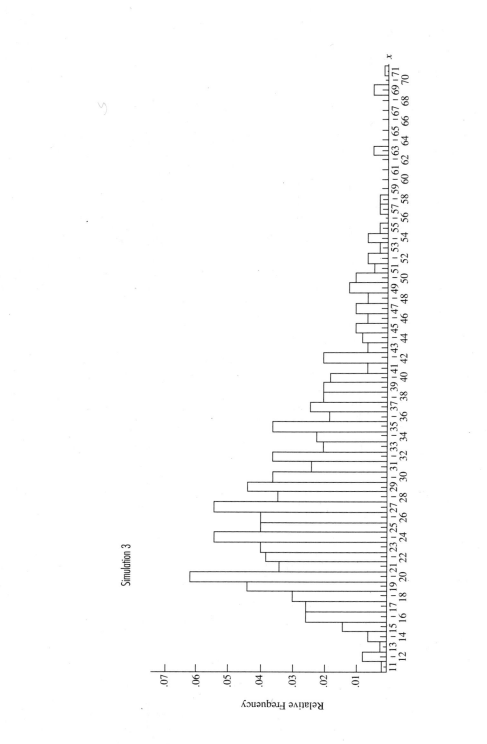

The sample mean in this case was computed to be 11.06, with a sample standard deviation of 4.65. Suppose that one is interested in finding $P(X \leq 10)$. Using the results of the simulations, the relative frequency probability is given as 0.555. It might also be noted that, from this simulation, the largest number of coupons needed to obtain the complete set was 31.

The second histogram represents a simulation where $n = 10$. The sample mean was computed to be 29.07, with a sample standard deviation of 10.16. The $P(X \leq 20) = 0.22$ from the simulated values. In this simulation, the largest number of coupons necessary to obtain the complete set was 71.

3.12.6   Binomial Proportions

Count data are often more conveniently discussed as a proportion—as in the proportion of inspected cars that have point defects, or the proportion of sampled voters who favor a certain candidate. For a binomial random variable X on n trials and a probability p of success on any one trial, the proportion of successes is X/n.

Generate 100 values of X/n and plot them on a line plot or histogram for each of the following cases:

1 $n = 20, \quad p = 0.2$

2 $n = 40, \quad p = 0.2$

3 $n = 20, \quad p = 0.5$

4 $n = 40, \quad p = 0.5$

Compare the four plots, and discuss any patterns you observe in their centering, variation, and symmetry. Can you propose a formula for the variance of X/n?

3.12.7   Waiting for Blood

A local blood bank knows that about 40% of its donors have A^+ blood. It needs $k = 3$ A^+ donors today. Generate a distribution of values for X, the number of donors that must be tested sequentially to find the three A^+ donors needed. What are the approximate expected value and standard deviation of X from your data? Do these values agree with the theory? What is the estimated probability that the blood bank must test 10 or more people to find the three A^+ donors? How will the answers to these questions change if $k = 4$?

3.13   Summary

The outcomes of interest in most investigations involving random events are numerical. The simplest numbered outcomes to model are counts, such as the number of nonconforming parts in a shipment, the number of sunny days in a month, or the number of water samples that contain a pollutant. One amazing result of probability theory is the fact that a small number of theoretical distributions can cover a wide array of applications. Six of the most useful discrete probability distributions are introduced in this chapter.

The *Bernoulli* random variable is simply an indicator random variable; it uses a numerical code to indicate the presence or absence of a characteristic.

The *binomial* random variable counts the number of "successes" among a fixed number n of independent events, each with the same probability of success.

The *geometric* random variable counts the number of events one needs to conduct sequentially until the first "success" is seen.

The *negative binomial* random variable counts the number of events one needs to conduct sequentially until the rth "success" is seen.

The *Poisson* random variable arises from counts in a restricted domain of time, area, or volume, and is most useful for counting fairly rare outcomes.

The *hypergeometric* random variable counts the number of "successes" in sampling from a finite population, which makes the sequential selections dependent on one another.

These theoretical distributions serve as models for real data that might arise in our quest to improve a process. Each involves assumptions that should be checked carefully before the distribution is applied.

Supplementary Exercises

3.95 Construct probability histograms for the binomial probability distribution for $n = 5$, and $p = 0.1, 0.5$, and 0.9. (Table 2 of the Appendix will reduce the amount of calculation you must perform.) Notice the symmetry for $p = 0.5$ and the direction of skewness for $p = 0.1$ and $p = 0.9$.

3.96 Use Table 2 of the Appendix to construct a probability histogram for the binomial probability distribution for $n = 20$ and $p = 0.5$. Notice that almost all of the probability falls in the interval $5 \leq y \leq 15$.

3.97 The probability that a single radar set will detect an enemy plane is 0.9. If we have five radar sets, what is the probability that exactly four sets will detect the plane? (Assume that the sets operate independently of each other.) At least one set?

3.98 Suppose that the four engines of a commercial aircraft were arranged to operate independently and that the probability of in-flight failure of a single engine is 0.01. What are the probabilities that, on a given flight, the following events occur?

a No failures are observed.

b No more than one failure is observed.

3.99 Sampling for defectives from among large lots of a manufactured product yields a number of defectives Y that follows a binomial probability distribution. A sampling plan consists of specifying the number of items to be included in a sample n and an acceptance number a. The lot is accepted if $Y \leq a$, and it is rejected if $Y > a$. Let p denote the proportion of defectives in the lot. For $n = 5$ and $a = 0$, calculate the probability of lot acceptance if the following lot proportions of defectives exist.

a $p = 0$ **b** $p = 0.1$

c $p = 0.3$ **d** $p = 0.5$

e $p = 1.0$

A graph showing the probability of lot acceptance as a function of lot fraction defective is called the *operating characteristics curve* for the sample plan. Construct this curve for the plan $n = 5, a = 0$. Notice that a sampling plan is an example of statistical inference. Accepting or rejecting a lot based on information contained in the sample is equivalent to concluding that the lot is either good or bad, respectively. "Good" implies that a low fraction of items are defective and, therefore, that the lot is suitable for shipment.

3.100 Refer to Exercise 3.99. Use Table 2 of the Appendix to construct the operating characteristic curve for a sampling plan with the following values.

a $n = 10, \quad a = 0$

b $n = 10, \quad a = 1$

c $n = 10, \quad a = 2$

For each, calculate P(Lot acceptance) for $p = 0, 0.05, 0.1, 0.3, 0.5$, and 1.0. Our intuition suggests that sampling plan (a) would be much less likely to accept bad lots than would plan (b) or plan (c). A visual comparison of the operating characteristic curves will confirm this supposition.

3.101 A quality control engineer wishes to study two alternative sample plans: $n = 5, a = 1$; and $n = 25, a = 5$. On a sheet of graph paper, construct the operating characteristic curves for both plans; make use of acceptance probabilities at $p = 0.05, 0.10, 0.20, 0.30$, and 0.40 in each case.

a If you were a seller producing lots whose fraction of defective items ranged from $p = 0$ to $p = 0.10$, which of the two sampling plans would you prefer?

b If you were a buyer wishing to be protected against accepting lots with a fraction defective exceeding $p = 0.30$, which of the two sampling plans would you prefer?

3.102 For a certain section of a pine forest, the number of diseased trees per acre Y has a Poisson distribution, with mean $\lambda = 10$. The diseased trees are sprayed with an insecticide at a cost of \$3 per tree, plus a fixed overhead cost for equipment rental of \$50. Letting C denote the total spraying cost for a randomly selected acre, find the expected value and the standard deviation for C. Within what interval would you expect C to lie with a probability of at least 0.75?

3.103 In checking river water samples for bacteria, a researcher places water in a culture medium to grow colonies of certain bacteria, if present. The number of colonies per dish averages twelve for water samples from a certain river.

a Find the probability that the next dish observed will have at least ten colonies.

b Find the mean and the standard deviation of the number of colonies per dish.

c Without calculating exact Poisson probabilities, find an interval in which at least 75% of the colony count measurements should lie.

3.104 The number of vehicles passing a specified point on a highway averages ten per minute.

a Find the probability that at least 15 vehicles will pass this point in the next minute.

b Find the probability that at least 15 vehicles will pass this point in the next two minutes.

c What assumptions must you make for your answers in (a) and (b) to be valid?

3.105 A production line often produces a variable number N of items each day. Suppose that each item produced has the same probability p of not conforming to manufacturing standards. If N has a Poisson distribution with mean λ, then the number of nonconforming items in one day's production Y has a Poisson distribution with mean λp.

The average number of resistors produced by a facility in one day has a Poisson distribution, with a mean of 100. Typically, 5% of the resistors produced do not meet specifications.

a Find the expected number of resistors that will not meet specifications on a given day.

b Find the probability that all resistors will meet the specifications on a given day.

c Find the probability that more than five resistors will fail to meet specifications on a given day.

3.106 A certain type of bacteria cell divides at a constant rate λ over time. Thus, the probability that a particular cell will divide in a small interval of time t is approximately λt. Given that a population starts out at time zero with k cells of this type, and cell divisions are independent of one another, the size of the population at time t, $Y(t)$, has the probability distribution

$$P[Y(t) = n] = \binom{N-1}{k-1} e^{-\lambda k t} (1 - e^{-\lambda t})^{n-k}$$

a Find the expected value of $Y(t)$ in terms of λ and t.

b If, for a certain type of bacteria cell, $\lambda = 0.1$ per second, and the population starts out with two cells at time zero, find the expected population size after 5 seconds.

3.107 The probability that any one vehicle will turn left at a particular intersection is 0.2. The left-turn lane at this intersection has room for three vehicles. If five vehicles arrive at this intersection while the light is red, find the probability that the left-turn lane will hold all of the vehicles that want to turn left.

3.108 Referring to Exercise 3.107, find the probability that six cars must arrive at the intersection while the light is red to fill up the left-turn lane.

3.109 For any probability $p(y)$, $\sum\limits_{y} p(y) = 1$ if the sum is taken over all possible values y that the random variable in question can assume. Show that this is true for the following distributions.

a the binomial distribution

b the geometric distribution

c the Poisson distribution

3.110 The supply office for a large construction firm has three welding units of Brand A in stock. If a welding unit is requested, the probability is 0.7 that the request will be for this particular brand. On a typical day five requests for welding units come to the office. Find the probability that all three Brand A units will be in use on that day.

3.111 Refer to Exercise 3.110. If the supply office also stocks three welding units that are not Brand A, find the probability that exactly one of these units will be left immediately after the third Brand A unit is requested.

3.112 The probability of a customer's arriving at a grocery service counter in any 1 second equals 0.1. Assume that customers arrive in a random stream and, hence, that the arrival at any 1 second is independent of any other arrival.

a Find the probability that the first arrival will occur during the third 1-second interval.

b Find the probability that the first arrival will not occur until at least the third 1-second interval.

3.113 Of a population of consumers, 60% are reputed to prefer a particular brand, A, of toothpaste. If a group of consumers are interviewed, what is the probability that exactly five people have to be interviewed before encountering a consumer who prefers brand A? At least five people?

3.114 The mean number of automobiles entering a mountain tunnel per 2-minute period is one. If an excessive number of cars enter the tunnel during a brief period of time, the result is a hazardous situation.

a Find the probability that the number of autos entering the tunnel during a 2-minute period exceeds three.

b Assume that the tunnel is observed during ten 2-minute intervals, thus giving ten independent observations, $Y_1, Y_2, \ldots, Y_{10}$, on a Poisson random variable. Find the probability that $Y > 3$ during at least one of the ten 2-minute intervals.

3.115 Suppose that 10% of a brand of microcomputers will fail before their guarantee has expired. If 1000 computers are sold this month, find the expected value and variance of Y, the number that have not failed during the guarantee period. Within what limit would Y be expected to fall? (*Hint:* Use Tchebysheff's Theorem.)

3.116 **a** Consider a binomial experiment for $n = 20$, $p = .05$. Use Table 2 of the Appendix to calculate the binomial probabilities for $Y = 0, 1, 2, 3, 4$.

b Calculate the same probabilities, but this time use the Poisson approximation with $\lambda = np$. Compare the two results.

3.117 The manufacturer of a low-calorie dairy drink wishes to compare the taste appeal of a new formula (B) with that of the standard formula (A). Each of four judges is given three glasses in random order, two containing formula A and the other containing formula B. Each judge is asked to choose which glass he most enjoyed. Suppose that the two formulas are equally attractive. Let Y be the number of judges stating a preference for the new formula.

a Find the probability function for Y.

b What is the probability that at least three of the four judges will state a preference for the new formula?

c Find the expected value of Y.

d Find the variance of Y.

3.118 Show that the hypergeometric probability function approaches the binomial in the limit as $N \to \infty$ and as $p = r/N$ remains constant; that is, show that

$$\lim_{\substack{N \to \infty \\ r \to \infty}} \frac{\binom{r}{y}\binom{N-r}{n-y}}{\binom{N}{n}} = \binom{n}{y} p^y q^{n-y}.$$

for constant $p = r/N$.

3.119 A lot of $N = 100$ industrial products contains 40 defectives. Let Y be the number of defectives in a random sample of size 20. Find $p(10)$ using the following distributions.

a the hypergeometric probability distribution

b the binomial probability distribution

Is N large enough so that the binomial probability function provides a good approximation to the hypergeometric probability function?

3.120 For simplicity, let us assume that there are two kinds of drivers. The safe drivers, who constitute 70% of the population, have a probability of 0.1 of causing an accident in a year. The rest of the population are accident makers, who have a probability of 0.5 of causing an accident in a year. The insurance premium is $400 times one's probability of causing an accident in the following year. A new subscriber has caused an accident during the first year. What should her insurance premium be for the next year?

3.121 A merchant stocks a certain perishable item. He knows that on any given day he will have a demand for two, three, or four of these items, with probabilities 0.1, 0.4, and 0.5, respectively. He buys the items for $1.00 each and sells them for $1.20 each. Any items left at the end of the day represent a total loss. How many items should the merchant stock to maximize his expected daily profit?

3.122 It is known that 5% of a population have disease A, which can be discovered by means of a blood test. Suppose that N (a large number) people are tested. This can be done in two ways:

1. Each person is tested separately.

2. The blood samples of k people are pooled together and analyzed. (Assume that $N = nk$, with n an integer.) If the test is negative, all of the persons in the pool are healthy (that is, just this one test is needed). If the test is positive, each of the k persons must be tested separately (that is, a total of $k + 1$ tests are needed).

a For fixed k, what is the expected number of tests needed in method (2)?

b Find the value for k that will minimize the expected number of tests in method (2).

c How many tests does part (b) save in comparison with part (a)?

3.123 Four possible winning numbers for a lottery—AB-4536, NH-7812, SQ-7855, and ZY-3221—are given to you. You will win a prize if one of your numbers matches one of the winning numbers. You are told that there is one first prize of $100,000; two second prizes of $50,000 each; and ten third prizes of $1000 each. All you have to do is mail the coupon back; no purchase is required. From the structure of the numbers you have received, it is obvious that the entire list consists of all the permutations of two alphabets, followed by four digits. Is the coupon worth mailing back for 29 cents postage?

3.124 For a discrete random variable X taking on values 0, 1, 2, ..., show that $E(X) = \sum_{n=0}^{\infty} P(x > n)$.

4

Continuous Probability Distributions

4.1 Continuous Random Variables and Their Probability Distributions

All of the random variables discussed in Chapter 3 were discrete, assuming only a finite number or countable infinity of values. However, many of the random variables seen in practice have more than a countable collection of possible values. Weights of adult patients coming into a clinic may be anywhere from, say, 80 to 300 pounds. Diameters of machined rods from a certain industrial process may be anywhere from 1.2 to 1.5 centimeters. Proportions of impurities in ore samples may run from 0.10 to 0.80. These random variables can take on any value in an interval of real numbers. That is not to say that every value in the interval can be found in the sample data if one looks long enough; one may never observe a patient weighing exactly 172.38 pounds. Yet, no value can be ruled out as a possible observation; one might encounter a patient weighing 172.38 pounds, so this number must be considered in the set of possible outcomes. Since random variables of this type have a continuum of possible values, they are called *continuous random variables*. Probability distributions for continuous random variables are developed in this chapter, and the basic ideas are presented in the context of an experiment on life lengths.

An experimenter is measuring the life length X of a transistor. In this case, X can assume an infinite number of possible values. We cannot assign a positive probability to each possible outcome of the experiment because, no matter how small we might make the individual probabilities, they would sum to a value greater than one when accumulated over the entire sample space. However, we can assign positive probabilities to *intervals* of real numbers in a manner consistent with the axioms of probability. To introduce the basic ideas involved here, let us consider a specific example in some detail.

Suppose that we have measured the life lengths of 50 batteries of a certain type, selected from a larger population of such batteries. The observed life lengths are as given in Table 4.1. The relative frequency histogram for these data (Figure 4.1) shows clearly that most of the life lengths are near zero, and the frequency drops off rather smoothly as we look at longer life lengths. Here, 32% of the 50 observations fall into the first subinterval (0–1), and another 22% fall into the second (1–2). There is a decline in frequency as we proceed across the subintervals, until the last subinterval (8–9) contains a single observation.

T A B L E **4.1**
Life Lengths of Batteries
(in hundreds of hours)

0.406	0.685	4.778	1.725	8.223
2.343	1.401	1.507	0.294	2.230
0.538	0.234	4.025	3.323	2.920
5.088	1.458	1.064	0.774	0.761
5.587	0.517	3.246	2.330	1.064
2.563	0.511	2.782	6.426	0.836
0.023	0.225	1.514	3.214	3.810
3.334	2.325	0.333	7.514	0.968
3.491	2.921	1.624	0.334	4.490
1.267	1.702	2.634	1.849	0.186

F I G U R E **4.1**
Relative frequency histogram of
data from Table 4.1.

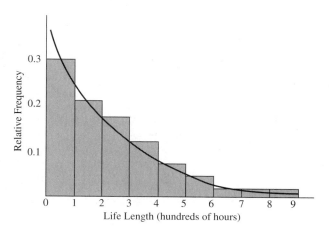

Not only does this sample relative frequency histogram allow us to picture how the sample behaves, but it also gives us some insight into a possible probabilistic model for the random variable X. This histogram of Figure 4.1 looks as though it could be approximated quite closely by a negative exponential curve. The particular function

$$f(x) = \frac{1}{2}e^{-x/2}, \qquad x > 0$$

is sketched through the histogram in Figure 4.1 and seems to fit reasonably well. Thus, we could take this function as a mathematical model for the behavior of the random variable X. If we want to use a battery of this type in the future, we might want to know the probability that it will last longer than 400 hours. This probability can be approximated by the area under the curve to the right of the value 4—that is, by

$$\int_4^\infty \frac{1}{2}e^{-x/2}dx = 0.135$$

Notice that this figure is quite close to the observed sample fraction of lifetimes that exceed 4—namely, $(8/50) = 0.16$. One might suggest that, because the sample fraction 0.16 is available, we do not really need the model. However, the model would give more satisfactory answers for other questions than could otherwise be obtained. For example, suppose that we are interested in the probability that X is greater than 9. Then, the model suggests the answer

$$\int_9^\infty \frac{1}{2}e^{-x/2}dx = 0.011$$

whereas the sample shows no observations in excess of 9. These are quite simple examples, in fact, and we shall see many examples of more involved questions for which a model is essential.

Why did we choose the exponential function as a model here? Would some others not do just as well? The choice of a model is a fundamental problem, and we shall spend considerable time in later sections delving into theoretical and practical reasons for these choices. Here, in this preliminary discussion, we merely examine some models that look as though they might do the job.

The function $f(x)$, which models the relative frequency behavior of X, is called the *probability density function*.

D E F I N I T I O N   **4.1**

A random variable X is said to be **continuous** if there is a function $f(x)$, called the **probability density function**, such that

1   $f(x) \geq 0$,   for all x
2   $\int_{-\infty}^{\infty} f(x)dx = 1$
3   $P(a \leq X \leq b) = \int_a^b f(x)dx$ ∎

Notice that, for a continuous random variable X,

$$P(X = a) = \int_a^a f(x)dx = 0$$

for any specific value a. The need to assign zero probability to any specific value should not disturb us, because X can assume an infinite number of possible values. For example, given all the possible lengths of life of a transistor, what is the probability that the transistor we are using will last exactly 497.392 hours? Assigning probability zero to this event does not rule out 497.392 as a possible length, but it does imply that the chance of observing this particular length is extremely small.

E X A M P L E   **4.1**   The random variable X of the life length example is associated with a probability density function of the form

$$f(x) = \begin{cases} \frac{1}{2}e^{-x/2} & \text{for } x > 0 \\ 0 & \text{elsewhere} \end{cases}$$

Find the probability that the life of a particular battery of this type is less than 200 or greater than 400 hours.

Solution   Let A denote the event that X is less than 2, and let B denote the event that X is greater than 4. Then, because A and B are mutually exclusive,

$$\begin{aligned} P(A \cup B) &= P(A) + P(B) \\ &= \int_0^2 \frac{1}{2}e^{-x/2}dx + \int_4^\infty \frac{1}{2}e^{-x/2}dx \\ &= (1 - e^{-1}) + (e^{-2}) \\ &= 1 - 0.368 + 0.135 \\ &= 0.767 \quad \blacksquare \end{aligned}$$

E X A M P L E   **4.2**   Refer to Example 4.1. Find the probability that a battery of this type lasts more than 300 hours, given that it already has been in use for more than 200 hours.

Solution   We are interested in $P(X > 3 | X > 2)$; and by the definition of conditional probability,

$$P(X > 3 | X > 2) = \frac{P(X > 3)}{P(X > 2)}$$

because the intersection of the events $(X > 3)$ and $(X > 2)$ is the event $(X > 3)$. Now

$$\frac{P(X > 3)}{P(X > 2)} = \frac{\int_3^\infty \frac{1}{2}e^{-x/2}dx}{\int_2^\infty \frac{1}{2}e^{-x/2}dx} = \frac{e^{-3/2}}{e^{-1}} = e^{-1/2} = 0.606 \quad \blacksquare$$

Sometimes it is convenient to look at cumulative probabilities of the form $P(X \le b)$. To do this, we can make use of what is called the *distribution function*.

D E F I N I T I O N   **4.2**   The **distribution function** for a random variable X is defined as

$$F(b) = P(X \le b)$$

If X is continuous, with probability density function $f(x)$, then

$$F(b) = \int_{-\infty}^{b} f(x)dx$$

Notice that $F'(x) = f(x)$. ∎

In the battery example, X has a probability density function given by

$$f(x) = \begin{cases} \frac{1}{2}e^{-x/2} & \text{for } x > 0 \\ 0 & \text{elsewhere} \end{cases}$$

Thus,

$$\begin{aligned} F(b) &= P(X \leq b) \\ &= \int_0^b \frac{1}{2}e^{-x/2}dx \\ &= -e^{-x/2}|_0^b \\ &= 1 - e^{-b/2}, \qquad b > 0 \\ &= 0, \qquad\qquad\quad b \leq 0 \end{aligned}$$

The function is shown graphically in Figure 4.2.

FIGURE **4.2**
Distribution function for a
continuous random variable.

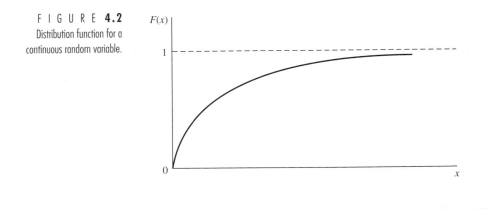

EXAMPLE **4.3** A supplier of kerosene has a 200-gallon tank filled at the beginning of each week. His weekly demands show a relative frequency behavior that increases steadily up to 100 gallons, and then levels off between 100 and 200 gallons. Letting X denote weekly demand in hundreds of gallons, suppose that the relative frequencies for demand are modeled adequately by

$$f(x) = \begin{cases} 0, & x < 0 \\ x, & 0 \leq x \leq 1 \\ 1/2, & 1 < x \leq 2 \\ 0, & x > 2 \end{cases}$$

This function has the graphical form shown in Figure 4.3. Find $F(b)$ for this random variable. Then use $F(b)$ to find the probability that demand will exceed 150 gallons on a given week.

F I G U R E **4.3**
$f(x)$ for Example 4.3.

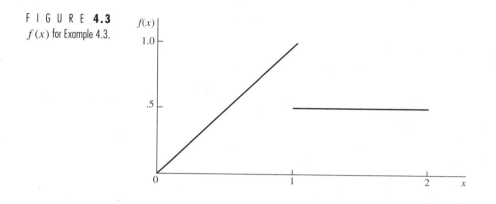

Solution From the definition,

$$
\begin{aligned}
F(b) &= \int_{-\infty}^{b} f(x)\,dx \\
&= 0, && b < 0 \\
&= \int_{0}^{b} x\,dx = \frac{b^2}{2}, && 0 \le b \le 1 \\
&= \frac{1}{2} + \int_{1}^{b} \frac{1}{2}\,dx \\
&= \frac{1}{2} + \frac{b-1}{2} = \frac{b}{2}, && 1 < b \le 2 \\
&= 1, && b > 2
\end{aligned}
$$

This function is graphed in Figure 4.4. Notice that $F(b)$ is continuous over the whole real line, even though $f(b)$ has two discontinuities. The probability that demand will exceed 150 gallons is given by

$$
\begin{aligned}
P(X > 1.5) &= 1 - P(X \le 1.5) \\
&= 1 - F(1.5) \\
&= 1 - \frac{1.5}{2} \\
&= .25 \quad \blacksquare
\end{aligned}
$$

FIGURE **4.4**
$F(b)$ for Example 4.3.

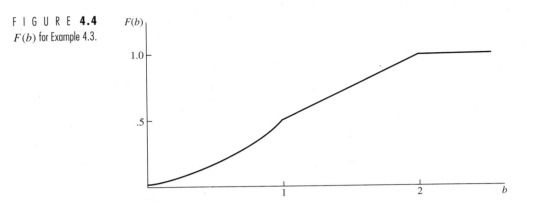

FIGURE **4.4**
$F(b)$ for Example 4.3.

Exercises

4.1 For each of the following situations, define an appropriate random variable and state whether it is continuous or discrete.

  a An environmental engineer is looking at ten field plots to determine whether they contain a certain type of insect.

  b A quality control technician samples a continuously produced fabric in square yard sections, and counts the number of defects observed in each sampled section.

  c A metallurgist counts the number of grains seen in a cross-sectional sample of aluminum.

  d The metallurgist of part (c) measures the area proportion covered by grains of a certain size, rather than simply counting them.

4.2 Suppose that a random variable X has a probability density function given by

$$f(x) = \begin{cases} kx(1-x) & \text{for } 0 \le x \le 1 \\ 0 & \text{elsewhere} \end{cases}$$

  a Find the value of k that makes this a probability density function.

  b Find $P(0.4 \le X \le 1)$.

  c Find $P(X \le 0.4 | X \le 0.8)$.

  d Find $F(b) = P(X \le b)$, and sketch the graph of this function.

4.3 The effectiveness of solar-energy heating units depends on the amount of radiation available from the sun. During a typical October, daily total solar radiation in Tampa, Florida, approximately follows the following probability density function (units are hundreds of calories):

$$f(x) = \begin{cases} \frac{3}{32}(x-2)(6-x) & \text{for } 2 \le x \le 6 \\ 0 & \text{elsewhere} \end{cases}$$

  a Find the probability that solar radiation will exceed 300 calories on a typical October day.

  b What amount of solar radiation is exceeded on exactly 50% of the October days, according to this model?

4.4 An accounting firm that does not have its own computing facilities rents time from a consulting company. The accounting firm must plan its computing budget carefully; hence, it has studied the weekly use of CPU time quite thoroughly. The weekly use of CPU time approximately conforms to the following probability density function (measurements in hours):

$$f(x) = \begin{cases} \frac{3}{64}x^2(4-x) & \text{for } 0 \le x \le 4 \\ 0 & \text{elsewhere} \end{cases}$$

a Find the distribution function $F(x)$ for weekly CPU time X.

b Find the probability that CPU time used by the firm will exceed 2 hours for a selected week.

c The current budget of the firm covers only 3 hours of CPU time per week. How often will the budgeted figure be exceeded?

d How much CPU time should be budgeted per week if this figure is to be exceeded with a probability of only 0.10?

4.5 The pH level, a measure of acidity, is important in studies of acid rain. For a certain Florida lake, baseline measurements of acidity are made so that any changes caused by acid rain can be noted. The pH for water samples from the lake is a random variable X, with probability density function

$$f(x) = \begin{cases} \frac{3}{8}(7-x)^2 & \text{for } 5 \le x \le 7 \\ 0 & \text{elsewhere} \end{cases}$$

a Sketch the curve of $f(x)$.

b Find the distribution function $F(x)$ for X.

c Find the probability that the pH for a water sample from this lake will be less than 6.

d Find the probability that the pH of a water sample from this lake will be less than 5.5, given that it is known to be less than 6.

4.6 The "on" temperature of a thermostatically controlled switch for an air conditioning system is set at $60°$, but the actual temperature X at which the switch turns on is a random variable having the probability density function

$$f(x) = \begin{cases} \frac{1}{2} & \text{for } 59 \le x \le 61 \\ 0 & \text{elsewhere} \end{cases}$$

a Find the probability that a temperature in excess of $60°$ is required to turn the switch on.

b If two such switches are used independently, find the probability that a temperature in excess of $60°$ is required to turn both on.

4.7 The proportion of time, during a 40-hour work week, that an industrial robot was in operation was measured for a large number of weeks. The measurements can be modeled by the probability density function

$$f(x) = \begin{cases} 2x & \text{for } 0 \le x \le 1 \\ 0 & \text{elsewhere} \end{cases}$$

If X denotes the proportion of time this robot will be in operation during a coming week, find the following values.

a $P(X > 1/2)$

b $P(X > 1/2 | X > 1/4)$

c $P(X > 1/4 | X > 1/2)$

d $F(x)$. Graph this function. Is $F(x)$ continuous?

4.8 The proportion of impurities by weight X in certain copper ore samples is a random variable having a probability density function of

$$f(x) = \begin{cases} 12x^2(1-x) & \text{for } 0 \le x \le 1 \\ 0 & \text{elsewhere} \end{cases}$$

If four such samples are independently selected, find the probabilities of the following events.

a   Exactly one sample has a proportion of impurities exceeding 0.5.

b   At least one sample has a proportion of impurities exceeding 0.5.

4.2 Expected Values of Continuous Random Variables

As in the discrete case, we often want to summarize the information contained in a continuous variable's probability distribution by calculating expected values for the random variable and for certain functions of the random variable.

DEFINITION **4.3**

> The **expected value** of a continuous random variable X that has a probability density function $f(x)$ is given by*
>
> $$E(X) = \int_{-\infty}^{\infty} x f(x) dx$$ ∎

For functions of random variables we have the following theorem.

THEOREM **4.1**

> If X is a continuous random variable with probability distribution $f(x)$, and if $g(x)$ is any real-valued function of X, then
>
> $$E[g(X)] = \int_{-\infty}^{+\infty} g(x) f(x) dx$$
>
> The proof of Theorem 4.1 will not be given here.   ∎

The definitions of *variance* and of *standard deviation* and the properties given in Theorems 3.2 and 3.3 hold for the continuous case, as well.

> For a random variable X with probability density function $f(x)$, the variance of X is given by
>
> $$V(X) = E(X - \mu)^2$$
> $$= \int_{-\infty}^{+\infty} (x - \mu)^2 f(x) dx$$
> $$= E(X^2) - \mu^2$$

*We assume absolute convergence of the integrals.

where $\mu = E(X)$. For constants a and b,

$$E(aX + b) = aE(X) + b$$

and

$$V(aX + b) = a^2 V(X)$$

We illustrate the expectations of continuous random variables in the following two examples.

EXAMPLE **4.4**   For a lathe in a machine shop, let X denote the percentage of time out of a 40-hour work week during which the lathe is actually in use. Suppose that X has a probability density function given by

$$f(x) = \begin{cases} 3x^2 & \text{for } 0 \le x \le 1 \\ 0 & \text{elsewhere} \end{cases}$$

Find the mean and the variance of X.

Solution   From Definition 4.3,

$$\begin{aligned}
E(X) &= \int_{-\infty}^{\infty} x f(x) \, dx \\
&= \int_{0}^{1} x(3x^2) \, dx \\
&= \int_{0}^{1} 3x^3 \, dx \\
&= 3 \left[\frac{x^4}{4} \right]_{0}^{1} \\
&= \frac{3}{4} \\
&= 0.75
\end{aligned}$$

Thus, on the average, the lathe is in use 75% of the time.
To compute $V(X)$, we first find $E(X^2)$:

$$\begin{aligned}
E(X^2) &= \int_{-\infty}^{\infty} x^2 f(x) \, dx \\
&= \int_{0}^{1} x^2 (3x^2) \, dx \\
&= \int_{0}^{1} 3x^4 \, dx
\end{aligned}$$

$$= 3 \left[\frac{x^5}{5} \right]_0^1$$

$$= \frac{3}{5}$$

$$= 0.60$$

Then,

$$V(X) = E(X^2) - \mu^2$$
$$= 0.60 - (0.75)^2$$
$$= 0.60 - 0.5625$$
$$= 0.0375 \quad \blacksquare$$

E X A M P L E   **4.5**   The weekly demand X for kerosene at a certain supply station has a density function given by

$$f(x) = \begin{cases} x & \text{for } 0 \le x \le 1 \\ \frac{1}{2} & \text{for } 1 < x \le 2 \\ 0 & \text{elsewhere} \end{cases}$$

Find the expected weekly demand.

Solution   Using Definition 4.3 to find $E(X)$, we must now carefully observe that $f(x)$ has different nonzero forms over two disjoint regions. Thus,

$$E(X) = \int_{-\infty}^{\infty} x f(x) dx$$

$$= \int_0^1 x(x) dx + \int_1^2 x \left(\frac{1}{2} \right) dx$$

$$= \int_0^1 x^2 dx + \frac{1}{2} \int_1^2 x dx$$

$$= \left[\frac{x^3}{3} \right]_0^1 + \frac{1}{2} \left[\frac{x^2}{2} \right]_1^2$$

$$= \frac{1}{3} + \frac{1}{2} \left[2 - \frac{1}{2} \right]$$

$$= \frac{1}{3} + \frac{3}{4}$$

$$= \frac{13}{12}$$

$$= 1.08$$

In other words, the expected weekly demand is for 108 gallons of kerosene.   ■

Tchebysheff's Theorem (Theorem 3.4) holds for continuous random variables, just as it does for discrete ones. Thus, if X is continuous, with mean μ and standard deviation σ, then

$$P(|X - \mu| < k\sigma) \geq 1 - \frac{1}{k^2}$$

for any positive number k. We illustrate the use of this result in the next example.

E X A M P L E   **4.6**   The weekly amount Y spent for chemicals in a certain firm has a mean of $445 and a variance of $236. Within what interval should these weekly costs for chemicals be expected to lie at least 75% of the time?

Solution   To find an interval guaranteed to contain at least 75% of the probability mass for Y, we specify

$$1 - \frac{1}{k^2} = 0.75$$

which yields

$$\frac{1}{k^2} = 0.25$$
$$k^2 = \frac{1}{0.25}$$
$$= 4$$
$$k = 2$$

Thus, the interval $\mu - 2\sigma$ to $\mu + 2\sigma$ will contain at least 75% of the probability. This interval is given by

$$445 - 2\sqrt{236} \text{ to } 445 + 2\sqrt{236}$$
$$445 - 30.72 \text{ to } 445 + 30.72$$
$$414.28 \text{ to } 475.72 \quad \blacksquare$$

Exercises

4.9   The temperature X at which a thermostatically controlled switch turns on has a probability density function of

$$f(x) = \begin{cases} \frac{1}{2} & \text{for } 59 \leq x \leq 61 \\ 0 & \text{elsewhere} \end{cases}$$

Find $E(X)$ and $V(X)$.

4.10   The proportion of time X during a 40-hour workweek that an industrial robot is in operation is a random variable with a probability density function of

$$f(x) = \begin{cases} 2x & \text{for } 0 \leq x \leq 1 \\ 0 & \text{elsewhere} \end{cases}$$

   a   Find $E(X)$ and $V(X)$.

   b   For the robot under study, the profit Y for a week is given by $Y = 200X - 60$. Find $E(Y)$ and $V(Y)$.

   c   Find an interval in which the profit should lie for at least 75% of the weeks that the robot is in use.

4.11   Daily total solar radiation for a certain location in Florida in October has a probability density function of

$$f(x) = \begin{cases} \frac{3}{32}(x - 2)(6 - x) & \text{for } 2 \leq x \leq 6 \\ 0 & \text{elsewhere} \end{cases}$$

   with measurements in hundreds of calories. Find the expected daily solar radiation for October.

4.12   Weekly CPU time used by an accounting firm has a probability density function (measured in hours) of

$$f(x) = \begin{cases} \frac{3}{64}x^2(4 - x) & \text{for } 0 \leq x \leq 4 \\ 0 & \text{elsewhere} \end{cases}$$

   a   Find the expected value and the variance of weekly CPU time.

   b   The CPU time costs the firm $200 per hour. Find the expected value and the variance of the weekly cost for CPU time.

   c   Would you expect the weekly cost to exceed $600 very often? Why?

4.13   The pH of water samples from a specific lake is a random variable X, with a probability density function of

$$f(x) = \begin{cases} \frac{3}{8}(7 - x)^2 & \text{for } 5 \leq x \leq 7 \\ 0 & \text{elsewhere} \end{cases}$$

   a   Find $E(X)$ and $V(X)$.

   b   Find an interval shorter than $(5, 7)$ within which at least 3/4 of the pH measurements must lie.

   c   Would you expect to see a pH measurement of less than 5.5 very often? Why?

4.14   A retail grocer has a daily demand X for a certain food sold by the pound, such that X (measured in hundreds of pounds) has a probability density function of

$$f(x) = \begin{cases} 3x^2 & \text{for } 0 \leq x \leq 1 \\ 0 & \text{elsewhere} \end{cases}$$

   The grocer, who cannot stock more than 100 pounds, wants to order $100k$ pounds of food on a certain day. He buys the food at 6 cents per pound and sells it at 10 cents per pound. What value of k will maximize his expected daily profit? (There is no salvage value for food not sold.)

4.3   The Uniform Distribution

4.3.1   Probability Density Function

We now move from a general discussion of continuous random variables to discussions of specific models that have been found useful in practice. Consider an experiment that consists of observing events in a certain time frame, such as buses arriving at a bus stop or telephone calls coming into a switchboard during a specified period. Suppose that we know that one such event has occurred in the time interval (a, b): a bus arrived between 8:00 and 8:10. It may then be of interest to place a probability distribution on the actual time of occurrence of the event under observation, which we will denote by X. A very simple model assumes that X is equally likely to lie in any small subinterval—say, of length d—no matter where that subinterval lies within (a, b). This assumption leads to the *uniform* probability distribution, which has the probability density function given by

$$f(x) = \begin{cases} \frac{1}{b-a} & \text{for } a \leq x \leq b \\ 0 & \text{elsewhere} \end{cases}$$

This density function is graphed in Figure 4.5.

F I G U R E **4.5**
Uniform probability density function.

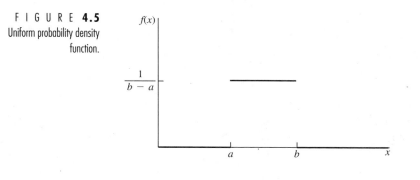

The distribution function for a uniformly distributed X is given by

$$F(x) = 0, \qquad\qquad\qquad x < a$$

$$F(x) = \int_a^x \frac{1}{b-a}\,dx = \frac{x-a}{b-a}, \qquad a \leq x \leq b$$

$$F(x) = 1, \qquad\qquad\qquad x > b$$

If we consider a subinterval $(c, c + d)$ contained entirely within (a, b), we have

$$P(c \leq X \leq c + d) = F(c + d) - F(c)$$
$$= \frac{(c + d) - a}{b - a} - \frac{c - a}{b - a}$$
$$= \frac{d}{b - a}$$

Notice that this probability does not depend on the subinterval's location c, but only on its length d.

A relationship exists between the uniform distribution and the Poisson distribution, which was introduced in Section 3.7. Suppose that the number of events that occur in an interval—say, $(0, t)$—has a Poisson distribution. If exactly one of these events is known to have occurred in the interval (a, b), with $a \geq 0$ and $b \leq t$, then the conditional probability distribution of the actual time of occurrence for this event is uniform over (a, b).

4.3.2 Mean and Variance

Paralleling our approach in Chapter 3, we now look at the mean and the variance of the uniform distribution. From Definition 4.3,

$$E(X) = \int_{-\infty}^{\infty} x f(x) dx$$
$$= \int_{a}^{b} x \left(\frac{1}{b - a} \right) dx$$
$$= \left(\frac{1}{b - a} \right) \left(\frac{b^2 - a^2}{2} \right)$$
$$= \frac{a + b}{2}$$

It is intuitively reasonable that the mean value of a uniformly distributed random variable should lie at the midpoint of the interval.

Recalling from Theorem 3.3 that $V(X) = E(X - \mu)^2 = E(X^2) - \mu^2$, we have, for the uniform case,

$$E(X^2) = \int_{-\infty}^{\infty} x^2 f(x) dx$$
$$= \int_{a}^{b} x^2 \left(\frac{1}{b - a} \right) dx$$
$$= \left(\frac{1}{b - a} \right) \left(\frac{b^3 - a^3}{3} \right)$$
$$= \frac{b^2 + ab + a^2}{3}.$$

Thus,

$$V(X) = \frac{b^2 + ab + a^2}{3} - \left(\frac{a + b}{2} \right)^2$$

$$= \frac{1}{12}[4(b^2 + ab + a^2) - 3(a + b)^2]$$
$$= \frac{1}{12}(b - a)^2$$

This result may not be intuitive, but we see that the variance depends only on the length of the interval (a, b).

E X A M P L E **4.7** The failure of a circuit board interrupts work by a computing system until a new board is delivered. Delivery time X is uniformly distributed over the interval of from one to five days. The cost C of this failure and interruption consists of a fixed cost c_0 for the new part and a cost that increases proportionally to X^2, so that

$$C = c_0 + c_1 X^2$$

1 Find the probability that the delivery time is two or more days.
2 Find the expected cost of a single failure, in terms of c_0 and c_1.

Solution 1 The delivery time X is distributed uniformly over the interval of from one to five days, which gives

$$f(x) = \begin{cases} \frac{1}{4} & \text{for } 1 \le x \le 5 \\ 0 & \text{elsewhere} \end{cases}$$

Thus,

$$P(X \ge 2) = \int_2^5 \left(\frac{1}{4}\right) dx$$
$$= \frac{1}{4}(5 - 2)$$
$$= \frac{3}{4}$$

2 We know that

$$E(C) = c_0 + c_1 E(X^2)$$

so it remains for us to find $E(X^2)$. This value could be found directly from the definition or by using the variance and the fact that

$$E(X^2) = V(X) + \mu^2$$

Using the latter approach, we find

$$E(X^2) = \frac{(b - a)^2}{12} + \left(\frac{a + b}{2}\right)^2$$
$$= \frac{(5 - 1)^2}{12} + \left(\frac{1 + 5}{2}\right)^2$$
$$= \frac{31}{3}$$

Thus,

$$E(C) = c_0 + c_1 \left(\frac{31}{3} \right) \quad \blacksquare$$

We now review the properties of the uniform distribution.

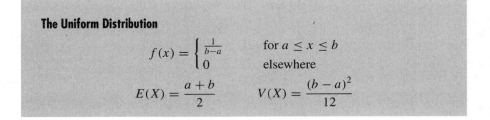

The Uniform Distribution

$$f(x) = \begin{cases} \frac{1}{b-a} & \text{for } a \le x \le b \\ 0 & \text{elsewhere} \end{cases}$$

$$E(X) = \frac{a+b}{2} \qquad V(X) = \frac{(b-a)^2}{12}$$

Exercises

4.15   Suppose X has a uniform distribution over the interval (a, b).

   a   Find $F(x)$.

   b   Find $P(X > c)$ for some point c between a and b.

   c   If $a \le c \le d \le b$, find $P(X > d | X > c)$.

4.16   Upon studying low bids for shipping contracts, a microcomputer manufacturing firm finds that intrastate contracts have low bids uniformly distributed between 20 and 25, in units of thousands of dollars. Find the probability that the low bid on the next intrastate shipping contract is as follows.

   a   below \$22,000

   b   in excess of \$24,000

   Now find the average cost of low bids on contracts of this type.

4.17   If a point is *randomly* located in an interval (a, b), and if X denotes the distance of the point from a, then X is assumed to have a uniform distribution over (a, b). A plant efficiency expert randomly picks a spot along a 500-foot assembly line from which to observe work habits. Find the probability that the point she selects is located as follows.

   a   within 25 feet of the end of the line

   b   within 25 feet of the beginning of the line

   c   closer to the beginning of the line than to the end of the line.

4.18   A bomb is to be dropped along a 1-mile-long line that stretches across a practice target zone. The target zone's center is at the midpoint of the line. The target will be destroyed if the bomb falls within $\frac{1}{10}$ mile on either side of the center. Find the probability that the target will be destroyed, given that the bomb falls at a random location along the line.

4.19   A telephone call arrived at a switchboard at a random instant within a 1-minute interval. The switchboard was fully busy for 15 seconds into this 1-minute period. Find the probability that the call arrived when the switchboard was not fully occupied.

4.20 Beginning at 12:00 midnight, a computer center is up for 1 hour and down for 2 hours on a regular cycle. A person who does not know the schedule dials the center at a random time between 12:00 midnight and 5:00 A.M. What is the probability that the center will be operating when this call comes in?

4.21 The number of defective circuit boards among those coming out of a soldering machine follows a Poisson distribution. For a particular 8-hour day, one defective board is found.

    a   Find the probability that it was produced during the first hour of operation for that day.

    b   Find the probability that it was produced during the last hour of operation for that day.

    c   Given that no defective boards were seen during the first 4 hours of operation, find the probability that the defective board was produced during the fifth hour.

4.22 In determining the range of an acoustic source by triangulation, one must accurately measure the time at which the spherical wave front arrives at a receiving sensor. According to Perruzzi and Hilliard (1984), errors in measuring these arrival times can be modeled as having uniform distributions. Suppose that measurement errors are uniformly distributed from -0.05 to $+0.05$ microseconds.

    a   Find the probability that a particular arrival time measurement will be in error by less than 0.01 microsecond.

    b   Find the mean and the variance of these measurement errors.

4.23 In the setting of Exercise 4.22, suppose that the measurement errors are distributed uniformly from -0.02 to $+0.05$ microseconds.

    a   Find the probability that a particular arrival time measurement will be in error by less than 0.01 microsecond.

    b   Find the mean and the variance of these measurement errors.

4.24 According to Y. Zimmels (1983), the sizes of particles used in sedimentation experiments often have uniform distributions. It is important to study both the mean and the variance of particle sizes because, in sedimentation with mixtures of various-size particles, the larger particles hinder the movements of the smaller particles.

    Suppose that spherical particles have diameters uniformly distributed between 0.01 and 0.05 centimeters. Find the mean and the variance of the *volumes* of these particles. (Recall that the volume of a sphere is $(4/3)\pi r^3$.)

4.25 Arrivals of customers at a certain checkout counter follow a Poisson distribution. During a given 30-minute period, one customer arrives at the counter. Find the probability that he arrives during the last 5 minutes of the period.

4.26 Using the conditions identified in Exercise 4.25, find the conditional probability that the customer arrives during the last 5 minutes of the 30-minute period, given that no one arrives during the first 10 minutes of the period.

4.27 In tests of stopping distances for automobiles, cars traveling 30 miles per hour before the brakes were applied tended to travel distances that appeared to be uniformly distributed between two points a and b. Find the probabilities of the following events.

    a   One of these automobiles, selected at random, stops closer to a than to b.

    b   One of these automobiles, selected at random, stops at a point where the distance to a is more than three times the distance to b.

4.28 Suppose that three automobiles are used in a test of the type discussed in Exercise 4.27. Find the probability that exactly one of the three travels past the midpoint between a and b.

4.29 The cycle time for trucks hauling concrete to a highway construction site is uniformly distributed over the interval from 50 to 70 minutes.

    a   Find the expected value and the variance for these cycle times.

    b   How many trucks should you expect to have to schedule for this job so that a truckload of concrete can be dumped at the site every 15 minutes?

4.4 The Exponential Distribution

4.4.1 Probability Density Function

The life-length data of Section 4.1 displayed a nonuniform probabilistic behavior: the probability over intervals of constant length decreased as the intervals moved farther and farther to the right. We saw that an exponential curve seemed to fit these data rather well, and we now discuss the exponential probability distribution in more detail. In general the exponential density function is given by

$$f(x) = \begin{cases} \frac{1}{\theta} e^{-x/\theta} & \text{for } x \geq 0 \\ 0 & \text{elsewhere} \end{cases}$$

where the parameter θ is a constant ($\theta > 0$) that determines the rate at which the curve decreases.

An exponential density function with $\theta = 2$ was sketched in Figure 4.1; and in general, the exponential functions have the form shown in Figure 4.6. Many random variables in engineering and the sciences can be modeled appropriately as having exponential distributions. Figure 4.7 shows two examples of relative frequency distributions for times between arrivals (interarrival times) of vehicles at a fixed point on a one-directional roadway. Both of these relative frequency histograms can be modeled quite nicely by exponential functions. Notice that the higher traffic density causes shorter interarrival times to be more frequent.

The distribution function for the exponential case has the following simple form:

$$F(t) = 0 \qquad \qquad \text{for } t < 0$$
$$F(t) = P(X \leq t)$$
$$= \int_0^t \frac{1}{\theta} e^{-x/\theta} \, dx$$
$$= -e^{-x/\theta} \big|_0^t$$
$$= 1 - e^{-t/\theta} \qquad \text{for } t \geq 0$$

FIGURE **4.6**
Exponential probability density
function.

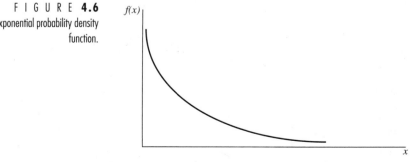

F I G U R E **4.7**
Interarrival times of vehicles on
a one-directional roadway.

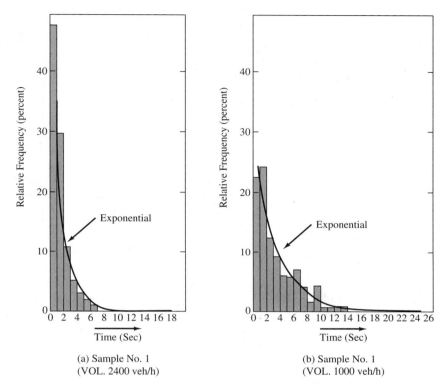

(a) Sample No. 1
(VOL. 2400 veh/h)

(b) Sample No. 1
(VOL. 1000 veh/h)

Source: Mahalel, D. and A.S. Hakkert, *Transportation Research,* 17A, No. 4, 1983, p. 267.

4.4.2   Mean and Variance

Finding expected values for the exponential distribution is simplified by understanding a certain type of integral called a *gamma* (Γ) *function.* The function $\Gamma(\alpha)$ is defined by

$$\Gamma(\alpha) = \int_0^\infty x^{\alpha-1} e^{-x} dx.$$

Integration by parts can be used to show that $\Gamma(\alpha + 1) = \alpha\Gamma(\alpha)$. It follows that $\Gamma(n) = (n - 1)!$, for any positive integer n. The integral

$$\int_0^\infty x^{\alpha-1} e^{-x/\beta} dx$$

for positive constants α and β, can be evaluated by making the transformation $y = x/\beta$, or $x = \beta y$, where $dx = \beta dy$. We have then

$$\int_0^\infty (\beta y)^{\alpha-1} e^{-y\beta} dy = \beta^\alpha \int_0^\infty y^{\alpha-1} e^{-y} dy = \beta^\alpha \Gamma(\alpha)$$

It is useful to note that $\Gamma(1/2) = \sqrt{\pi}$.

Using this result, we see that, for the exponential distribution,

$$E(X) = \int_{-\infty}^{\infty} xf(x)dx$$

$$= \int_0^{\infty} x\left(\frac{1}{\theta}\right)e^{-x/\theta}dx$$

$$= \frac{1}{\theta}\int_0^{\infty} xe^{-x/\theta}dx$$

$$= \frac{1}{\theta}\Gamma(2)\theta^2 = \theta$$

Thus, the parameter θ actually is the mean of the distribution.

To evaluate the variance of the exponential distribution, we start by finding

$$E(X^2) = \int_0^{\infty} x^2\left(\frac{1}{\theta}\right)e^{-x/\theta}dx$$

$$= \frac{1}{\theta}\Gamma(3)\theta^3$$

$$= 2\theta^2$$

It follows that

$$V(X) = E(X^2) - \mu^2$$
$$= 2\theta^2 - \theta^2 = \theta^2$$

and θ becomes the standard deviation as well as the mean.

E X A M P L E   **4.8**   A sugar refinery has three processing plants, all of which receive raw sugar in bulk. The amount of sugar that one plant can process in one day can be modeled as having an exponential distribution with a mean of 4 (measurements in tons) for each of the three plants. If the plants operate independently, find the probability that exactly two of the three plants will process more than 4 tons on a given day.

Solution   The probability that any given plant will process more than 4 tons on a given day, with X denoting the amount used, is

$$P(X > 4) = \int_4^{\infty} f(x)dx$$

$$= \int_4^{\infty} \frac{1}{4}e^{-x/4}dx$$

$$= -e^{-x/4}\big|_4^{\infty}$$

$$= e^{-1}$$

$$= 0.37$$

(*Note:* $P(X > 4) \neq 0.5$, even though the mean of X is 4.)

Knowledge of the distribution function would allow us to evaluate this immediately as

$$P(X > 4) = 1 - P(X \le 4)$$
$$= 1 - (1 - e^{-4/4})$$
$$= e^{-1} = .37$$

Assuming that the three plants operate independently, the problem is to find the probability of two successes out of three tries, where the probability of success is 0.37. This is a binomial problem, and the solution is

$$P(\text{Exactly two plants use more than 4 tons}) = \binom{3}{2}(0.37)^2(0.63)$$
$$= 3(0.37)^2(0.63)$$
$$= 0.26 \quad \blacksquare$$

E X A M P L E   **4.9**   Consider a particular plant in Example 4.8. How much raw sugar should be stocked for that plant each day so that the chance of running out of product is only 0.05?

Solution   Let a denote the amount to be stocked. Since the amount to be used X has an exponential distribution,

$$P(X > a) = \int_a^\infty \frac{1}{4} e^{-x/4} dx = e^{-a/4}$$

We want to choose a so that

$$P(X > a) = e^{-a/4} = 0.05$$

and solving this equation yields

$$a = 11.98 \quad \blacksquare$$

As in the uniform case, a relationship exists between the exponential distribution and the Poisson distribution. Suppose that events are occurring in time according to a Poisson distribution with a rate of λ events per hour. Thus, in t hours, the number of events—say, Y—will have a Poisson distribution with mean value λt. Suppose that we start at time zero and ask the question, How long do I have to wait to see the first event occur? Let X denote the length of time until this first event. Then,

$$P(X > t) = P[Y = 0 \text{ on the interval } (0, t)]$$
$$= \frac{(\lambda t)^0 e^{-\lambda t}}{0!}$$
$$= e^{-\lambda t}$$

and

$$P(X \le t) = 1 - P(X > t) = 1 - e^{-\lambda t}$$

We see that $P(X \le t) = F(t)$, the distribution function for X, has the form of an exponential distribution function, with $\lambda = (1/\theta)$. Upon differentiating, we see that the probability density function of X is given by

$$
\begin{aligned}
f(t) &= \frac{dF(t)}{dt} \\
&= \frac{d(1 - e^{-\lambda t})}{dt} \\
&= \lambda e^{-\lambda t} \\
&= \frac{1}{\theta} e^{-t/\theta}, \qquad t > 0
\end{aligned}
$$

and X has an exponential distribution. Actually, we need not start at time zero, because it can be shown that the waiting time from the occurrence of any one event until the occurrence of the next has an exponential distribution for events occurring according to a Poisson distribution. The properties of the exponential distribution are summarized next.

The Exponential Distribution

$$f(x) = \begin{cases} \frac{1}{\theta} e^{-x/\theta} & \text{for } x > 0 \\ 0 & \text{elsewhere} \end{cases}$$

$$E(X) = \theta \qquad V(X) = \theta^2$$

Exercises

4.30   Suppose that Y has an exponential density function with mean θ. Show that $P(Y > a + b | Y > a) = P(Y > b)$. This is referred to as the *memoryless property* of the exponential distribution. [This is a unique property of the exponential distribution.]

4.31   The magnitudes of earthquakes recorded in a region of North America can be modeled by an exponential distribution with a mean of 2.4, as measured on the Richter scale. Find the probabilities that the next earthquake to strike this region will have the following characteristics:

a   It will exceed 3.0 on the Richter scale.

b   It will fall between 2.0 and 3.0 on the Richter scale.

4.32   Referring to Exercise 4.31, find the probability that, of the next ten earthquakes to strike this region, at least one will exceed 5.0 on the Richter scale.

4.33   A pumping station operator observes that the demand for water at a certain hour of the day can be modeled as an exponential random variable with a mean of 100 cfs (cubic feet per second).

a   Find the probability that the demand will exceed 200 cfs on a randomly selected day.

b What is the maximum water-producing capacity that the station should keep on line for this hour so that the demand will have a probability of only 0.01 of exceeding this production capacity?

4.34 Suppose that customers arrive at a certain checkout counter at a rate of two every minute.

a Find the mean and the variance of the waiting time between successive customer arrivals.

b If a clerk takes 3 minutes to serve the first customer arriving at the counter, what is the probability that at least one more customer will be waiting when the service provided to the first customer is completed?

4.35 The length of time X required to complete a certain key task in house construction is an exponentially distributed random variable, with a mean of 10 hours. The cost C of completing this task is related to the square of the time required for completion by the formula

$$C = 100 + 40X + 3X^2$$

a Find the expected value and the variance of C.

b Would you expect C to exceed 2000 very often?

4.36 The interaccident times (times between accidents) for all fatal accidents on scheduled American domestic passenger airplane flights for the period 1948–1961 were found to follow an exponential distribution, with a mean of approximately 44 days (Pyke 1968, p. 426).

a If one of those accidents occurred on July 1, find the probability that another one occurred in that same (31-day) month.

b Find the variance of the interaccident times.

c What does this information suggest about the clumping of airline accidents?

4.37 Under average driving conditions, the life lengths of automobile tires of a certain brand are found to follow an exponential distribution, with a mean of 30,000 miles. Find the probability that one of these tires, bought today, will last the following numbers of miles.

a over 30,000 miles

b over 30,000 miles, given that it already has gone 15,000 miles

4.38 The dial-up connections from remote terminals come into a computing center at the rate of four per minute. The calls follow a Poisson distribution. If a call arrives at the beginning of a 1-minute period, find the probability that a second call will not arrive within the next 20 seconds.

4.39 The breakdowns of an industrial robot follow a Poisson distribution, with an average of 0.5 breakdowns per 8-hour workday. If this robot is placed in service at the beginning of the day, find the probabilities of the following events.

a It will not break down during the day.

b It will work for at least 4 hours without breaking down.

Does what happened the day before have any effect on your answers? Why?

4.40 Air samples from a large city are found to have 1-hour carbon monoxide concentrations in an exponential distribution, with a mean of 3.6 ppm (Zamurs 1984, p. 637).

a Find the probability that a concentration will exceed 9 parts per million (ppm).

b A traffic control strategy reduced the mean to 2.5 ppm. Now find the probability that a concentration will exceed 9 ppm.

4.41 The weekly rainfall totals for a section of the midwestern United States follow an exponential distribution, with a mean of 1.6 inches.

a Find the probability that a randomly chosen weekly rainfall total in this section will exceed 2 inches.

b Find the probability that the weekly rainfall totals will not exceed 2 inches in either of the next two weeks.

4.42 The service times at teller windows in a bank were found to follow an exponential distribution, with a mean of 3.2 minutes. A customer arrives at a window at 4:00 P.M.

  a Find the probability that he will still be there at 4:02 P.M.

  b Find the probability that he will still be there at 4:04 P.M. given that he was there at 4:02.

4.43 In deciding how many customer service representatives to hire and in planning their schedules, a firm that markets electronic typewriters studies repair times for the machines. One such study revealed that repair times have an approximately exponential distribution, with a mean of 22 minutes.

  a Find the probability that a randomly selected repair time will be less than 10 minutes.

  b The charge for typewriter repairs is $50 for each half hour (or part thereof) for labor. What is the probability that a repair job will result in a charge for labor of $100?

  c In planning schedules, how much time should the firm allow for each repair to ensure that the chance of any one repair time's exceeding this allowed time is only 0.01?

4.44 Explosive devices used in a mining operation cause nearly circular craters to form in a rocky surface. The radii of these craters are exponentially distributed, with a mean of 10 feet. Find the mean and the variance of the area covered by such a crater.

4.5 The Gamma Distribution

4.5.1 Probability Density Function

Many sets of data, of course, do not have relative frequency curves with the smooth decreasing trend found in the exponential model. It is perhaps more common to see distributions that have low probabilities for intervals close to zero, with the probability increasing for a while as the interval moves to the right (in the positive direction) and then decreasing as the interval moves out even farther; that is, the relative frequency curves follow the pattern graphed as in Figure 4.8. In the case of electronic components, for example, few have very short life lengths, many have something close to an average life length, and very few have extraordinarily long life lengths.

FIGURE 4.8
Common relative frequency curve.

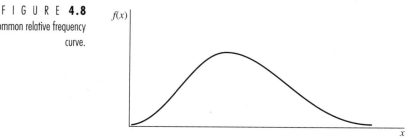

A class of functions that serve as good models for this type of behavior is the *gamma* class. The gamma probability density function is given by

$$f(x) = \begin{cases} \frac{1}{\Gamma(\alpha)\beta^\alpha} x^{\alpha-1} e^{-x/\beta} & \text{for } x \geq 0 \\ 0 & \text{elsewhere} \end{cases}$$

where α and β are parameters that determine the specific shape of the curve. Notice immediately that the gamma density reduces to the exponential density when $\alpha = 1$. The parameters α and β must be positive, but they need not be integers. The symbol $\Gamma(\alpha)$ is defined by

$$\Gamma(\alpha) = \int_0^\infty x^{\alpha-1} e^{-x} dx$$

Since we have already seen that

$$\int_0^\infty x^{\alpha-1} e^{-x/\beta} dx = \beta^\alpha \Gamma(\alpha)$$

it follows that

$$\int_0^\infty (\beta y)^{\alpha-1} e^{-y} \beta dy = \beta^\alpha \int_0^\infty y^{\alpha-1} e^{-y} dy = \beta^\alpha \Gamma(\alpha)$$

which makes $\int_0^\infty f(x) dx = 1$. Some typical gamma densities are shown in Figure 4.9.

F I G U R E **4.9**
Gamma density function,
$\beta = 1$.

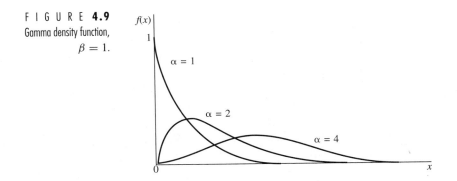

An example of a real data set that closely follows a gamma distribution is shown in Figure 4.10. The data consist of 6-week summer rainfall totals for Ames, Iowa. Notice that many totals fall in the range of from 2 to 8 inches, but occasionally a rainfall total goes well beyond 8 inches. Of course, no rainfall measurements can be negative.

4.5.2   Mean and Variance

The derivation of expectations here is very similar to the exponential case of Section 4.4. We have

$$E(X) = \int_{-\infty}^\infty x f(x) dx$$

$$= \int_0^\infty x \frac{1}{\Gamma(\alpha)\beta^\alpha} x^{\alpha-1} e^{-x/\beta} dx$$

$$= \frac{1}{\Gamma(\alpha)\beta^\alpha} \int_0^\infty x^\alpha e^{-x/\beta} dx$$

F I G U R E **4.10**
Summer rainfall (6-week totals)
for Ames, Iowa.

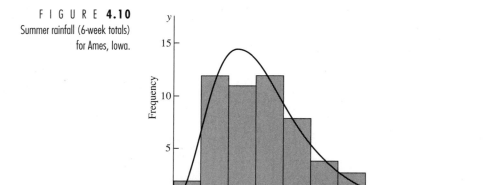

Source: G. L. Barger and H. C. S. Thom, *Agronomy Journal,* 41, 1949, p.521.

$$= \frac{1}{\Gamma(\alpha)\beta^\alpha} \Gamma(\alpha + 1)\beta^{\alpha+1}$$
$$= \alpha\beta$$

Similar manipulations yield $E(X^2) = \alpha(\alpha + 1)\beta^2$; and hence,

$$V(X) = E(X^2) - \mu^2$$
$$= \alpha(\alpha + 1)\beta^2 - \alpha^2\beta^2$$
$$= \alpha\beta^2$$

A simple and often-used property of sums of identically distributed, independent gamma random variables will be stated, but not proved, at this point. Suppose that $X_1, X_2, \ldots, X_n$ represent independent gamma random variables with parameters α and β, as just used. If

$$Y = \sum_{i=1}^n X_i.$$

then Y also has a gamma distribution with parameters $n\alpha$ and β. Thus, one can immediately see that

$$E(Y) = n\alpha\beta$$

and

$$V(Y) = n\alpha\beta^2$$

E X A M P L E   **4.10**   A certain electronic system that has a life length of X_1, with an exponential distribution and a mean of 400 hours, is supported by an identical backup system that has a life length of X_2. The backup system takes over immediately when the primary system fails. If the systems operate independently, find the probability distribution and expected value for the total life length of the primary and backup systems.

Solution Letting Y denote the total life length, we have $Y = X_1 + X_2$, where X_1 and X_2 are independent exponential random variables, each with a mean $\beta = 400$. By the results stated earlier, Y has a gamma distribution with $\alpha = 2$ and $\beta = 400$; that is,

$$f_Y(y) = \frac{1}{\Gamma(2)(400)^2} y e^{-y/400}, \qquad y > 0$$

The mean value is given by

$$E(Y) = \alpha\beta = 2(400) = 800$$

which is intuitively reasonable. ∎

EXAMPLE **4.11** Suppose that the length of time Y needed to conduct a periodic maintenance check on a dictating machine (known from previous experience) follows a gamma distribution, with $\alpha = 3$ and $\beta = 2$ (minutes). Suppose that a new repair person requires 20 minutes to check a particular machine. Does this time required to perform a maintenance check seem out-of-line with prior experience?

Solution The mean and the variance for the length of maintenance time (prior experience) are

$$\mu = \alpha\beta \quad \text{and} \quad \sigma^2 = \alpha\beta^2$$

Then, for our example

$$\mu = \alpha\beta = (3)(2) = 6$$
$$\sigma^2 = \alpha\beta^2 = (3)(2)^2 = 12$$
$$\sigma = \sqrt{12} = 3.46$$

and the observed deviation $(Y - \mu)$ is $20 - 6 = 14$ minutes.

For our example, $y = 20$ minutes exceeds the mean $\mu = 6$ by $k = 14/3.46$ standard deviations. Then, from Tchebysheff's Theorem,

$$P(|Y - \mu| \geq k\sigma) \leq \frac{1}{k^2}$$

or

$$P(|Y - 6| \geq 14) \leq \frac{1}{k^2} = \frac{(3.46)^2}{(14)^2} = 0.06$$

Notice that this probability is based on the assumption that the distribution of maintenance times has not changed from prior experience. Then, observing that $P(Y \geq 20$ minutes) is small, we must conclude either that our new maintenance person has encountered a machine needing an unusually lengthy maintenance time (which occurs with low probability) or that the person is somewhat slower than previous repairers.

Noting the low probability for $P(Y \geq 20)$, we would be inclined to favor the latter view. ■

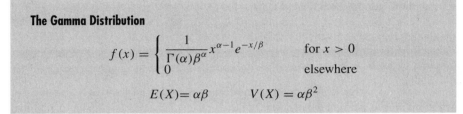

The Gamma Distribution

$$f(x) = \begin{cases} \dfrac{1}{\Gamma(\alpha)\beta^{\alpha}} x^{\alpha-1} e^{-x/\beta} & \text{for } x > 0 \\ 0 & \text{elsewhere} \end{cases}$$

$$E(X) = \alpha\beta \qquad V(X) = \alpha\beta^2$$

Exercises

4.45 Four-week summer rainfall totals in a certain section of the midwestern United States have a relative frequency histogram that appears to fit closely to a gamma distribution, with $\alpha = 1.6$ and $\beta = 2.0$.

    a Find the mean and the variance of this distribution of 4-week rainfall totals.

    b Find an interval that will include the rainfall total for a randomly selected 4-week period with probability at least 0.75.

4.46 Annual incomes for engineers in a certain industry have approximately a gamma distribution, with $\alpha = 600$ and $\beta = 50$.

    a Find the mean and the variance of these incomes.

    b Would you expect to find many engineers in this industry whose annual incomes exceed $35,000?

4.47 The weekly downtime Y (in hours) for a certain industrial machine has approximately a gamma distribution, with $\alpha = 3$ and $\beta = 2$. The loss L (in dollars) to the industrial operation as a result of this downtime is given by

$$L = 30Y + 2Y^2$$

    a Find the expected value and the variance of L.

    b Find an interval that will contain L on approximately 89% of the weeks that the machine is in use.

4.48 Customers arrive at a checkout counter according to a Poisson process, at a rate of two per minute. Find the mean, the variance, and the probability density function of the waiting time between the opening of the counter and the following events.

    a the arrival of the second customer

    b the arrival of the third customer

4.49 Suppose that two houses are to be built, each involving the completion of a certain key task. Completion of the task has an exponentially distributed time, with a mean of 10 hours. Assuming that the completion times are independent for the two houses, find the expected value and the variance of the following times.

a   the total time to complete both tasks

b   the average time to complete the two tasks

4.50   The total sustained load on the concrete footing of a planned building is the sum of the dead load plus the occupancy load. Suppose that the dead load X_1 has a gamma distribution, with $\alpha_1 = 50$ and $\beta_1 = 2$; whereas the occupancy load X_2 also has a gamma distribution, but with $\alpha_2 = 20$ and $\beta_2 = 2$. (Units are in kips, or thousands of pounds.)

a   Find the mean, the variance, and the probability density function of the total sustained load on the footing.

b   Find a value for the sustained load that should be exceeded only with a probability of less than 1/16.

4.51   A 40-year history of annual maximum river flows for a certain small river in the United States shows a relative frequency histogram that can be modeled by a gamma density function, with $\alpha = 1.6$ and $\beta = 150$ (measurements in cubic feet per second).

a   Find the mean and the standard deviation of the annual maximum river flows.

b   Within what interval should the maximum annual flow be contained with a probability of at least 8/9?

4.52   The time intervals between dial-up connections to a computer center from remote terminals are exponentially distributed, with a mean of 15 seconds. Find the mean, the variance, and the probability distribution of the waiting time from the opening of the computer center until the fourth dial-up connection from a remote terminal.

4.53   If service times at a teller window of a bank are exponentially distributed, with a mean of 3.2 minutes, find the probability distribution, the mean, and the variance of the time taken to serve three waiting customers.

4.54   The response times at an on-line terminal have, approximately, a gamma distribution, with a mean of 4 seconds and a variance of 8. Write the probability density function for these response times.

4.6   The Normal Distribution

4.6.1   Normal Probability Density Function

The most widely used continuous probability distribution is referred to as the *normal distribution*. The normal probability density function has the familiar symmetric "bell" shape, as indicated in the two graphs shown in Figure 4.11. The curve is centered at the mean value μ, and its spread is measured by the standard deviation σ. These two parameters, μ and σ^2, completely determine the shape and location of the normal density function, whose functional form is given by

$$f(x) = \frac{1}{\sqrt{2\pi}\,\sigma} e^{-(x-\mu)^2/2\sigma^2}, \qquad -\infty < x < \infty.$$

The basic reason that the normal distribution works well as a model for many different types of measurements generated in real experiments will be discussed in some detail in Chapter 7. For now it suffices to say that, any time responses tend to be averages of independent qualities, the normal distribution quite likely will provide a reasonably good model for their relative frequency behavior. Many naturally occurring measurements tend to have relative frequency distributions that closely resemble the normal curve, probably because nature tends to "average out" the

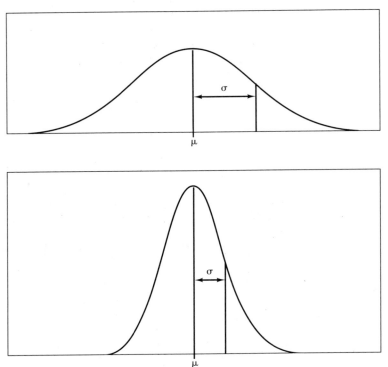

effects of the many variables that relate to a particular response. For example, heights of adult American men tend to have a distribution that shows many measurements clumped closely about a mean height, with relatively few very short or very tall men in the population. In other words, the relative frequency distribution is close to normal.

In contrast, life lengths of biological organisms or electronic components tend to have relative frequency distributions that are not normal or even close to normal. This often is because life length measurements are a product of "extreme" behavior, not "average" behavior. A component may fail because of one extremely severe shock rather than the average effect of many shocks. Thus, the normal distribution is not often used to model life lengths, and we will not discuss the failure rate function for this distribution.

A naturally occurring example of the normal distribution is seen in Michelson's measurements of the speed of light. A histogram of these measurements is given in Figure 4.12. The distribution is not perfectly symmetrical, but it still exhibits an approximately normal shape.

4.6.2 Mean and Variance

A very important property of the normal distribution (which will be proved in Section 4.9) is that any linear function of a normally distributed random variable also is normally distributed; that is, if X has a normal distribution with mean μ and variance

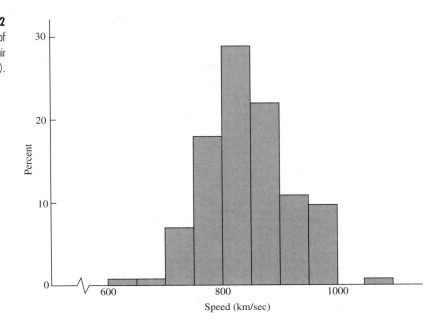

FIGURE **4.12**
Michelson's 100 measures of the speed of light in air ($-299,000$ km/sec).

σ^2, and if $Y = aX + b$ for constants a and b, then Y also is normally distributed. It can easily be seen that

$$E(Y) = a\mu + b \qquad V(Y) = a^2\sigma^2$$

Suppose that Z has a normal distribution, with $\mu = 0$ and $\sigma = 1$. This random variable Z is said to have a *standard normal distribution*. Direct integration will show that $E(Z) = 0$ and $V(Z) = 1$. We have

$$E(Z) = \int_{-\infty}^{\infty} z \frac{1}{\sqrt{2\pi}} e^{-z^2/2} dz$$
$$= \frac{1}{\sqrt{2\pi}} \int_{-\infty}^{\infty} e^{-z^2/2} z \, dz$$
$$= \frac{1}{\sqrt{2\pi}} [-e^{-z^2/2}]_{-\infty}^{\infty}$$
$$= 0$$

Similarly,

$$E(Z^2) = \int_{-\infty}^{\infty} \frac{1}{\sqrt{2\pi}} z^2 e^{-z^2/2} dz$$
$$= \frac{1}{\sqrt{2\pi}} (2) \int_{0}^{\infty} z^2 e^{-z^2/2} dz$$

On our making the transformation $u = z^2$, the integral becomes

$$\frac{1}{\sqrt{2\pi}} \int_{0}^{\infty} u^{1/2} e^{-u/2} du = \frac{1}{\sqrt{2\pi}} \Gamma\left(\frac{3}{2}\right) (2)^{2/3}$$

$$= \frac{1}{\sqrt{\pi}} (2)^{3/2} \left(\frac{1}{2} \right) \Gamma \left(\frac{1}{2} \right) = 1$$

since $\Gamma(1/2) = \sqrt{\pi}$. Therefore, $E(Z) = 0$ and $V(Z) = E(Z^2) - \mu^2 = E(Z^2) = 1$. For any normally distributed random variable X, with parameters μ and σ^2,

$$Z = \frac{X - \mu}{\sigma}$$

will have a standard normal distribution. Then

$$X = Z\sigma + \mu,$$
$$E(X) = \sigma E(Z) + \mu = \mu$$

and

$$V(X) = \sigma^2 V(Z) = \sigma^2$$

This shows that the parameters μ and σ^2 do, indeed, measure the mean and the variance of the distribution.

4.6.3 Calculating Normal Probabilities

Since any normally distributed random variable can be transformed into standard normal form, probabilities can be evaluated for any normal distribution simply by having a table of standard normal integrals available. Such a table is given in Table 4 of the Appendix. Table 4 gives numerical values for

$$P(0 \le Z \le z) = \int_0^z \frac{1}{\sqrt{2\pi}} e^{-z^2/2} dx$$

Values of the integral are given for values of z between 0.00 and 3.09.

We shall now use Table 4 to find $P(-0.5 \le Z \le 1.5)$ for a standard normal variable Z. Figure 4.13 will help us visualize the necessary areas.

F I G U R E **4.13**
Standard normal density
function.

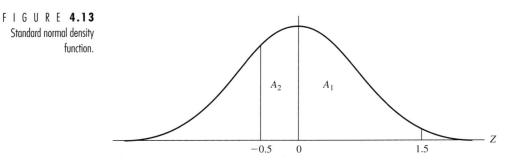

We first must write the probability in terms of intervals to the left and to the right of zero (the mean of the distribution). This produces

$$P(-0.5 \le Z \le 1.5) = P(0 \le Z \le 1.5) + P(-0.5 \le Z \le 0)$$

Now $(0 \leq Z \leq 1.5) = A_1$ in Figure 4.13 and this value can be found by looking up $z = 1.5$ in Table 4. The result is $A_1 = 0.4332$. Similarly, $P(-0.5 \leq Z \leq 0) = A_2$ in Figure 4.13, and this value can be found by looking up $z = 0.5$ in Table 4. Areas under the standard normal curve for negative z-values are equal to those for corresponding positive z-values, since the curve is symmetric around zero. We find $A_2 = 0.1915$. It follows that

$$P(-0.5 \leq Z \leq 1.5) = A_1 + A_2 = 0.4332 + 0.1915 = 0.6247$$

E X A M P L E   **4.12**   If Z denotes a standard normal variable, find the following probabilities.

1 $P(Z \leq 1)$
2 $P(Z > 1)$
3 $P(Z < -1.5)$
4 $P(-1.5 \leq Z \leq 0.5)$

Also find a value of z—say z_0—such that $P(0 \leq Z \leq z_0) = 0.49$.

Solution   This example provides practice in reading Table 4. We see from the table that the following values are correct.

1
$$P(Z \leq 1) = P(Z \leq 0) + P(0 \leq Z \leq 1)$$
$$= 0.5 + 0.3413$$
$$= 0.8413$$

2
$$P(Z > 1) = 0.5 - P(0 \leq Z \leq 1)$$
$$= 0.5 - 0.3413$$
$$= 0.1587$$

3
$$P(Z < -1.5) = P(Z > 1.5)$$
$$= 0.5 - P(0 \leq Z \leq 1.5)$$
$$= 0.5 - 0.4332$$
$$= 0.0668$$

4
$$P(1.5 \leq Z \leq 0.5) = P(-1.5 \leq Z \leq 0) + P(0 \leq Z \leq 0.5)$$
$$= P(0 \leq Z \leq 1.5) + P(0 \leq Z \leq 0.5)$$
$$= 0.4332 + 0.1915$$
$$= 0.6247.$$

To find the value of z_0, we must look for the given probability of 0.49 on the area side of Table 4. The closest we can come to 0.49 is 0.4901, which corresponds to a z-value of 2.33. Hence, $z_0 = 2.33$.   ∎

Studying the table of normal curve areas for z-scores of 1, 2, and 3 reveals how the percentages used in what is commonly called the Empirical Rule were determined. These percentages actually represent areas under the standard normal curve, as depicted in Figure 4.14. The next example illustrates how the standardization works to allow Table 4 to be used for any normally distributed random variable.

FIGURE **4.14**
Justification of the empirical rule.

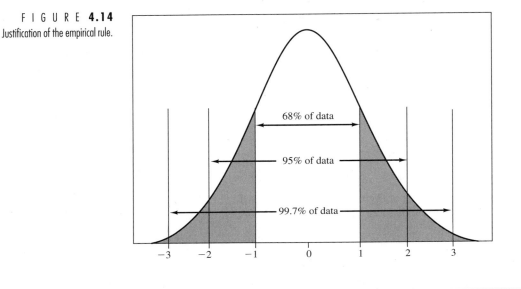

68% of data

95% of data

99.7% of data

-3 -2 -1 0   1   2   3

EXAMPLE **4.13**

A firm that manufactures and bottles apple juice has a machine that automatically fills 16-ounce bottles. There is some variation, however, in the amount of liquid dispensed into each bottle by the machine. Over a long period of time, the average amount dispensed into the bottles was 16 ounces, but the underlying measurements have a standard deviation of 1 ounce. If the ounces of fill per bottle can be assumed to be normally distributed, find the probability that the machine will dispense more than 17 ounces of liquid in any one bottle.

Solution

Let X denote the amount of liquid (in ounces) dispensed into one bottle by the filling machine. Then X is assumed to be normally distributed, with a mean of 16 and a standard deviation of 1. Hence,

$$P(X > 17) = P\left(\frac{X - \mu}{\sigma} > \frac{17 - \mu}{\sigma}\right) = P\left(Z > \frac{17 - 16}{1}\right)$$
$$= P(Z > 1) = 0.1587$$

The answer can be found from Table 4, since $Z = (X - \mu)/\sigma$ has a *standard* normal distribution. ▪

EXAMPLE **4.14**

Suppose that another machine, similar to the one described in Example 4.13, operates in such a way that the ounces of fill have a mean value equal to the dial setting for "amount of liquid" but also have a standard deviation of 1.2 ounces. Find the proper setting for the dial so that 17- ounce bottles will overflow only 5% of the time. Assume that the amounts dispensed have a normal distribution.

Solution   Letting X denote the amount of liquid dispensed, we now look for a value of μ such that

$$P(X > 17) = 0.05.$$

This is represented graphically in Figure 4.15. Now

$$P(X > 17) = P\left(\frac{X - \mu}{\sigma} > \frac{17 - \mu}{\sigma}\right)$$
$$= P\left(Z > \frac{17 - \mu}{1.2}\right)$$

From Table 4, we know that if

$$P(Z > z_0) = 0.05,$$

then $z_0 = 1.645$. Thus, it must be that

$$\frac{17 - \mu}{1.2} = 1.645$$

and

$$\mu = 17 - 1.2(1.645) = 15.026 \quad \blacksquare$$

F I G U R E **4.15**
Diagram for Example 4.14.

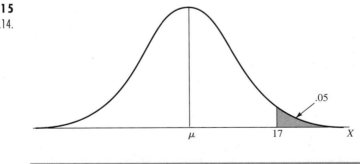

4.6.4   Applications to Real Data

The practical value of probability distributions in relation to data analysis is that the probability models help explain key features of the data succinctly and aid in the constructive use of data to help predict future outcomes. Patterns that appeared regularly in the past are expected to appear again in the future. If they do not, something of importance may have happened to disturb the process under study.

Describing data patterns through their probability distribution is a very useful tool for data analysis.

EXAMPLE **4.15** The SAT and ACT college entrance exams are taken by thousands of students each year. The scores on the exams for any one year produce a histogram that looks very much like a normal curve. Thus, we can say that the scores are approximately normally distributed. In recent years, the SAT mathematics scores have averaged around 480, with a standard deviation of 100. The ACT mathematics scores average around 18, with a standard deviation of 6.

1 An engineering school sets 550 as the minimum SAT math score for new students. What percentage of students would score less than 550 in a typical year? (This percentage is called the **percentile** score equivalent for 550.)

2 What score should the engineering school set as a comparable standard on the ACT math test?

3 What is the probability that a randomly selected student will score over 700 on the SAT math test?

Solution 1 The percentile score corresponding to a raw score of 550 is the area under the normal curve to the left of 550, as shown in Figure 4.16. The area marked A can be found from Table 4 (on your computer). The area to the left of the mean is 0.5. To find A, observe that the z-score for 550 is

$$z = \frac{x - \mu}{\sigma} = \frac{550 - 480}{100} = 0.7$$

This gives $A = 0.258$, and the percentile score is $0.5 + 0.258 = 0.758$. The score of 550 is almost the 76th percentile. About 75.8% of all students taking the SAT should have scores below this value.

FIGURE **4.16**
Diagram for part 1 of Example 4.15.

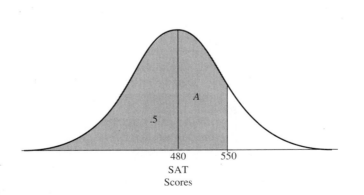

2 The question now is "What is the 75.8th percentile of the ACT math score distribution?" From the preceding information, we know that percentile 75.8 for

any normal distribution is 0.7 standard deviations above the mean. Therefore, using the ACT distribution,

$$x = \mu + z\sigma$$
$$= 18 + (0.7)6$$
$$= 22.2$$

3   A score of 22.2 on the ACT should be equivalent to a score of 550 on the SAT. The score of 700 is transformed into a z-score of

$$z = \frac{x - \mu}{\sigma} = \frac{700 - 480}{100} = 2.2$$

which, in turn, gives an area of $A = 0.4861$ from Table 4 (see Figure 4.17). Thus, the area above 700 is $0.50 - 0.4861 = 0.0139$. The chance of a randomly selected student's scoring above 700 is quite small (around 0.014).   ∎

FIGURE **4.17**
Diagram for part 3 of Example 4.15.

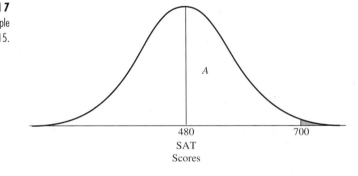

480
SAT
Scores

700

EXAMPLE   **4.16**   The batting averages of American League batting champions for the years from 1941 through 1991 are graphed on the histogram in Figure 4.18. This graph looks somewhat normal in shape, but it has a little skewness toward the high values. The mean is .344 and the standard deviation is .022 for these data.

1   Ted Williams batted .406 in 1941, and George Brett batted .390 in 1980. How would you compare these performances?

2   Is there a good chance that anyone in the American League will hit over .400 in any one year?

Solution   **1**   Obviously .406 is better than .390, but how much better? One way to describe how these numbers compare to each other—and how they compare to the remaining data points in the distribution—is to look at z-scores and percentile scores. For 1941, Ted Williams had a z-score of

$$z = \frac{.406 - .344}{.022} = 2.82$$

F I G U R E **4.18** Batting averages: American League batting champions, 1941–1991.

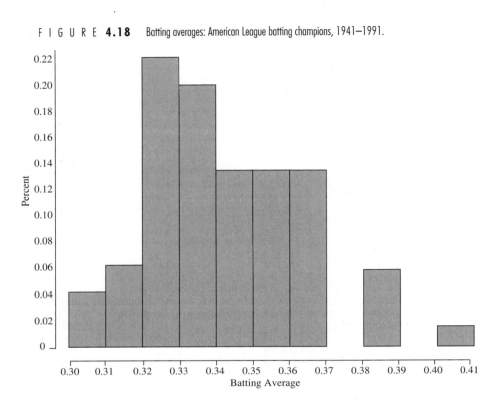

and a percentile score of

$$0.50 + 0.4976 = 0.9976$$

For 1980, George Brett had a z-score of

$$z = \frac{.390 - .344}{.022} = 2.09$$

and a percentile score of

$$0.50 + 0.4817 = 0.9817$$

Both are above the 98th percentile and are far above average. In that sense, both performances are outstanding and, perhaps, not very far apart.

2 The chance of the league leader's hitting over .400 in a given year can be approximated by looking at a z-score of

$$z = \frac{.400 - .344}{.022} = 2.54$$

This translates into a probability of

$$0.50 - 0.4945 = 0.0055$$

or about 6 chances out of 1000. (This the probability for the league leader; what would happen to this chance if *any* player were eligible for consideration?)   ∎

What happens when the normal model is used to describe skewed distributions? To answer this, let's look at two data sets involving comparable variables and see how good the normal approximations are. Figure 4.19 shows dotplots of cancer mortality rates for white males during the entire decade of the 1970s. (These rates are expressed as deaths per 100,000 people.) The top graph shows data for the 67 counties of Florida, and the bottom graph shows data for the 93 counties of Nebraska. Summary statistics are as follows:

State	Mean	Standard Deviation
Florida	199.3	21.7
Nebraska	192.5	95.7

F I G U R E **4.19**   County-by-county cancer mortality rates (per 100,000) among white males in Florida and in Nebraska, 1970s.

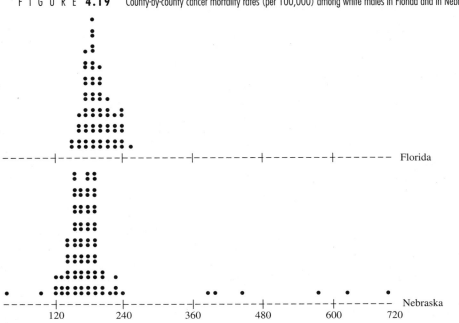

Note: Each dot represents the results of one county.
Source: National Cancer Institute.

On average, the states perform about the same, but the distributions are quite different. Key features of the difference can be observed by looking at empirical versus theoretical relative frequencies, as shown in Table 4.2.

Interval	Observed Proportion		Theoretical Proportion
	Florida	Nebraska	
$\bar{x} \pm s$	$\frac{46}{67} = 0.686$	$\frac{85}{93} = 0.914$	0.68
$\bar{x} \pm 2s$	$\frac{66}{67} = 0.985$	$\frac{87}{93} = 0.935$	0.95

For Florida, observation and theory are close together, although the two-standard-deviation interval does pick up a few too many data points. For Nebraska, observation and theory are not close. In this case, the large outliers inflate the standard deviation so that the one-standard-deviation interval is far too long to agree with normal theory. The two-standard-deviation interval still doesn't reach the outliers, so the observed relative frequency is still smaller than expected. This is typical of performance of relative frequencies in highly skewed situations; be careful in interpreting standard deviation under skewed conditions.

4.6.5   Quantile-Quantile (Q-Q) Plots

The normal model is popular for describing data distributions; but as we have just seen, it often does *not* work well in instances where the data distribution is skewed. How can we recognize such instances and avoid using the normal model in situations where it would be inappropriate? One way, as we have seen, is to look carefully at histograms, dotplots, and stemplots to gauge symmetry and outliers visually. Another way, presented in this subsection, is to take advantage of the unique properties of z-scores for the normal case.

If X has a normal (μ, σ) distribution, then

$$X = \mu + \sigma Z$$

and there is a perfect linear relationship between X and Z. Now, suppose that we observe n measurements and order them so that $x_1 \leq x_2 \leq x_3 \leq \cdots \leq x_n$. The value x_k has (k/n) values that are less than or equal to it, so it is the (k/n)th sample percentile. If the observations come from a normal distribution, x_k should approximate the (k/n)th percentile from the normal distribution and, therefore, should be linearly related to z_k (the corresponding z-score). The cancer data used in the preceding subsection (Figure 4.19) will serve to illustrate this point. The sample percentiles corresponding to each state were determined for the 25th, 50th, 75th, 95th, and 99th percentiles as shown in Table 4.3. The z-scores corresponding to these percentiles for the normal distribution are also listed in the table.

TABLE **4.3**
Percentile and z-Score Values
for Cancer Data

	Mortality Rate		
	Florida	Nebraska	
Percentile	Sample	Sample	z-score
25th	185	155	−0.680
50th	195	174	0.000
75th	214	192	0.680
95th	237	230	1.645
99th	242	622	2.330

In other words, 25% of the Florida data fell on or below 185, and 95% of the Nebraska data fell on or below 230. For the normal distribution, 25% of the area will fall below a point with a z-score −0.68.

Plots of the sample percentiles against the z-scores are shown in Figure 4.20. Such plots are called quantile-quantile or Q-Q plots. For Florida, the sample percentiles and z-scores fall nearly on a straight line, indicating a good fit of the data to a normal distribution. The slope of the line is around 20, which is close to the standard deviation for these data, and the intercept at $z = 0$ is around 195, which is close to the sample mean.

For Nebraska, four points fall on a line, or nearly so, but the fifth point is far off the line. The observed 99th percentile is much *larger* than would have been expected from the normal distribution, indicating a skewness toward the larger data points. Figure 4.20(c) and (d) show the plots for all the sample data, with the same general result. (Integers in the plots indicate the number of points at that position.)

For small samples, it is better to think of x_k as the $(k/(n + 1))$th sample percentile. The reason for this is that $x_1 \leq x_2 \leq \cdots \leq x_n$ actually divide the population distribution into $(n + 1)$ segments, all of which are expected to process roughly equal probability masses.

EXAMPLE **4.17**   In the interest of predicting future peak particulate matter values, 12 observations of this variable were obtained from various sites across the United States. The ordered measurements, in $\mu g/m^3$, are as follows:

$$22, 24, 24, 28, 30, 31, 33, 45, 45, 48, 51, 79$$

Should the normal probability model be used to anticipate peak particulate matter values in the future?

Solution   Table 4.4 provides the key components of the analysis. Recall that the z-scores are the standard normal values corresponding to the percentiles listed in the $i/(n + 1)$ column. For example, about 69% of the normal curve's area falls below a z-score of 0.50. The Q-Q plot of x_i versus z_i in Figure 4.21 shows that the data departs from normality at both ends: the lower values have too short a tail, and the higher values have too long a tail. The data points appear to come from a highly skewed distribution; it would be unwise to predict future values for this variable by using a normal distribution.   ∎

F I G U R E **4.20**   Q-Q plot: cancer mortality rates.

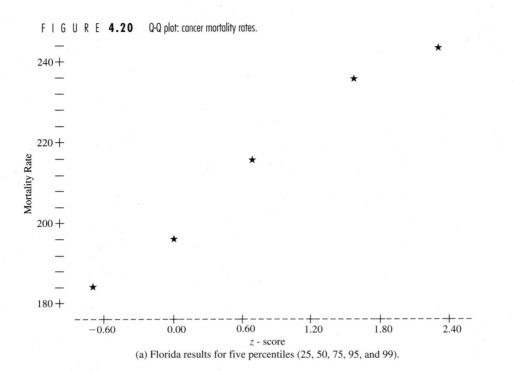

(a) Florida results for five percentiles (25, 50, 75, 95, and 99).

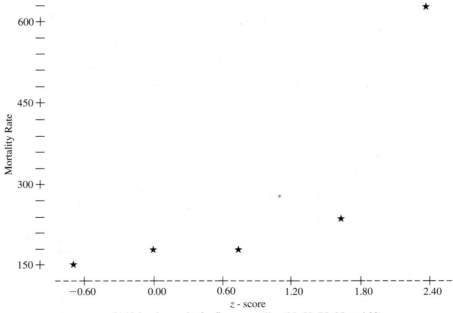

(b) Nebraska results for five percentiles (25, 50, 75, 95, and 99).

F I G U R E **4.20** (continued)

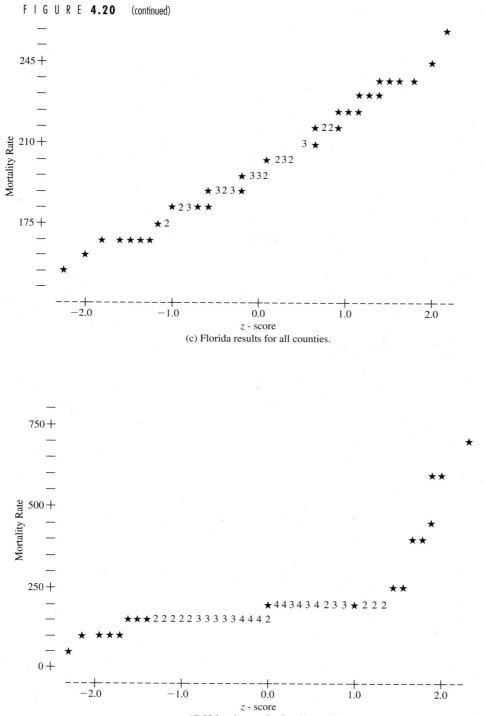

(c) Florida results for all counties.

(d) Nebraska results for all counties.

T A B L E **4.4**	i	x_i	$i/(n+1)$	z-**Score**
Key Values for Analysis of	1	22	0.077	−1.43
Particulate Matter Data	2	24	0.153	−1.02
	3	24	0.231	−0.73
	4	28	0.308	−0.50
	5	30	0.385	−0.29
	6	31	0.462	−0.10
	7	33	0.538	0.10
	8	45	0.615	0.29
	9	45	0.692	0.50
	10	48	0.769	0.74
	11	51	0.486	1.02
	12	79	0.923	1.43

Q-Q plots are cumbersome to construct by hand, especially if the data set is large; but most computer programs for statistical analysis can generate the essential parts quite easily. In Minitab, for example, issuing the NSCORES (for normal scores) command prompts Minitab to produce the z-scores corresponding to a set of sample data.

F I G U R E **4.21**   Q-Q plot: peak particulate matter values, 1990.

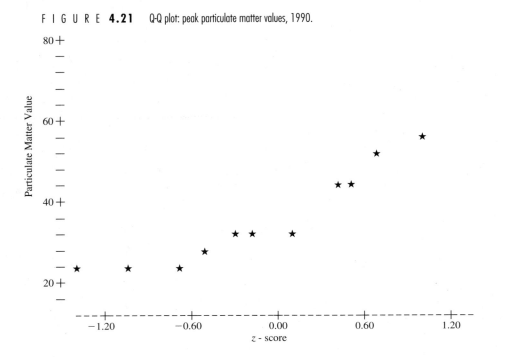

EXAMPLE   **4.18**   The heights of males between the ages of 18 and 24 have the *cumulative* relative frequency distribution given in Table 4.5.

TABLE **4.5**
Cumulative Relative Frequency
Distribution for Heights of Males,
Ages 18 to 24

Height (inches)	Cumulative Percentage
61	0.18%
62	0.34
63	0.61
64	2.37
65	3.85
66	8.24
67	16.18
68	26.68
69	38.89
70	53.66
71	68.25
72	80.14
73	88.54
74	92.74
75	96.17
76	98.40

Source: Statistical Abstract of the United States, 1991.

Should the normal distribution be used to model the male height distribution? If so, what mean and what standard deviation should be used?

Solution   Here, the percentiles can easily be derived from the cumulative percentages, simply by dividing by 100. Using a normal curve area table—or an inverse normal probability function (going from probabilities to z-scores) on a computer—we find that the corresponding z-scores are as follows:

$$-3.29, -2.63, -2.35, -2.01, -1.72, -1.33, -0.92,$$
$$-0.55, -0.21, 0.20, 0.67, 1.05, 1.42, 1.78, 2.13, 2.50$$

The Q-Q plot of heights versus z-scores appears in Figure 4.22.

The plot shows a nearly straight line; thus, the normal distribution can legitimately be used to model these data.

Marking a line with a straight edge through these points to approximate the line of best fit for the data yields a slope of approximately 2.9 and a y-intercept (at $z = 0$) of about 70. Thus, male heights can be modeled by a normal distribution with a mean of 70 inches and a standard deviation of 2.9 inches. (Can you verify the mean and the standard deviation approximations by another method?)   ∎

F I G U R E **4.22**   Q-Q plot: male heights (18–24 years old).

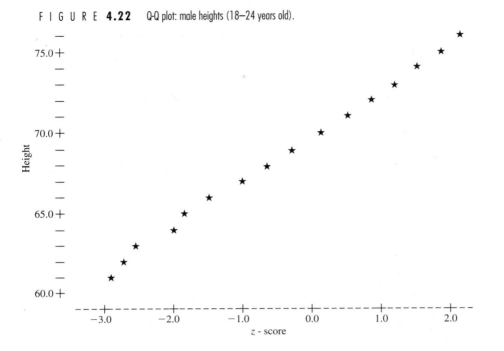

As was mentioned earlier, we shall make much more use of the normal distribution in later chapters. The properties of the normal distribution are summarized next.

The Normal Distribution

$$f(x) = \frac{1}{\sqrt{2\pi}\,\sigma} e^{-(x-\mu)^2/2\sigma^2}, \qquad -\infty < x < \infty$$

$$E(X) = \mu \qquad\qquad V(X) = \sigma^2$$

Exercises

4.55   Use Table 4 of the Appendix to find the following probabilities for a standard normal random variable Z.

a $P(0 \le Z \le 1.2)$

b $P(-0.9 \le Z \le 0)$

c $P(0.3 \le Z \le 1.56)$

d $P(-0.2 \le Z \le 0.2)$

e $P(-2.00 \leq Z \leq 1.56)$

4.56   For a standard normal random variable Z, use Table 4 of the Appendix to find a number z_0 such that the following probabilities are obtained.

  a $P(Z \leq z_0) = 0.5$

  b $P(Z \leq z_0) = 0.8749$

  c $P(Z \geq z_0) = 0.117$

  d $P(Z \geq z_0) = 0.617$

  e $P(-z_0 \leq Z \leq z_0) = 0.90$

  f $P(-z_0 \leq Z \leq z_0) = 0.95$

4.57   The weekly amount spent for maintenance and repairs in a certain company has an approximately normal distribution, with a mean of $400 and a standard deviation of $20. If $450 is budgeted to cover repairs for next week, what is the probability that the actual costs will exceed the budgeted amount?

4.58   In the setting of Exercise 4.57, how much should be budgeted weekly for maintenance and repairs to ensure that the probability that the budgeted amount will be exceeded in any given week is only 0.1?

4.59   A machining operation produces steel shafts where diameters have a normal distribution, with a mean of 1.005 inches and a standard deviation of 0.01 inch. Specifications call for diameters to fall within the interval 1.00 ± 0.02 inches. What percentage of the output of this operation will fail to meet specifications?

4.60   Referring to Exercise 4.59, what should be the mean diameter of the shafts produced to minimize the fraction that fail to meet specifications?

4.61   Wires manufactured for a certain computer system are specified to have a resistance of between 0.12 and 0.14 ohms. The actual measured resistances of the wires produced by Company A have a normal probability distribution, with a mean of 0.13 ohms and a standard deviation of 0.005 ohms.

  a   What is the probability that a randomly selected wire from Company A's production lot will meet the specifications?

  b   If four such wires are used in a single system and all are selected from Company A, what is the probability that all four will meet the specifications?

4.62   At a temperature of 25°C, the resistances of a type of thermistor are normally distributed, with a mean of 10,000 ohms and a standard deviation of 4000 ohms. The thermistors are to be sorted, and those having resistances between 8000 and 15,000 ohms are to be shipped to a vendor. What fraction of these thermistors will actually be shipped?

4.63   A vehicle driver gauges the relative speed of the next vehicle ahead by observing the speed with which the image of the width of that vehicle varies. This speed is proportional to X, the speed of variation of the angle at which the eye subtends this width. According to P. Ferrani and others (1984, pp. 50, 51), a study of many drivers revealed X to be normally distributed, with a mean of 0 and a standard deviation of 10 (10^{-4} radian per second). What fraction of these measurements is more than 5 units away from 0? What fraction is more than 10 units away from 0?

4.64   A type of capacitor has resistances that vary according to a normal distribution, with a mean of 800 megohms and a standard deviation of 200 megohms (see Nelson 1967, pp. 261–268, for a more thorough discussion). A certain application specifies capacitors with resistances of between 900 and 1000 megohms.

  a   What proportion of these capacitors will meet this specification?

  b   If two capacitors are randomly chosen from a lot of capacitors of this type, what is the probability that both will satisfy the specification?

4.65   Sick-leave time used by employees of a firm in the course of one month has approximately a normal distribution, with a mean of 200 hours and a variance of 400 hours.

    a   Find the probability that total sick leave for next month will be less than 150 hours.

    b   In planning schedules for next month, how much time should be budgeted for sick leave if that amount is to be exceeded with a probability of only 0.10?

4.66   The time until first failure of a brand of ink-jet printers is approximately normally distributed, with a mean of 1500 hours and a standard deviation of 200 hours.

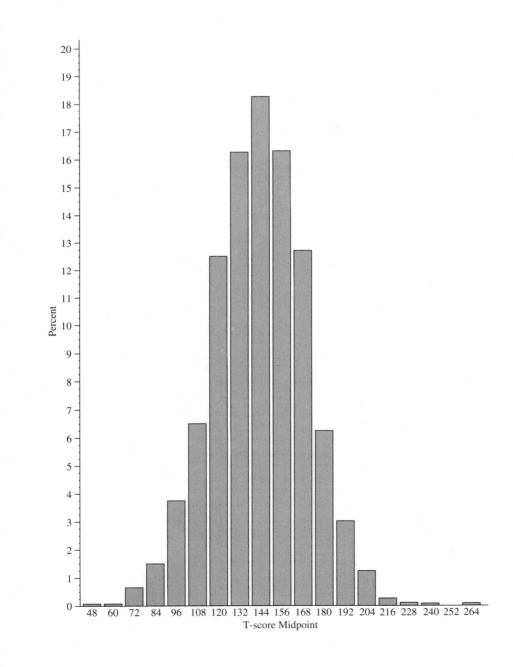

    a   What fraction of these printers will fail before 1000 hours?

    b   What should be the guarantee time for these printers if the manufacturer wants no more than 5% to fail within the guarantee period?

4.67   A machine for filling cereal boxes has a standard deviation of 1 ounce in relation to ounces of fill per box. What setting of the mean ounces of fill per box will allow 16-ounce boxes to overflow only 1% of the time? (Assume that the ounces of fill per box are normally distributed.)

4.68   Referring to Exercise 4.67, suppose the standard deviation σ is not known but can be fixed at certain levels by carefully adjusting the machine. What is the largest value of σ that will allow the actual value dispensed to fall within 1 ounce of the mean with a probability at least 0.95?

4.69   The histogram on the facing page depicts the total points scored per game for every NCAA tournament basketball game between 1939 and 1992 ($n = 1521$). The mean is 143, and the standard deviation is 26.

    a   Does it appear that total points can be modeled as a normally distributed random variable?

    b   Show that the empirical rule holds for these data.

    c   Would you expect a total score to top 200 very often? 250? Why?

    d   Sixty-four games will be played in this year's tournament. How many of these would you expect to record total scores of less than 100 points?

4.70   One major contributor to air pollution is sulfur dioxide. The following are twelve peak sulfur dioxide measurements (in ppm) for 1990 from randomly selected U.S. sites:

$$0.003, 0.010, 0.014, 0.024, 0.024, 0.032, 0.038, 0.042, 0.043, 0.044, 0.047, 0.061$$

    a   Should the normal distribution be used as a model for these measurements?

    b   From a Q-Q plot for these data, approximate the mean and the standard deviation. Check the approximation by direct calculation.

4.71   The heights of females of ages 18 to 24 are distributed according to the cumulative percentages listed in the accompanying table.

Cumulative Relative Frequency Distribution for Heights of Females, Ages 18 to 24

Height (inches)	Cumulative Percentage
56	0.05
57	0.43
58	0.94
59	2.22
60	4.22
61	9.13
62	17.75
63	29.06
64	41.81
65	58.09
66	74.76
67	85.37
68	92.30
69	96.23
70	98.34
71	99.38

Source: Statistical Abstract of the United States, 1991.

a Produce a Q-Q plot for these data; discuss their goodness of fit to a normal distribution.

b Approximate the mean and the standard deviation of these heights, based on the Q-Q plot.

4.72 The cumulative proportions of U.S. residents of various ages are shown in the accompanying table for the years 1900 and 2000 (projected):

Cumulative Age	Cumulative Proportion 1900	2000
0–5	0.121	0.066
0–15	0.344	0.209
0–25	0.540	0.344
0–35	0.700	0.480
0–45	0.822	0.643
0–55	0.906	0.781
0–65	0.959	0.870
0–100	0.990	0.990

a Construct a Q-Q plot for each year. Use these plots as a basis for discussing key differences between the two age distributions.

b Each of the Q-Q plots should show some departures from normality. Explain the nature of these departures.

4.7 The Beta Distribution

4.7.1 Probability Density Function

Except for the uniform distribution of Section 4.3, the continuous distributions discussed thus far have had density functions that are positive over an infinite interval. It is useful to have another class of distributions to model phenomena constrained to a finite interval of possible values. One such class, the beta distributions, is valuable for modeling the probabilistic behavior of certain random variables (such as proportions) constrained to fall in the interval (0, 1). [Actually, any finite interval can be transformed into (0, 1).]

The beta distribution has the functional form

$$f(x) = \begin{cases} \frac{\Gamma(\alpha+\beta)}{\Gamma(\alpha)\Gamma(\beta)} x^{\alpha-1}(1-x)^{\beta-1} & \text{for } 0 < x < 1 \\ 0 & \text{elsewhere} \end{cases}$$

where α and β are positive constants. The constant term in $f(x)$ is necessary so that

$$\int_0^1 f(x)\mathrm{d}x = 1$$

In other words,

$$\int_0^1 x^{\alpha-1}(1-x)^{\beta-1}\mathrm{d}x = \frac{\Gamma(\alpha)\Gamma(\beta)}{\Gamma(\alpha+\beta)}$$

FIGURE **4.23**
Beta density function.

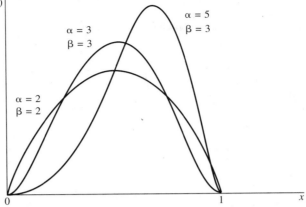

for positive α and β. This is a handy result to keep in mind. The graphs of some common beta density functions are shown in Figure 4.23.

One important measurement in the process of sintering copper relates to the proportion of the volume that is solid, rather than made up of voids. (The proportion due to voids is sometimes called the *porosity of the solid*.) Figure 4.24 shows a relative frequency histogram of proportions of solid copper in samples drawn from a sintering process. This distribution could be modeled with a beta distribution that has a large α and a small β.

FIGURE **4.24**
Solid mass in sintered linde copper.

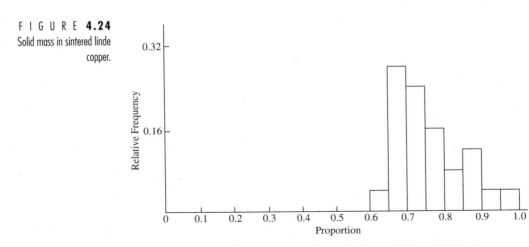

Source: Department of Materials Science, University of Florida.

4.7.2 Mean and Variance

The expected value of a beta random variable can easily be found because

$$E(X) = \int_0^1 x \frac{\Gamma(\alpha + \beta)}{\Gamma(\alpha)\Gamma(\beta)} x^{\alpha-1}(1 - x)^{\beta-1} \mathrm{d}x$$

$$= \frac{\Gamma(\alpha + \beta)}{\Gamma(\alpha)\Gamma(\beta)} \int_0^1 x^\alpha (1 - x)^{\beta-1} dx$$

$$= \frac{\Gamma(\alpha + \beta)}{\Gamma(\alpha)\Gamma(\beta)} \times \frac{\Gamma(\alpha + 1)\Gamma(\beta)}{\Gamma(\alpha + \beta + 1)}$$

$$= \frac{\alpha}{\alpha + \beta}$$

[Recall that $\Gamma(n + 1) = n\Gamma(n)$.] Similar manipulations reveal that

$$V(X) = \frac{\alpha\beta}{(\alpha + \beta)^2(\alpha + \beta + 1)}$$

We illustrate the use of this density function in an example.

E X A M P L E   **4.19**   A gasoline wholesale distributor uses bulk storage tanks to hold a fixed supply. The tanks are filled every Monday. Of interest to the wholesaler is the proportion of this supply sold during the week. Over many weeks, this proportion has been observed to match fairly well a beta distribution with $\alpha = 4$ and $\beta = 2$.

1   Find the expected value of this proportion.

2   Is it highly likely that the wholesaler will sell at least 90% of the stock in a given week?

Solution   1   By the results given earlier, with X denoting the proportion of the total supply sold in a given week,

$$E(X) = \frac{\alpha}{\alpha + \beta} = \frac{4}{6} = \frac{2}{3}$$

2   Now we are interested in

$$P(X > 0.9) = \int_{0.9}^1 \frac{\Gamma(4 + 2)}{\Gamma(4)\Gamma(2)} x^3 (1 - x) dx$$

$$= 20 \int_{0.9}^1 (x^3 - x^4) dx$$

$$= 20(0.004) = 0.08$$

Therefore, it is not very likely that 90% of the stock will be sold in a given week.   ▪

The basic properties of the beta distribution are summarized next.

The Beta Distribution

$$f(x) = \begin{cases} \dfrac{\Gamma(\alpha + \beta)}{\Gamma(\alpha)\Gamma(\beta)} x^{\alpha-1}(1-x)^{\beta-1} & \text{for } 0 < x < 1 \\ 0 & \text{elsewhere} \end{cases}$$

$$E(X) = \frac{\alpha}{\alpha + \beta} \qquad V(X) = \frac{\alpha\beta}{(\alpha + \beta)^2(\alpha + \beta + 1)}$$

Exercises

4.73   Suppose that X has a probability density function given by

$$f(x) = \begin{cases} kx^3(1-x)^2 & \text{for } 0 \le x \le 1 \\ 0 & \text{elsewhere} \end{cases}$$

  a   Find the value of k that makes this a probability density function.
  b   Find $E(X)$ and $V(X)$.

4.74   If X has a beta distribution with parameters α and β, show that

$$V(X) = \frac{\alpha\beta}{(\alpha + \beta)^2(\alpha + \beta + 1)}$$

4.75   During an 8-hour shift, the proportion of time X that a sheet-metal stamping machine is down for maintenance or repairs has a beta distribution, with $\alpha = 1$ and $\beta = 2$; that is,

$$f(x) = \begin{cases} 2(1-x) & \text{for } 0 \le x \le 1 \\ 0 & \text{elsewhere} \end{cases}$$

The cost (in \$100s) of this downtime, in lost production and repair expenses, is given by

$$C = 10 + 20X + 4X^2$$

  a   Find the mean and the variance of C.
  b   Find an interval within which C has a probability of at least 0.75 of lying.

4.76   The percentage of impurities per batch in a certain type of industrial chemical is a random variable X that has the probability density function

$$f(x) = \begin{cases} 12x^2(1-x) & \text{for } 0 \le x \le 1 \\ 0 & \text{elsewhere} \end{cases}$$

  a   Suppose that a batch with more than 40% impurities cannot be sold. What is the probability that a randomly selected batch cannot be sold for this reason?
  b   Suppose that the dollar value of each batch is given by $V = 5 - 0.5X$. Find the expected value and the variance of V.

4.77   To study the dispersal of pollutants emerging from a power plant, researchers measured the prevailing wind direction for a large number of days. The direction is measured on a scale of $0°$ to $360°$; but by dividing each daily direction by 360, one can rescale the measurements to the interval $(0, 1)$. These rescaled measurements X prove to follow a beta distribution, with $\alpha = 4$ and $\beta = 2$. Find $E(X)$. To what angle does this mean correspond?

4.78 Errors in measuring the arrival time (in microseconds) of a wavefront from an acoustic source can sometimes be modeled by a beta distribution (Perruzzi and Hilliard 1984, p. 197). Suppose that these errors have a beta distribution, with $\alpha = 1$ and $\beta = 2$.

a Find the probability that a particular measurement error will be less than 0.5 microseconds.

b Find the mean and the standard deviation of these error measurements.

4.79 The proper blending of fine and coarse powders in copper sintering is essential for uniformity in the finished product. One way to check the blending is to select many small samples of the blended powders and to measure the weight fractions of the fine particles. These measurements should be relatively constant if good blending has been achieved.

a Suppose that the weight fractions have a beta distribution, with $\alpha = \beta = 3$. Find the mean and the variance of these fractions.

b Repeat part (a) for $\alpha = \beta = 2$.

c Repeat part (a) for $\alpha = \beta = 1$.

d Which case—(a), (b), or (c)—would exemplify the best blending?

4.80 The proportion of pure iron in certain ore samples has a beta distribution, with $\alpha = 3$ and $\beta = 1$.

a Find the probability that one of these samples will have more than 50% pure iron.

b Find the probability that two out of three samples will have less than 30% pure iron.

4.8 The Weibull Distribution

4.8.1 Probability Density Function

We have observed that the gamma distribution often can serve as a probabilistic model for life lengths of components or systems. However, the failure rate function for the gamma distribution has an upper bound that limits its applicability to real systems. For this and other reasons, other distributions often provide better models for life-length data. One such distribution is the *Weibull*, which is explored in this section.

A Weibull density function has the form

$$f(x) = \begin{cases} \frac{\gamma}{\theta} x^{\gamma-1} e^{-x^\gamma/\theta} & \text{for } x > 0 \\ 0 & \text{elsewhere} \end{cases}$$

for positive parameters θ and γ. For $\gamma = 1$, this becomes an exponential density. For $\gamma > 1$, the functions look something like the gamma functions of Section 4.5, but they have somewhat different mathematical properties. We can integrate directly to see that

$$F(x) = \int_0^x \frac{\gamma}{\theta} t^{\gamma-1} e^{-t^\gamma/\theta} dt$$
$$= -e^{-t^\gamma/\theta} \big|_0^x$$
$$= 1 - e^{-x^\gamma/\theta}, \qquad x > 0$$

A convenient way to look at properties of the Weibull density is to use the transformation $Y = X^\gamma$. Then,

$$F_Y(y) = P(Y \le y)$$

$$
\begin{aligned}
&= P(X^\gamma \le y)\\
&= P(X \le y^{1/\gamma})\\
&= F_X(y^{1/\gamma})\\
&= 1 - e^{-(y^{1/\gamma})^\gamma / \theta}\\
&= 1 - e^{-y/\theta}, \qquad y > 0
\end{aligned}
$$

Hence,

$$
f_Y(y) = \frac{\mathrm{d} F_Y(y)}{\mathrm{d} y} = \frac{1}{\theta} e^{-y/\theta}, \qquad y > 0
$$

and Y has the familiar exponential density.

4.8.2   Mean and Variance

If we want to find $E(X)$ for a variable X that has the Weibull distribution, then

$$
\begin{aligned}
E(X) &= E(Y^{1/\gamma})\\
&= \int_0^\infty y^{1/\gamma} \frac{1}{\theta} e^{-y/\theta} \mathrm{d} y\\
&= \frac{1}{\theta} \int_0^\infty y^{1/\gamma} e^{-y/\theta} \mathrm{d} y\\
&= \frac{1}{\theta} \Gamma\left(1 + \frac{1}{\gamma}\right) \theta^{(1+1/\gamma)}\\
&= \theta^{1/\gamma} \Gamma\left(1 + \frac{1}{\gamma}\right)
\end{aligned}
$$

This result follows from recognizing the integral to be of the gamma type.

If we let $\gamma = 2$ in the Weibull density, we see that $Y = X^2$ has an exponential distribution. To reverse the idea just outlined, if we start with an exponentially distributed random variable Y, then the square root of Y will have a Weibull distribution, with $\gamma = 2$. We can illustrate this empirically by taking the square roots of the data from an exponential distribution that were given in Table 4.1. These square roots are given in Table 4.6, and a relative frequency histogram for these data is given in Figure 4.25. Notice that the exponential form has now disappeared and that the curve given by the Weibull density, with $\gamma = 2$ and $\theta = 2$ (shown in Figure 4.26) is a much more plausible model for these observations.

0.637	0.828	2.186	1.313	2.868
1.531	1.184	1.223	0.542	1.459
0.733	0.484	2.006	1.823	1.709
2.256	1.207	1.032	0.880	0.872
2.364	1.305	1.623	1.360	0.431
1.601	0.719	1.802	1.526	1.032
0.152	0.715	1.668	2.535	0.914
1.826	0.474	1.230	1.793	0.617
1.868	1.525	0.577	2.746	0.984
1.126	1.709	1.274	0.578	2.119

T A B L E **4.6**
Square Roots of the Battery Life
Lengths of Table 4.1

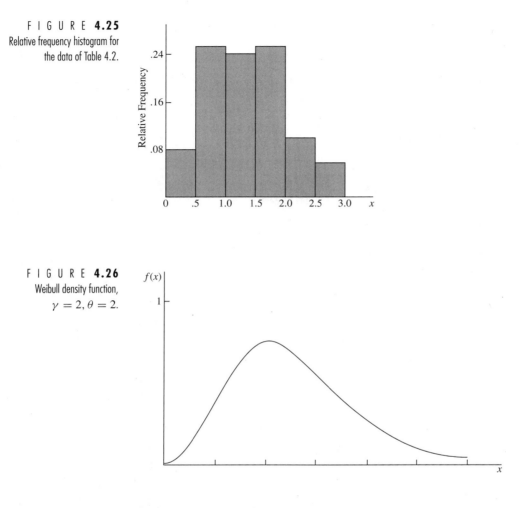

F I G U R E **4.25**
Relative frequency histogram for
the data of Table 4.2.

We illustrate the use of the Weibull distribution in the following example.

E X A M P L E   **4.20**   The lengths of time in service during which a certain type of thermistor produces re-sistances within its specifications have been observed to follow a Weibull distribution, with $\theta = 50$ and $\gamma = 2$ (measurements in thousands of hours).

1   Find the probability that one of these thermistors, which is to be installed in a system today, will function properly for more than 10,000 hours.

2   Find the expected life length for thermistors of this type.

Solution   **1**   The Weibull distribution has a closed-form expression for $F(x)$. Thus, if X represents the life length of the thermistor in question, then

$$P(X > 10) = 1 - F(10)$$
$$= 1 - [1 - e^{-(10)^2/50}]$$

$$= e^{-(10)^2/50}$$
$$= e^{-2}$$
$$= 0.14$$

because $\theta = 50$ and $\gamma = 2$.

2   We know that

$$E(X) = \theta^{1/\gamma} \Gamma \left(1 + \frac{1}{\gamma} \right)$$
$$= (50)^{1/2} \Gamma \left(\frac{3}{2} \right)$$
$$= (50)^{1/2} \frac{1}{2} \Gamma \left(\frac{1}{2} \right)$$
$$= (50)^{1/2} \left(\frac{1}{2} \right) \sqrt{\pi}$$
$$= 6.27$$

Thus, the average service time for these thermistors is 6270 hours.   ∎

4.8.3   Applications to Real Data

The Weibull distribution is versatile enough to be a good model for many distributions of data that are mound-shaped but skewed. Another advantage of the Weibull distribution over, say, the gamma is that the number of relatively straightforward techniques exist for actually fitting the model to data. One such technique is illustrated here. For the Weibull distribution,

$$1 - F(x) = e^{-x^{\gamma}/\theta}, \qquad x > 0$$

and therefore (with ln denoting natural logarithm),

$$\ln \left[\frac{1}{1 - F(x)} \right] = x^{\gamma} \theta$$

and

$$\ln\ln \left[\frac{1}{1 - F(x)} \right] = \gamma \ln(x) - \ln(\theta)$$

For simplicity, let's call the double ln expression on the left $LF(x)$. Plotting $LF(x)$ as a function of $\ln(x)$ produces a straight line with slope γ and intercept $\ln(\theta)$.

Now let's see how this works with real data. Peak particulate matter counts from locations across the United States did not appear to fit the normal distribution. Do they fit the Weibull? In ascending order the counts (in $\mu g/m^3$) used earlier (see Table 4.4) were as follows:

$$22, 24, 24, 28, 30, 31, 33, 45, 45, 48, 51, 79$$

Checking for outliers (by using a boxplot, for example) reveals that 79 is an extreme outlier. No smooth distribution will do a good job of picking it, so we remove it from the data set. Proceeding then, as in the normal case, to produce sample percentiles as approximations to $F(x)$, we obtain the data listed in Table 4.7.

TABLE **4.7**
Analysis of Particulate Matter Data

i	x	$\ln(x)$	$i/(n+1)$	Approximate LF(x)
1	22	3.09104	0.083333	−2.44172
2	24	3.17805	0.166667	−1.70198
3	24	3.17805	0.250000	−1.24590
4	28	3.33220	0.333333	−0.90272
5	30	3.40120	0.416667	−0.61805
6	31	3.43399	0.500000	−0.36651
7	33	3.49651	0.583333	−0.13300
8	45	3.80666	0.666667	0.09405
9	45	3.80666	0.750000	0.32663
10	48	3.87120	0.833333	0.58320
11	51	3.93183	0.916667	0.91024

Here, $i/(n+1)$ is used to approximate $F(x_i)$ in LF(x). Figure 4.27 shows the plot of LF(x) versus $\ln(x)$. The points lie rather close to a straight line that has a slope of about 3.2 and a y-intercept of about -11.6. (This line can be approximated by drawing a straight line with a straightedge through the middle of the scatterplot.) Thus, the particulate counts appear to be adequately modeled by a Weibull distribution, with $\gamma \approx 3.2$ and $\ln(\theta) \approx 11.6$.

Notice that fitting an exponential distribution to data could follow a similar procedure, with $\gamma = 1$. In this case,

$$\ln\left[\frac{1}{1 - F(x)}\right] = x/\theta$$

so a plot of an estimate of the left side against x should reveal a straight line through the origin, with a slope of $1/\theta$.

The Weibull Distribution

$$f(x) = \begin{cases} \frac{\gamma}{\theta} x^{\gamma-1} e^{-x^{\gamma}/\theta} & \text{for } x > 0 \\ 0 & \text{elsewhere} \end{cases}$$

$$E(X) = \theta^{1/\gamma} \Gamma\left(1 + \frac{1}{\gamma}\right)$$

$$V(X) = \theta^{2/\gamma} \left\{ \Gamma\left(1 + \frac{2}{\gamma}\right) - \left[\Gamma\left(1 + \frac{1}{\gamma}\right)\right]^2 \right\}$$

FIGURE **4.27**   Plot of LF(x) versus ln(x).

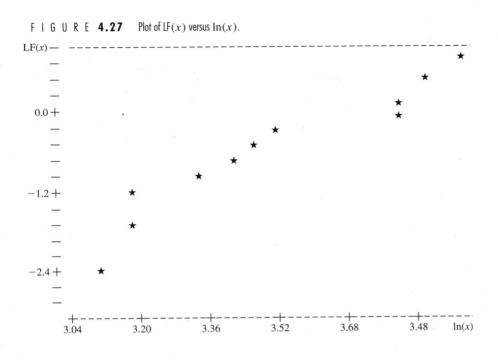

Exercises

4.81   Fatigue life (in hundreds of hours) for a certain type of bearings has approximately a Weibull distribution, with $\gamma = 2$ and $\theta = 4$.

   a   Find the probability that a randomly selected bearing of this type will fail in less than 200 hours.

   b   Find the expected value of the fatigue life for these bearings.

4.82   The yearly maximum flood levels (in millions of cubic feet per second) for a certain United States river have a Weibull distribution, with $\gamma = 1.5$ and $\theta = 0.6$ (see Cohen, Whitten, and Ding 1984, p. 165, for more details). Find the probabilities that the maximum flood level for next year will have the following characteristics.

   a   It will exceed 0.5.

   b   It will be less than 0.8.

4.83   The times necessary to achieve proper blending of copper powders before sintering were found to have a Weibull distribution, with $\gamma = 1.1$ and $\theta = 2$ (measurements in minutes). Find the probability that proper blending in a particular case will take less than 2 minutes.

4.84   The ultimate tensile strength of steel wire used to wrap concrete pipe was found to have a Weibull distribution, with $\gamma = 1.2$ and $\theta = 270$ (measurements in thousands of pounds). Pressure in the pipe at a certain point may require an ultimate tensile strength of a least 300,000 pounds. What is the probability that a randomly selected wire will possess this strength?

4.85   The yield strengths of certain steel beams have a Weibull distribution, with $\gamma = 2$ and $\theta = 3600$ (measurement in thousands of pounds per square inch). Two such beams are used in a construction project, which calls for yield strengths in excess of 70,000 psi. Find the probability that both beams will meet the specifications for the project.

4.86   The pressure (in thousands of psi) exerted on the tank of a steam boiler has a Weibull distribution, with $\gamma = 1.8$ and $\theta = 1.5$. The tank is built to withstand pressures of up to 2000 psi. Find the probability that this limit will be exceeded.

4.87   Resistors used in an aircraft guidance system have life lengths that follow a Weibull distribution, with $\gamma = 2$ and $\theta = 10$ (measurements in thousands of hours).

    a   Find the probability that a randomly selected resistor of this type has a life length that exceeds 5000 hours.

    b   If three resistors of this type operate independently, find the probability that exactly one of the three will burn out prior to 5000 hours of use.

    c   Find the mean and the variance of the life length of such a resistor.

4.88   Failures in the bleed systems of jet engines were causing some concern at an air base. It was decided to model the failure-time distribution for these systems so that future failures could be anticipated better. The following randomly identified failure times (in operating hours since installation) were observed (from *Weibull Analysis Handbook, Pratt & Whitney Aircraft*, 1983):

$$1198, 884, 1251, 1249, 708, 1082, 884, 1105, 828, 1013$$

Does a Weibull distribution appear to be a good model for these data? If so, what values should be used for λ and θ?

4.89   Maximum wind-gust velocities in summer thunderstorms were found to follow a Weibull distribution, with $\gamma = 2$ and $\theta = 400$ (measurements in feet per second). Engineers who design structures in areas where these thunderstorms occur are interested in finding a gust velocity that will be exceeded only with a probability of 0.01. Find such a value.

4.90   The velocities of gas particles can be modeled by the Maxwell distribution, with the probability density function given by

$$f(v) = 4\pi \left(\frac{m}{2\pi K T} \right)^{3/2} v^2 e^{-v^2(m/2KT)}, \qquad v > 0$$

where m is the mass of the particle, K is Boltzmann's constant, and T is the absolute temperature.

    a   Find the mean velocity of these particles.

    b   The kinetic energy of a particle is given by $(1/2)mV^2$. Find the mean kinetic energy of these particles.

4.9   Reliability

One important measure of the quality of products is their **reliability**, or probability of working for a specified period of time. We want products—whether they be cars, television sets, or shoes—that do not break down or wear our for some definite period of time, and we want to know how long this time period can be expected to last. The study of relaibility is a probabilistic exercise because data or models on component lifetimes are used to predict future behavior: how long a process will operate before it fails.

In reliability studies, the underlying random variable of interest, X, is usually lifetime.

DEFINITION   **4.4**

> If a component has lifetime X with distribution function F, then the **reliability** of the component is
>
> $$R(t) = P(X > t) = 1 - F(t) \quad \blacksquare$$

It follows that, for exponentially distributed lifetimes,

$$R(t) = e^{-t/\theta}, \qquad t \geq 0$$

and for lifetimes following a Weibull distribution,

$$R(t) = e^{-t^{\gamma}/\theta}, \qquad t \geq 0,$$

Reliability functions for gamma and normal distributions do not exist in closed form. In fact, the normal model is not often used to model life-length data since length of life tends to exhibit positively skewed behavior.

4.9.1   Failure Rate Function

Besides probability density and distribution functions and the reliability function, another function is useful in work with life-length data. Suppose that X denotes the life-length of a component with density function $f(x)$ and distribution function $F(x)$. The *failure rate function $r(t)$* is defined as

$$r(t) = \frac{f(t)}{1 - F(t)}, \qquad t > 0, \qquad F(t) < 1$$

For an intuitive look at what $r(t)$ is measuring, suppose that dt denotes a very small interval around the point t. Then $f(t)dt$ is approximately the probability that X will take on a value in $(t, t + dt)$. Further, $1 - F(t) = P(X > t)$. Thus,

$$r(t)dt = \frac{f(t)dt}{1 - F(t)}$$
$$\approx P[X \in (t, t + dt)|X > t]$$

In other words, $r(t)dt$ represents the probability of failure during the time interval $(t, t + dt)$, given that the component has survived up to time t.

For the exponential case,

$$r(t) = \frac{f(t)}{1 - F(t)} = \frac{\frac{1}{\theta}e^{-t/\theta}}{e^{-t/\theta}} = \frac{1}{\theta}$$

Thus, X has a *constant* failure rate. It is unlikely that many individual components have a constant failure rate over time (most fail more frequently as they age), but it may be true of some systems that undergo regular preventive maintenance.

The failure rate function $r(t)$ for the gamma case is not easily displayed, since $F(t)$ does not have a simple closed form. For $\alpha > 1$, however, this function will

FIGURE **4.28**
Failure rate function for the
gamma distribution ($\alpha > 1$).

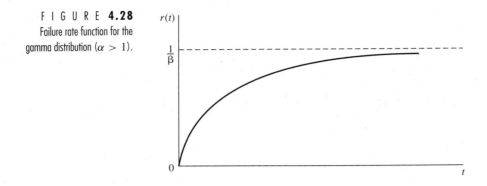

increase but always remain bounded above by $1/\beta$. A typical form is shown in Figure 4.28.

4.9.2   Series and Parallel Systems

A system is made up of a number of components, such as relays in an electrical system or check valves in a water system. The reliability of system depends critically on how the components are networked into the system. A **series** system (Figure 4.29(a)) fails as soon as any one component fails. A **parallel** system (Figure 4.29(b)) fails only when all components have failed.

FIGURE **4.29**
Series and parallel systems.

(a)

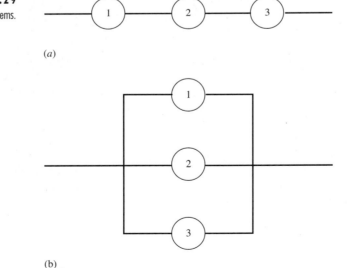

(b)

Suppose that the system components in Figure 4.29 each have a reliability function of $R(t)$, and suppose that the components operate independently of one another. What are the system reliabilities $R_s(t)$? For the series system, life length will exceed t only if each component has a life length exceeding t. Thus,

$$R_s(t) = R(t)R(t)R(t) = [R(t)]^3$$

For the parallel system, life length will be less than t only if all components fail before t. Thus,

$$1 - R_s(t) = [1 - R(t)][1 - R(t)][1 - R(t)]$$
$$R_s(t) = 1 - [1 - R(t)]^3$$

Of course, most systems are combinations of components in series and components in parallel. The rules of probability must be used to evaluate system reliability in each special case, but breaking the system into series and parallel subsystems often simplifies this process.

4.9.3   Redundancy

What happens as we add components to a system? For a series of n independent components,

$$R_s(t) = [R(t)]^n$$

And since $R(t) \leq 1$, adding more components in series will just make things worse! For n components operating in parallel, however,

$$R_s(t) = 1 - [1 - R(t)]^n$$

which will increase with n. Thus, system reliability can be improved by adding backup components in parallel—a practice called *redundancy*.

How many components must we use to achieve a specified level of system reliability? If we have a fixed value for $R_s(t)$ in mind, and if we have independently operating parallel components, then

$$1 - R_s(t) = [1 - R(t)]^n$$

Letting ln denote a natural logarithm, we can calcualte

$$\ln[1 - R_s(t)] = n\ln[1 - R(t)]$$

and

$$n = \frac{\ln[1 - R_s(t)]}{\ln 1 - R(t)]}$$

This is but a brief introduction to the interesting and important area of reliability. But the key to all reliability problems is a firm understanding of basic probability.

Exercises

4.91 For a component with an exponentially distributed lifetime, find the reliability up to time $t_1 + t_2$, given that the component has already lived past t_1. Why is the constant failure rate referred to as the "memoryless" property?

4.92 For a series of n components operating independently—each with the same exponential life distribution—find an expression for $R_s(t)$. What is the mean lifetime of the system?

4.93 For independently operating components with identical life distributions, find the system reliability for each of the following:

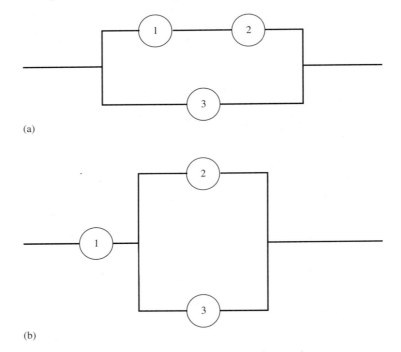

(a)

(b)

Which has the higher reliability?

4.94 Suppose that each relay in an electrical circuit has a reliability of 0.9 for a specified operating period of t hours. How could you configure a system of such relays to bring $R_s(t)$ up to 0.999?

4.10 Moment-generating Functions for Continuous Random Variables

As in the case of discrete distributions, the moment-generating functions of continuous random variables help us find expected values and identify certain properties of probability distributions. We now undertake a short discussion of moment-generating functions for continuous random variables.

The moment-generating function of a continuous random variable X with a probability density function of $f(x)$ is given by

$$M(t) = E(e^{tX}) = \int_{-\infty}^{\infty} e^{tx} f(x) dx$$

when the integral exists.

For the exponential distribution, this becomes

$$\begin{aligned}
M(t) &= \int_{0}^{\infty} e^{tx} \frac{1}{\theta} e^{-x/\theta} dx \\
&= \frac{1}{\theta} \int_{0}^{\infty} e^{-x(1/\theta - t)} dx \\
&= \frac{1}{\theta} \int_{0}^{\infty} e^{-x(1-\theta t)/\theta} dx \\
&= \frac{1}{\theta} \Gamma(1) \left(\frac{\theta}{1 - \theta t} \right) = (1 - \theta t)^{-1}
\end{aligned}$$

We can now use $M(t)$ to find $E(X)$, because

$$M'(0) = E(X)$$

and for the exponential distribution we have

$$\begin{aligned}
E(X) &= M'(0) \\
&= [-(1 - \theta t)^{-2}(-\theta)]_{t=0} \\
&= \theta
\end{aligned}$$

Similarly, we could find $E(X^2)$ and then $V(X)$ by using the moment-generating function.

An argument analogous to the one used for the exponential distribution shows that, for the gamma distribution,

$$M(t) = (1 - \beta t)^{-\alpha}$$

From this we can see that, if X has a gamma distribution,

$$\begin{aligned}
E(X) &= M'(0) \\
&= [-\alpha(1 - \beta t)^{-\alpha-1}(-\beta)]_{t=0} \\
&= \alpha\beta \\
E(X^2) &= M^{(2)}(0) \\
&= [\alpha\beta(-\alpha - 1)(1 - \beta t)^{-\alpha-2}(-\beta)]_{t=0} \\
&= \alpha(\alpha + 1)\beta^2 \\
E(X^3) &= M^{(3)}(0) \\
&= [\alpha(\alpha + 1)\beta^2(-\alpha - 2)(1 - \beta t)^{-\alpha-3}(-\beta)]_{t=0} \\
&= \alpha(\alpha + 1)(\alpha + 2)\beta^3
\end{aligned}$$

and so on.

EXAMPLE   **4.21**   The force h exerted by a mass m moving at velocity v is

$$h = \frac{mv^3}{2}$$

Consider a device that fires a serrated nail into concrete at a mean velocity of 500 feet per second, where the random velocity V possesses a density function of

$$f(v) = \frac{v^2 e^{-v/b}}{b^4 \Gamma(4)} \qquad b = 500, \quad v \geq 0$$

If each nail possesses mass m, find the expected force exerted by a nail.

Solution
$$E(H) = E\left(\frac{mV^2}{2}\right) = \frac{m}{2} E(V^2)$$

The density function for V is a gamma-type function, with $\alpha = 4$ and $\beta = 500$. Therefore,

$$\begin{aligned} E(V^2) &= \alpha(\alpha + 1)\beta^2 \\ &= (4)(5)(500)^2 \\ &= 5{,}000{,}000 \end{aligned}$$

and

$$\begin{aligned} E(H) &= \frac{m}{2} E(V^2) \\ &= \frac{m}{2}(5{,}000{,}000) \\ &= 2{,}500{,}000m \quad \blacksquare \end{aligned}$$

Moment-generating functions have two important properties:

1   If a random variable X has the moment-generating function $M_X(t)$, then $Y = aX + b$, for constants a and b, has the moment-generating function

$$M_Y(t) = e^{tb} M_X(at)$$

2   Moment-generating functions are unique; that is, two random variables that have the same moment-generating function have the same probability distribution as well.

Property 1 is easily shown (see Exercise 3.85). The proof of property 2 is beyond the scope of this textbook; but we make use of it in identifying distributions, as will be demonstrated later.

The normal random variable is used so commonly that we ought to investigate its moment-generating function. We can start to do so by finding the moment-generating function of $X - \mu$, where X is normally distributed with a mean of μ and a variance of σ^2. Now,

$$E(e^{t(X-\mu)}) = \int_{-\infty}^{\infty} e^{t(x-\mu)} \frac{1}{\sigma\sqrt{2\pi}} e^{-(x-\mu)^2/2\sigma^2} dx$$

Letting $y = x - \mu$, we find that the integral becomes

$$\frac{1}{\sigma\sqrt{2\pi}} \int_{-\infty}^{\infty} e^{ty-y^2/2\sigma^2} dy = \frac{1}{\sigma\sqrt{2\pi}} \int_{-\infty}^{\infty} \exp\left[-\frac{1}{2\sigma^2}(y^2 - 2\sigma^2 ty)\right] dy$$

Upon completion of the square in the exponent, we have

$$-\frac{1}{2\sigma^2}(y^2 - 2\sigma^2 ty) = -\frac{1}{2\sigma^2}(y^2 - 2\sigma^2 ty + \sigma^4 t^2) + \frac{1}{2}t^2\sigma^2$$

$$= -\frac{1}{2\sigma^2}(y - \sigma^2 t)^2 + \frac{1}{2}t^2\sigma^2$$

so the integral becomes

$$e^{t^2\sigma^2/2} \frac{1}{\sigma\sqrt{2\pi}} \int_{-\infty}^{\infty} \exp\left[-\frac{1}{2\sigma^2}(y - \sigma^2 t)^2\right] dy = e^{t^2\sigma^2/2}$$

because the remaining integrand forms a normal probability density that integrates to unity.

If $Y = X - \mu$ has the moment-generating function given by

$$M_Y(t) = e^{t^2\sigma^2/2}$$

then

$$X = Y + \mu$$

has the moment-generating function given by

$$M_X(t) = e^{t\mu} M_Y(t)$$
$$= e^{t\mu} e^{t^2\sigma^2/2}$$
$$= e^{t\mu + t^2\sigma^2/2}$$

by property 1 (stated earlier).

For this same normal random variable X, let

$$Z = \frac{X - \mu}{\sigma} = \frac{1}{\sigma}X - \frac{\mu}{\sigma}$$

Then, by property 1,

$$M_Z(t) = e^{-\mu t/\sigma} e^{\mu t/\sigma + t^2\sigma^2/2\sigma^2}$$
$$= e^{t^2/2}$$

The normal generating function of Z has the form of a moment-generating function for a normal random variable with a mean of 0 and a variance of 1. Thus, by property 2, Z must have that distribution.

Exercises

4.95 Show that a gamma distribution with parameters α and β has the moment-generating function

$$M(t) = (1 - \beta t)^{-\alpha}$$

4.96 Using the moment-generating function for the exponential distribution with mean θ, find $E(X^2)$. Use this result to show that $V(X) = \theta^2$.

4.97 Let Z denote a standard normal random variable. Find the moment-generating function of Z directly from the definition.

4.98 Let Z denote a standard normal random variable. Find the moment-generating function of Z^2. What does the uniqueness property of the moment-generating function tell you about the distribution of Z^2?

4.11 Expectations of Discontinuous Functions and Mixed Probability Distributions

Problems in probability and statistics frequently involve functions that are partly continuous and partly discrete in one of two ways. First, we may be interested in the properties—perhaps the expectation—of a random variable $g(X)$ that is a discontinuous function of a discrete or continuous random variable X. Second, the random variable of interest may itself have a probability distribution made up of isolated points having discrete probabilities and of intervals having continuous probability. The first of these two situations is illustrated by the following example.

EXAMPLE **4.22** A certain retailer for a petroleum product sells a random amount X each day. Suppose that X (measured in hundreds of gallons) has the probability density function

$$f(x) = \begin{cases} \left(\frac{3}{8}\right) x^2 & \text{for } 0 \leq x \leq 2 \\ 0 & \text{elsewhere} \end{cases}$$

The retailer's profit turns out to be \$5 for each 100 gallons sold (5 cents per gallon) if $X \leq 1$, and \$8 per 100 gallons if $X > 1$. Find the retailer's expected profit for any given day.

Solution Let $g(X)$ denote the retailer's daily profit. Then,

$$g(X) = \begin{cases} 5X & \text{if } 0 \leq X \leq 1 \\ 8X & \text{if } 1 < X \leq 2 \end{cases}$$

We want to find expected profit, and

$$E[g(X)] = \int_{-\infty}^{\infty} g(x) f(x) dx$$

$$= \int_0^1 5x \left[\left(\frac{3}{8}\right) x^2 \right] dx + \int_1^2 8x \left[\left(\frac{3}{8}\right) x^2 \right] dx$$

$$= \frac{15}{(8)(4)}[x^4]_0^1 + \frac{24}{(8)(4)}[x^4]_1^2$$

$$= \frac{15}{32}(1) + \frac{24}{32}(15)$$

$$= \frac{(15)(25)}{32}$$

$$= 11.72$$

Thus, the retailer can expect to profit by \$11.72 on the daily sale of this particular product. ∎

A random variable X that has some of its probability at discrete points and the remainder spread over intervals is said to have a *mixed distribution* . Let $F(x)$ denote a distribution function representing a mixed distribution. For practical purposes, any mixed distribution function $F(x)$ can be written uniquely as

$$F(x) = c_1 F_1(x) + c_2 F_2(x)$$

where $F_1(x)$ is a step distribution function, $F_2(x)$ is a continuous distribution function, c_1 is the accumulated probability of all discrete points, and $c_2 = 1 - c_1$ is the accumulated probability of all continuous portions. The following example offers an instance of a mixed distribution.

E X A M P L E   **4.23**   Let X denote the life length (in hundreds of hours) of a certain type of electronic component. These components frequently fail immediately upon insertion into a system; the probability of immediate failure is 1/4. If a component does not fail immediately, its life-length distribution has the exponential density

$$f(x) = \begin{cases} e^{-x} & \text{for } x > 0 \\ 0 & \text{elsewhere} \end{cases}$$

Find the distribution function for X, and evaluate $P(X > 10)$.

Solution   There is only one discrete point, $X = 0$, and this point has probability 1/4. Hence, $c_1 = 1/4$ and $c_2 = 3/4$. It follows that X is a mixture of two random variables, X_1 and X_2, where X_1 has a probability of 1 at the point 0 and X_2 has the given exponential density; that is,

$$F_1(x) = \begin{cases} 0 & \text{if } x < 0 \\ 1 & \text{if } x \geq 0 \end{cases}$$

and

$$F_2(x) = \int_0^x e^{-y}\, dy$$

$$= 1 - e^{-x} \qquad \text{for } x > 0$$

Now,

$$F(x) = \left(\frac{1}{4}\right) F_1(x) + \left(\frac{3}{4}\right) F_2(x)$$

Hence,

$$
\begin{aligned}
P(X > 10) &= 1 - P(X \le 10) \\
&= 1 - F(10) \\
&= 1 - \left[\frac{1}{4} + \left(\frac{3}{4}\right)(1 - e^{-10})\right] \\
&= \left(\frac{3}{4}\right)[1 - (1 - e^{-10})] \\
&= \left(\frac{3}{4}\right) e^{-10} \quad \blacksquare
\end{aligned}
$$

An easy method for finding expectations of random variables that have mixed distributions is given in Definition 4.5.

DEFINITION   **4.5**

Let X have the mixed distribution function

$$F(x) = c_1 F_1(x) + c_2 F_2(x)$$

and suppose that X_1 is a discrete random variable having distribution function $F_1(x)$ and that X_2 is a continuous random variable having distribution function $F_2(x)$. Let $g(X)$ denote a function of X. Then

$$E[g(X)] = c_1 E[g(X_1)] + c_2 E[g(X_2)] \quad \blacksquare$$

EXAMPLE   **4.24**   Find the mean and the variance of the random variable defined in Example 4.23.

Solution   With all definitions remaining as given in Example 4.23, it follows that

$$E(X_1) = 0$$

and

$$E(X_2) = \int_0^\infty y e^{-y} dy = 1$$

Therefore,

$$
\begin{aligned}
\mu &= E(X) \\
&= \left(\frac{1}{4}\right) E(X_1) + \left(\frac{3}{4}\right) E(X_2)
\end{aligned}
$$

$$= \frac{3}{4}$$

Also,

$$E(X_1^2) = 0$$

and

$$E(X_2^2) = \int_0^\infty y^2 e^{-y} \, dy = 2$$

Therefore,

$$E(X^2) = \left(\frac{1}{4}\right) E(X_1^2) + \left(\frac{3}{4}\right) E(X_2^2)$$

$$= \left(\frac{1}{4}\right)(0) + \left(\frac{3}{4}\right)(2)$$

$$= \frac{3}{2}$$

Then,

$$V(X) = E(X^2) - \mu^2$$

$$= \frac{3}{2} - \left(\frac{3}{4}\right)^2$$

$$= \frac{15}{16} \quad \blacksquare$$

For any nonnegative random variable X, the mean can be expressed as

$$E(X) = \int_0^\infty [1 - F(t)] dt$$

where $F(t)$ is the distribution function for X. To see this, write

$$\int_0^\infty [1 - F(t)] dt = \int_0^\infty \left(\int_t^\infty f(x) dx \right) dt$$

Upon changing the order of integration, we find that the integral becomes

$$\int_0^\infty \left(\int_0^x f(T) dt \right) dX = \int_0^\infty x f(x) dx = E(X)$$

Employing this result for the mixed distribution of Examples 4.23 and 4.24, we see that

$$1 - F(x) = \frac{3}{4} e^{-x} \qquad \text{for } x > 0$$

and

$$E(X) = \int_0^\infty [1 - F(t)]dt$$

$$= \int_0^\infty \frac{3}{4}e^{-t}dt$$

$$= \frac{3}{4}[-e^t]_0^\infty$$

$$= \frac{3}{4}$$

Thus, the mean can be found without resorting to the two-part calculation of Definition 4.5.

Exercises

4.99   A retail grocer has daily demand X for a certain food that is sold by the pound wholesale. Food left over at the end of the day is a total loss. The grocer buys the food for $6 per pound and sells it for $10 per pound. If demand is uniformly distributed over the interval 0 to 1 pound, how much of this food should the grocer order to maximize his expected daily profit?

4.100   Suppose that a distribution function has the form

$$
\begin{aligned}
F(x) &= 0 & \text{for } x < 0 \\
&= x^2 + 0.1 & \text{for } 0 \le x < 0.5 \\
&= x & \text{for } 0.5 \le x < 1 \\
&= 1 & \text{for } 1 \le x
\end{aligned}
$$

a   Describe $F_1(x)$ and $F_2(x)$, the discrete and continuous components of $F(x)$.

b   Write $F(x)$ as

$$c_1 F_1(x) + c_2 F_2(x)$$

c   Sketch $F(x)$.

d   Find the expected value of the random variable whose distribution function is $F(x)$.

4.101   The duration X of telephone calls coming through a pay phone is a random variable whose distribution function is

$$
\begin{aligned}
F(X) &= 0 & \text{for } x \le 0 \\
&= 1 - \frac{2}{3}de^{-x/3} - \frac{1}{3}e^{-[x/3]} & \text{for } x > 0
\end{aligned}
$$

where $[y]$ denotes the greatest integer less than or equal to y.

a   Sketch $F(x)$.

b   Find $P(X \le 6)$.

c   Find $P(X > 4)$.

d   Describe the discrete and continuous parts that make up $F(x)$.

e   Find $E(X)$.

4.12 Activities for Students: Simulation

When generating random variables from a continuous distribution with a (cumulative) distribution function (CDF) of $F(x)$, we may wish to use the inverse transformation method or inverse CDF method. The procedure for this method is as follows:

1   Generate R_i, which is uniform over the interval (0, 1). This is denoted as $R_i \sim U(0, 1)$.
2   Let $R_i = F(X)$, where $F(x)$ is the CDF for the distribution of values of x.
3   Evaluate step 2 for X, which gives $X = F^{-1}(R)$.

4.12.1   Uniform Distribution

The probability density function (pdf) for uniform distribution is defined as $f(x) = 1/(b - a)$, where $a \leq x \leq b$. The CDF is given by $F(x) = (x - a)/(b - a)$. Using the inverse CDF method to generate uniform random numbers x_i, we let $r_i = (x - a)/(b - a)$. Solving for x, we obtain $x_i = a + (b - a)r_i$.

4.12.2   Exponential Distribution

The pdf for exponential distribution is defined as $f(x) = (1/\theta)e^{-x/\theta}$, for $x \geq 0$. The CDF is given by $F(x) = 1 - e^{-x/\theta}$. Using the inverse CDF method, we can generate exponential random numbers x_i, by letting $r_i = 1 - e^{-x/\theta}$. Solving for x, we get $e^{-x/\theta} = 1 - r_i$. Taking the ln on both sides, we obtain $x = -\theta \ln(1 - r_i)$. Because $(1 - r_i) \sim U(0, 1)$, we let $x = -\theta \ln r_i$.

4.12.3   Gamma Distribution

If X_i is a random variable from the gamma distribution, with parameters (α, β), and if α is an integer n, then X_i is the convolution (sum) of n exponential random variables Y_j, with parameter β. Thus, to simulate a gamma random variable X_i, generate n exponential random variables as described earlier. Then, $X_i = \sum Y_j$, where $j = 1, 2, \ldots, n$.

4.12.4   Poisson Distribution

We noted in Chapter 3 that the method of generating Poisson random variables would appear in this chapter. The procedure that follows requires us to make use of exponential random variables. If the number of events in an interval of a specified unit length follows the Poisson distribution, then the waiting time between successive events in the interval follows an exponential distribution with $\theta = 1/\lambda$, where λ is the average value of the Poisson random variable. Thus, we can generate exponential random variables as many times as are necessary until the sum of these exponential random variables exceeds the length of the interval (L) under consideration. Let n represent the number of exponential random variables generated. The Poisson random variable will be given by $Y_j = n - 1$. For example, suppose that we desire to generate Poisson random variables from a distribution with a mean rate of $\lambda = 2$ arrivals per day. Let $L = 1$ day. We proceed to generate exponential random variables with mean $\theta = 1/2$ until the sum of the exponentials exceeds $L = 1$. The Poisson random variable Y_j is given by $n - 1$. We can simplify this procedure. Let $X_i = -(1/\lambda a)\ln R_i$. The random variable Y_j is defined as the value of $(n - 1)$ such that $\sum_{i=1}^{n-1} X_i \leq L < \sum_{i=0}^{n} X_i$. Replacing X_i, we have $-\sum_{i=0}^{n-1} \ln R_i < \lambda(L) < -\sum_{i=0}^{n} \ln R_i$. Recall that $\sum \ln R_i = \ln \Pi R_i$. Multiplying by (-1), we have

$$\ln \prod_{i=0}^{n-1} R_i \geq -\lambda(L) > \ln \prod_{i=0}^{n} R_i$$

$$\prod_{i=0}^{n-1} R_i \geq e^{-\lambda(L)} > \prod_{i=0}^{n} R_i$$

Therefore, the method for generating Poisson random variables can be stated as follows:

1 Let $w = e^{(-\lambda(L))}$, with $z = 1$ and $i = 0$.
2 Generate $R_i \sim U(0, 1)$. Replace z by $z R_i$. If $z < w$, then output $Y = i$ and return to step 1; otherwise, proceed to step 3.
3 Replace i by $i + 1$ and return to step 2.

4.12.5 Normal Distribution

In attempting to use the inverse CDF method to generate normal random variables, we encounter the problem that there is no explicit closed-form expression for the CDF of x, given by $F(x)$, or of the inverse function $F^{-1}(R)$. Methods of approximation have been developed, however, that work well in simulating normal random variables. A few of these will be discussed briefly here.

Box–Muller Polar Method for Standard Normal Variables

This method can be used to generate a pair of standard normal random variables. The following steps are involved:

1 Generate R_1 and R_2, which are two uniform numbers on $(0, 1)$.
2 Evaluate $Z_1 = (\sqrt{-2 \ln R_i}) \cos(2\pi R_2)$ and $Z_2 = (\sqrt{-2 \ln R_1}) \sin(2\pi R_2)$.
3 The normal random variables X_1 and X_2, with mean μ_x and standard deviation σ_x, are given by $X_1 = \mu_x + Z_1 \sigma_x$ and $X_2 = \mu_x + Z_2 \sigma_x$.

Central Limit Theorem Method

Discussion of this method will be deferred until Chapter 7.

Let us now consider some situations where simulation in regard to the distributions mentioned earlier might be useful. First, we look at the discrete Poisson distribution.

Suppose that a local business firm is interested in comparing the number of local telephone calls to the number of long-distance calls coming into their switchboard. Assuming independence, we define X as representing the number of local calls received during a 15-minute period and Y as representing the number of long-distance calls during a 15-minute period. A random variable of interest would be $W = X/Y$, where X and Y represent Poisson random variables with parameters λ_1 and λ_2, respectively. How will the distribution of W behave, given the restriction that Y cannot take on the value of 0? The results of two simulations are presented. In Simulation 1, the random variable X has a mean of 6, and the random variable Y has a mean of 3; the histogram represents 200 random values of W. The sample mean of W is 2.76, with a sample standard deviation of 2.12.

Simulation 1

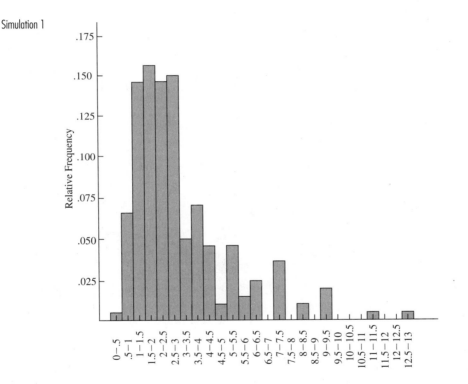

In Simulation 2, the mean of the random variable X is 2, and the mean of the random variable Y is 5; the histogram represents 200 random values of W. Here, the sample mean for W is 0.53, with a sample standard deviation of 0.57.

Let us now consider a situation involving exponential random variables. Engineers often are concerned with a parallel system. A system is defined as being *parallel* if it functions when at least one of its components is capable of working. Let us consider a parallel system that has two components. We define the random variable X as representing the time until failure for component 1, and we define the random variable Y as representing the time until failure for component 2. X and Y are independent

Simulation 2

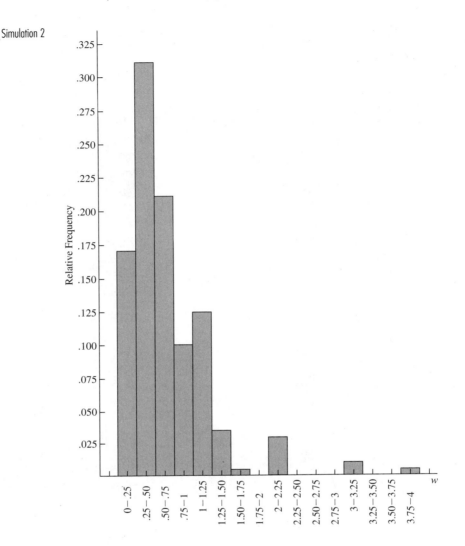

exponential random variables, with mean θ_x for X and mean θ_y for Y. A variable of interest is the maximum of X and Y; let $W = \max\{X, Y\}$. In Simulations 3 through 5, we can see how the distribution of W behaves.

As a final example, we consider a situation that involves normal random variables. A business is interested in looking at its profits versus its total income before expenses are paid. Let X represent the amount of profit, and let Y represent the amount of expenses. Suppose that both X and Y are normally distributed. A variable of interest is W where $W = X/(X + Y)$. Simulations 6 through 8 show the form of W for different settings of the parameters underlying X and Y.

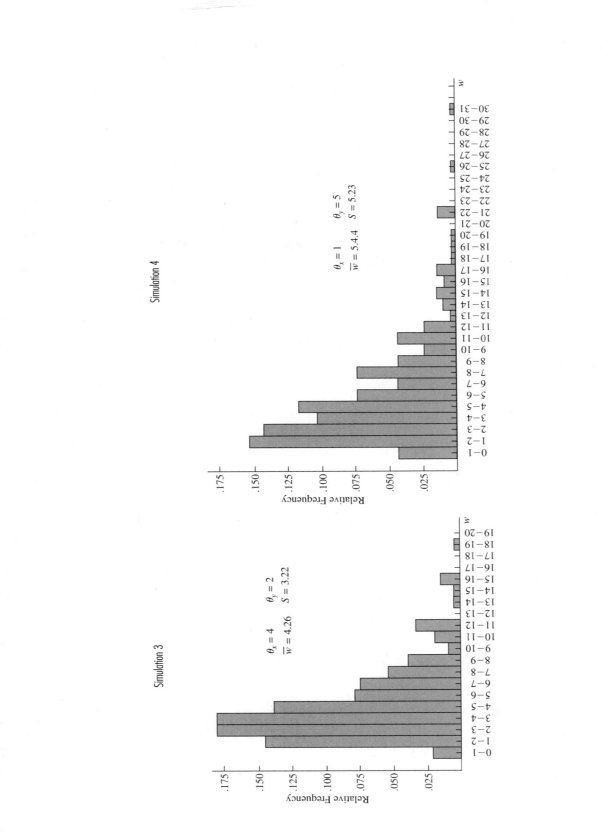

Simulation 4

$\theta_x = 1$ $\theta_y = 5$
$\overline{w} = 5.4.4$ $S = 5.23$

Simulation 3

$\theta_x = 4$ $\theta_y = 2$
$\overline{w} = 4.26$ $S = 3.22$

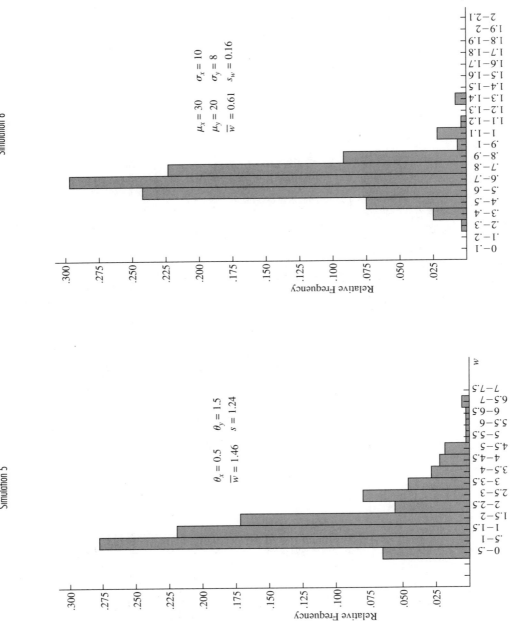

Simulation 6

$\mu_x = 30$ $\sigma_x = 10$
$\mu_y = 20$ $\sigma_y = 8$
$\overline{w} = 0.61$ $s_w = 0.16$

Simulation 5

$\theta_x = 0.5$ $\theta_y = 1.5$
$\overline{w} = 1.46$ $s = 1.24$

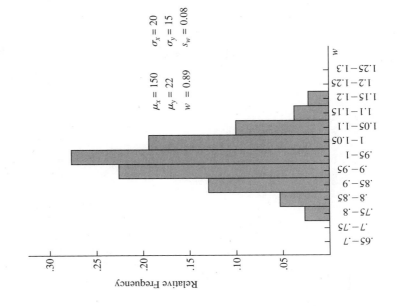

Simulation 8

Relative Frequency

$\mu_x = 150$ $\sigma_x = 20$
$\mu_y = 22$ $\sigma_y = 15$
$\overline{w} = 0.89$ $s_w = 0.08$

w

.65–.7
.7–.75
.75–.8
.8–.85
.85–.9
.9–.95
.95–1
1–1.05
1.05–1.1
1.1–1.15
1.15–1.2
1.2–1.25
1.25–1.3

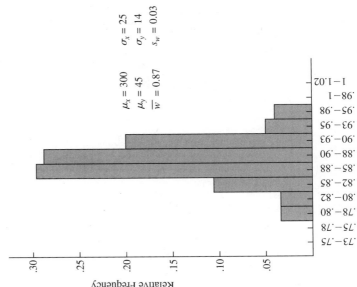

Simulation 7

Relative Frequency

$\mu_x = 300$ $\sigma_x = 25$
$\mu_y = 45$ $\sigma_y = 14$
$\overline{w} = 0.87$ $s_w = 0.03$

.73–.75
.75–.78
.78–.80
.80–.82
.82–.85
.85–.88
.88–.90
.90–.93
.93–.95
.95–.98
.98–1
1–1.02

It is time for you to try your hand at simulating the behavior of continuous random variables, as follows:

1 Suppose that a two-component system has its two components operating independently and in parallel. Each component has the same Weibull distribution of life lengths. Generate a simulated distribution for the life length of the system. (You may choose your own parameter values.)

2 Repeat part 1 for the same components operating in series.

4.13 Summary

Practical applications of measurement, such as to improve the quality of a system, often require careful study of the behavior of observations measured on a continuum (time, weight, distance, or the like). Probability distributions for such measurements are characterized by **probability density functions** and cumulative **distribution functions.** The **uniform** distribution models the waiting time for an event that is equally likely to occur anywhere on a finite interval. The **exponential** model is useful for modeling the lengths of time between occurrences of random events. The **gamma** model fits the behavior of sums of exponential variables and, therefore, models the total waiting time until a number of random events have occurred. The most widely used continuous probability model is the **normal**, which exhibits the symmetric mound-shaped behavior often seen in data. For data that is mound-shaped but skewed, the **Weibull** provides a useful and versatile model.

All models are approximations to reality, but the ones discussed here are relatively easy to use and yet provide good approximations in a variety of settings.

Supplementary Exercises

4.102 Let Y possess a density function

$$f(y) = \begin{cases} cy & \text{for } 0 \le y \le 2 \\ 0 & \text{elsewhere} \end{cases}$$

a Find c.

b Find $F(y)$.

c Graph $f(y)$ and $F(y)$.

d Use $F(y)$ from part (b) to find $P(1 \le Y \le 2)$.

e Use the geometric figure for $f(y)$ from part (c) to calculate $P(1 \le Y \le 2)$.

4.103 Let Y have the density function given by

$$f(y) = \begin{cases} cy^2 + y & \text{for } 0 \le y \le 2 \\ 0 & \text{elsewhere} \end{cases}$$

a Find c.

b Find $F(y)$.

c Graph $f(y)$ and $F(y)$.

d Use $F(y)$ from part (b) to find $F(-1)$, $F(0)$, and $F(1)$.

e   Find $P(0 \leq Y \leq 0.5)$.

f   Find the mean and the variance of Y.

4.104   Let Y have the density function given by

$$f(y) = \begin{cases} 0.2 & \text{for } -1 < y \leq 0 \\ 0.2 + cy & \text{for } 0 < y \leq 1 \\ 0 & \text{elsewhere} \end{cases}$$

Answer parts (a) through (f) of Exercise 4.103 for this density function.

4.105   The grade-point averages of a large population of college students are approximately normally distributed, with a mean equal to 2.4 and a standard deviation equal to 0.5. What fraction of the students possess a grade-point average in excess of 3.0?

4.106   Referring to Exercise 4.105, if students who possess a grade-point average equal to or less than 1.9 are dropped from college, what percentage of the students will be dropped?

4.107   Referring to Exercise 4.105, suppose that three students are selected at random from the student body. What is the probability that all three possess a grade-point average in excess of 3.0?

4.108   A machine operation produces bearings whose diameters are normally distributed, with mean and standard deviation of 3.005 and 0.001, respectively. Customer specifications require that the bearing diameters lie in the interval 3.000 ± 0.0020. Units falling outside the interval are considered scrap and must be remachined or used as stock for smaller bearings. At the current machine setting, what fraction of total production will be scrap?

4.109   Referring to Exercise 4.108, suppose that five bearings are drawn from production. What is the probability that at least one of these is defective?

4.110   Let Y have the density function

$$f(y) = \begin{cases} cye^{-2y} & \text{for } 0 \leq y \leq \infty \\ 0 & \text{elsewhere} \end{cases}$$

a   Give the mean and the variance for Y.

b   Give the moment-generating function for Y.

c   Find the value of c.

4.111   Find $E(X^k)$ for the beta random variable. Then find the mean and the variance for the beta random variable.

4.112   The yield force of a steel reinforcing bar of a certain type is found to be normally distributed, with a mean of 8500 pounds and a standard deviation of 80 pounds. If three such bars are to be used on a certain project, find the probability that all three will have yield forces in excess of 8700 pounds.

4.113   An engineer has observed that the gap times between vehicles passing a certain point on a highway have an exponential distribution, with a mean of 10 seconds.

a   Find the probability that the next gap observed will be no longer than 1 minute.

b   Find the probability density function for the sum of the next four gap times to be observed. What assumptions must be true for this answer to be correct?

4.114   The proportion of time, per day, that all checkout counters in a supermarket are busy is a random variable X that has the probability density function

$$f(x) = \begin{cases} kx^2(1-x)^4 & \text{for } 0 \leq x \leq 1 \\ 0 & \text{elsewhere} \end{cases}$$

a   Find the value of k that makes this a probability density function.

b   Find the mean and the variance of X.

4.115   If the life length X of a certain type of battery has a Weibull distribution, with $\gamma = 2$ and $\theta = 3$ (with measurements in years), find the probability that such a battery will last less than 4 years given that it is now 2 years old.

4.116 The time (in hours) a manager takes to interview an applicant has an exponential distribution, with $\theta = 1/2$. Three applicants arrive at 8:00 A.M., and the interviews begin. A fourth applicant arrives at 8:45 A.M. What is the probability that the latecomer must wait before seeing the manager?

4.117 The weekly repair cost, Y for a certain machine has a probability density function given by

$$f(y) = \begin{cases} 3(1-y)^2 & \text{for } 0 < y < 1 \\ 0 & \text{elsewhere} \end{cases}$$

with measurements in \$100s. How much money should be budgeted each week for repair cost to ensure that the actual cost should exceed the budgeted amount only 10% of the time?

4.118 A house builder must order some supplies that have a waiting time for delivery Y that is uniformly distributed over the interval 1 to 4 days. Because the builder can get by without them for 2 days, the cost of the delay is a fixed \$100 for any waiting time of up to 2 days. After 2 days, however, the cost of the delay becomes \$100 plus \$20 per day for each additional day. Thus, if the waiting time is 3.5 days, the cost of the delay is $\$100 + \$20(1.5) = \$130$. Find the expected value of the builder's cost due to waiting for supplies.

4.119 The relationship between incomplete gamma integrals and sums of Poisson probabilities is given by

$$\frac{1}{\Gamma(\alpha)} \int_\lambda^\infty y^{\alpha-1} e^{-y} dy = \sum_{y=0}^{\alpha-1} \frac{\lambda^y e^{-\lambda}}{y!}$$

for integer values of α.
    If Y has a gamma distribution, with $\alpha = 2$ and $\beta = 1$, find $P(Y > 1)$ by using this equality and Table 3 of the Appendix.

4.120 The weekly down time (in hours) for a certain production line has a gamma distribution, with $\alpha = 3$ and $\beta = 2$. Find the probability that the down time for a given week will not exceed 10 hours.

4.121 Suppose that plants of a certain species are randomly dispersed over a region, with a mean density of λ plants per unit area; that is, the number of plants in a region of area A has a Poisson distribution, with mean λA. For a randomly selected plant in this region, let R denote its distance from the nearest neighboring plant.

a Find the probability density function for R. [*Hint:* Notice that $P(R > r)$ is the same as the probability of seeing no plants within a circle of radius r.]

b Find $E(R)$.

4.122 A random variable X is said to have a log normal distribution if $Y = \ln(X)$ has a normal distribution. (The symbol ln denotes natural logarithm.) In this case, X must not be negative. The shape of the log normal probability density function is similar to that of the gamma distribution, with long tails to the right. The equation of the log normal density function is

$$f(x) = \begin{cases} \dfrac{1}{\sqrt{2\pi}\sigma x} e^{-(\ln(x)-\mu)^2/2\sigma^2} & \text{for } x > 0 \\ 0 & \text{elsewhere} \end{cases}$$

Since $\ln(x)$ is a monotonic function of x,

$$P(X \le x) = P[\ln(X) \le \ln(x)] = P[Y \le \ln(x)]$$

where Y has a normal distribution with mean μ and variance σ^2. Thus, probabilities in the log normal case can be found by transforming them in to probabilities in the normal case. If X has a log normal distribution, with $\mu = 4$ and $\sigma^2 = 1$, find the following probabilities.

a $P(X \le 4)$

b $P(X > 8)$

4.123   If X has a log normal distribution, with parameters μ and σ^2, it can be shown that

$$E(X) = e^{\mu + \sigma^2/2}$$

and

$$V(X) = e^{2\mu + \sigma^2}(e^{\sigma^2} - 1)$$

The grains composing polycrystalline metals tend to have weights that follow a log normal distribution. For a certain type of aluminum, grain weights have a log normal distribution with $\mu = 3$ and $\sigma = 4$ (in units of 10^{-2} gram).

a   Find the mean and the variance of the grain weights.

b   Find an interval within which at least 75% of the grain weights should lie (use Tchebysheff's Theorem).

c   than the mean grain weight.

4.124   Let Y denote a random variable whose probability density function is given by

$$f(y) = \left(\frac{1}{2}\right) e^{-|y|}, \qquad -\infty < y < \infty$$

Find the moment-generating function of Y, and use it to find $E(Y)$.

4.125   The life length Y of a certain component in a complex electronic system is known to have an exponential density, with a mean of 100 hours. The component is replaced at failure or at age 200 hours, whichever comes first.

a   Find the distribution function for X, the length of time that the component is in use.

b   Find $E(X)$.

4.126   We can show that the normal density function integrates to unity, by showing that

$$\frac{1}{\sqrt{2\pi}} \int_{-\infty}^{\infty} e^{-(1/2)uy^2} dy = \frac{1}{\sqrt{\mu}}$$

This, in turn, can be shown by considering the product of two such integrals,

$$\frac{1}{2\pi} \left(\int_{-\infty}^{\infty} e^{-(1/2)uy^2} dy \right) \left(\int_{-\infty}^{\infty} e^{-(1/2)ux^2} dx \right) = \frac{1}{2\pi} \int_{-\infty}^{\infty} \int_{-\infty}^{\infty} e^{-(1/2)u(x^2+y^2)} dx dy$$

By transforming the expression to polar coordinates, show that the double integral is equal to $1/u$.

4.127   The function $\Gamma(u)$ is defined by

$$\Gamma(u) = \int_0^\infty y^{u-1} e^{-y} dy$$

Integrate by parts to show that

$$\Gamma(u) = (u - 1)\Gamma(u - 1)$$

Hence, if n is a positive integer, then $\Gamma(n) = (n - 1)!$.

4.128   Show that $\Gamma(1/2) = \sqrt{\pi}$, by writing

$$\Gamma\left(\frac{1}{2}\right) = \int_0^\infty y^{-1/2} e^{-y} dy$$

making the transformation $y = (1/2)x^2$, and employing the result of Exercise 4.126.

4.129   The function $B(\alpha, \beta)$ is defined by

$$B(\alpha, \beta) = \int_0^1 y^{\alpha-1}(1-y)^{\beta-1}dy$$

a   Let $y = \sin^2 \theta$, and show that

$$B(\alpha, \beta) = 2 \int_0^{\pi/2} \sin^{2\alpha-1} \theta \cos^{2\beta-1} \theta d\theta$$

b   Write $\Gamma(\alpha)\Gamma(\beta)$ as a double integral, transform the integral to polar coordinates, and conclude that

$$B(\alpha, \beta) = \frac{\Gamma(\alpha)\Gamma(\beta)}{\Gamma(\alpha+\beta)}$$

4.130   The lifetime X of a certain electronic component is a random variable with density function

$$f(x) = \begin{cases} \left(\frac{1}{100}\right) e^{-x/100} & \text{for } x > 0 \\ 0 & \text{elsewhere} \end{cases}$$

Three of these components operate independently in a piece of equipment. The equipment fails if at least two of the components fail. Find the probability that the equipment will operate for at least 200 hours without failure.

Multivariate Probability Distributions

5.1 Bivariate and Marginal Probability Distributions

Chapters 3 and 4 dealt with experiments that produced a single numerical response, or random variable, of interest. We discussed, for example, the life lengths X of a battery or the strength Y of a steel casing. Often, however, we want to study the joint behavior of two random variables, such as the joint behavior of life length *and* casing strength for these batteries. Perhaps in such a study we can identify a region in which some combination of life length and casing strength is optimal in balancing the cost of manufacturing with customer satisfaction. To proceed with such a study, we must know how to handle a joint probability distribution. When only two random variables are involved, joint distributions are called *bivariate distributions*. We will discuss the bivariate case in some detail; extensions to more than two variables follow along similar lines.

Other situations in which bivariate probability distributions are important come to mind easily. A physician studies the joint behavior of pulse and exercise. An educator studies the joint behavior of grades and time devoted to study, or the interrelationship of pretest and posttest scores. An economist studies the joint behavior of business volume and profits. In fact, most real problems we come across have more than one underlying random variable of interest. The National Highway Traffic Safety Administration, among others, is interested in the effect of seat belt use on saving lives. One study reported statistics on children under the age of 5 who were involved in motor vehicle accidents in which at least one fatality occurred. For 7,060 such accidents between 1985 and 1989, the results were as shown in Table 5.1. As noted

Seat Belt Status	Survivors	Fatalities	Total
No belt	1129	509	1638
Adult belt	432	73	505
Children's carseat belt	733	139	872
Total	2294	721	3015

Source: Statistical Abstract of the United States, 1991.

in earlier chapters, it's more convenient to analyze such data in terms of meaningful random variables rather than in terms of described categories. For each child, we want to know whether or not he or she survived and what type of seat belt (if any) he or she was wearing. An X_1 defined as

$$X_1 = \begin{cases} 0 & \text{if child survived} \\ 1 & \text{if child did not survive} \end{cases}$$

will keep track of the number of fatalities per child. Now, children's carseats usually involve two belts: the regular adult belt in the vehicle; and the belt on the carseat itself. An X_2 defined as

$$X_2 = \begin{cases} 0 & \text{if no belt} \\ 1 & \text{if adult belt used} \\ 2 & \text{if child carseat belt used} \end{cases}$$

provides a random variable that keeps track of the number of belts (properly defined) restraining the child.

The frequencies from Table 5.1 are turned into the relative frequencies of Table 5.2 to produce the *joint probability distribution* of X_1 and X_2.

		X_1		
		0	1	
	0	0.38	0.17	0.55
X_2	1	0.14	0.02	0.16
	2	0.24	0.05	0.29
		0.76	0.24	1.00

In general, we write

$$P(X_1 = x_1, \ X_2 = x_2) = p(x_1, x_2)$$

and call $p(x_1, x_2)$ the *joint probability function* of (X_1, X_2). From Table 5.2, we can see that

$$P(X_1 = 0, \ X_2 = 2) = p(0, 2) = 0.24 \qquad \textbf{(5.1)}$$

represents the approximate probability that a child will both survive *and* be properly belted in a children's carseat when involved in an accident in which someone dies. The probability that a child will be in a children's carseat is

$$P(X_2 = 2) = P(X_1 = 0, \ X_2 = 2) + P(X_1 = 1, \ X_2 = 2)$$
$$= 0.24 + 0.05$$
$$= 0.29$$

which is one of the *marginal probabilities* for X_2. The univariate distribution along the right margin is the *marginal distribution* for X_2 alone, and the one along the bottom row is the marginal distribution for X_1 alone. The *conditional probability distribution* for X_1, given X_2, fixes a value of X_2 (a row of Table 5.2) and looks at the relative frequencies for values of X_1 in that row. For example, conditioning on $X_2 = 0$, produces

$$P(X_1 = 0 | X_2 = 0) = \frac{P(X_1 = 0, \ X_2 = 0)}{P(X_2 = 0)}$$
$$= \frac{0.38}{0.55}$$
$$= 0.69$$

and

$$P(X_1 = 1 | X_2 = 0) = 0.31$$

These two values show how survivorship behaves in the case where no seat belts were in use; 69% of children in such situations survived fatal accidents. That number may seem high, but compare it to $P(X_1 = 0 | X_2 = 2)$. Does seat belt use seem to be beneficial? Another set of conditional distributions here consists of those for X_2 conditioning on X_1. In this case,

$$P(X_2 = 0 | X_1 = 0) = \frac{P(X_1 = 0, \ X_2 = 0)}{P(X_1 = 0)}$$
$$= \frac{0.38}{0.76}$$
$$= 0.50$$

and

$$P(X_2 = 0 | X_1 = 1) = \frac{0.17}{0.24} = 0.71$$

What do these two probabilities tell you about the effectiveness of seat belts? The bivariate discrete case is summarized in Definition 5.1.

DEFINITION 5.1

Let X_1 and X_2 be discrete random variables. The **joint probability distribution** of X_1 and X_2 is given by

$$p(x_1, x_2) = P(X_1 = x_1, \ X_2 = x_2)$$

defined for all real numbers x_1 and x_2. The function $p(x_1, x_2)$ is called the *joint probability function* of X_1 and X_2.

The marginal probability functions of X_1 and X_2, respectively, are given by

$$p_1(x_1) = \sum_{x_2} p(x_1, x_2)$$

and

$$p_2(x_2) = \sum_{x_1} p(x_1, x_2) \quad \blacksquare$$

E X A M P L E   **5.1**   Three checkout counters are in operation at a local supermarket. Two customers arrive at the counters at different times, when the counters are serving no other customers. It is assumed that the customers then choose a checkout station at random and independent of one another. Let X_1 denote the number of times counter A is selected, and let X_2 denote the number of times counter B is selected by the two customers. Find the joint probability distribution of X_1 and X_2. Find the probability that one of the customers visits counter B, given that one of the customers is known to have visited counter A.

Solution   For convenience, let us introduce X_3—the number of customers who visit counter C. Now the event ($X_1 = 0$ and $X_2 = 0$) is equivalent to the event ($X_1 = 0$, $X_2 = 0$, and $X_3 = 2$). It follows that

$$
\begin{aligned}
P(X_1 = 0,\ X_2 = 0) &= P(X_1 = 0,\ X_2 = 0,\ X_3 = 2) \\
&= P(\text{customer I selects counter C and customer II selects counter C}) \\
&= P(\text{customer I selects counter C}) \times P(\text{customer II selects counter C}) \\
&= \frac{1}{3} \times \frac{1}{3} \\
&= \frac{1}{9}
\end{aligned}
$$

because customers' choices are independent and because each customer makes a random selection from among the three available counters.

It is slightly more complicated to calculate $P(X_1 = 1,\ X_2 = 0) = P(X_1 = 1,\ X_2 = 0,\ X_3 = 1)$. In this event, customer I could select counter A and customer II could select counter C, or I could select C and II could select A. Thus,

$$
\begin{aligned}
P(X_1 = 1,\ X_2 = 0) &= P(\text{I selects A})P(\text{II selects C}) + P(\text{I selects C})P(\text{II selects A}) \\
&= \left(\frac{1}{3} \times \frac{1}{3} \right) + \left(\frac{1}{3} \times \frac{1}{3} \right) \\
&= \frac{2}{9}
\end{aligned}
$$

Similar arguments allow one to derive the results given in Table 5.3. The second

T A B L E **5.3**
Joint Distribution of X_1 and X_2 for Example 5.1

		X_1			
		0	**1**	**2**	**Marginal probabilities for X_2**
	0	1/9	2/9	1/9	4/9
X_2	**1**	2/9	2/9	0	4/9
	2	1/9	0	0	1/9
Marginal probabilities for X_1		4/9	4/9	1/9	1

statement asks for

$$P(X_2 = 1 | X_1 = 1) = \frac{P(X_1 = 1, X_2 = 1)}{P(X_1 = 1)}$$
$$= \frac{2/9}{4/9}$$
$$= \frac{1}{2}$$

Does this answer (1/2), agree with your intuition?   ▪

Before we move to the continuous case, let us quickly review the situation in one dimension. If $f(x)$ denotes the probability density function of a random variable X, then $f(x)$ represents a relative frequency curve; and probabilities, such as $P(a \le X \le b)$, are represented as areas under this curve. In symbolic terms,

$$P(a \le X \le b) = \int_a^b f(x)dx$$

Now, suppose that we are interested in the joint behavior of two continuous random variables—say, X_1 and X_2—where X_1 and X_2 might represent the amounts of two different hydrocarbons found in an air sample taken for a pollution study, for example. The relative frequency of these two random variables can be modeled by a bivariate function, $f(x_1, x_2)$, which forms a probability, or relative frequency, surface in three dimensions. Figure 5.1 shows such a surface. The probability that X_1 lies in one interval and that X_2 lies in another interval is then represented as a volume under this surface. Thus,

$$P(a_1 \le X_1 \le a_2, \quad b_1 \le X_2 \le b_2) = \int_{b_1}^{b_2} \int_{a_1}^{a_2} f(x_1, x_2)dx_1 dx_2$$

Notice that this integral simply gives the volume under the surface and over the shaded region in Figure 5.1. We trace the actual computations involved in such a bivariate problem with a very simple example.

E X A M P L E   **5.2**   A certain process for producing an industrial chemical yields a product that contains two main types of impurities. For a certain volume of sample from this process,

F I G U R E **5.1**
A bivariate density function.

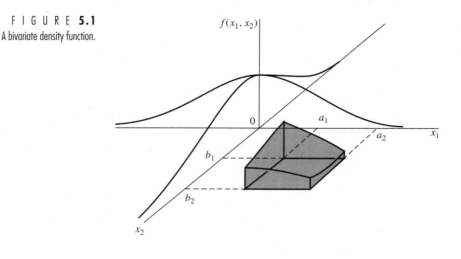

let X_1 denote the proportion of total impurities in the sample, and let X_2 denote the proportion of type I impurity among all impurities found. Suppose that, after investigation of many such samples, the joint distribution of X_1 and X_2 can be adequately modeled by the following function:

$$f(x_1, x_2) = \begin{cases} 2(1 - x_1) & \text{for } 0 \leq x_1 \leq 1, \text{ and } \quad 0 \leq x_2 \leq 1 \\ 0 & \text{elsewhere} \end{cases}$$

This function graphs as the surface given in Figure 5.2. Calculate the probability that X_1 is less than 0.5 and that X_2 is between 0.4 and 0.7.

Solution   From the preceding discussion, we see that

$$P(0 \leq x_1 \leq 0.5, \ 0.4 \leq x_2 \leq 0.7) = \int_{0.4}^{0.7} \int_0^{0.5} 2(1 - x_1)dx_1 dx_2$$
$$= \int_{0.4}^{0.7} [-(1 - x_1)^2]_0^{0.5} dx_2$$
$$= \int_{0.4}^{0.7} (0.75)dx_2$$

F I G U R E **5.2**
Probability density function for
Example 5.2.

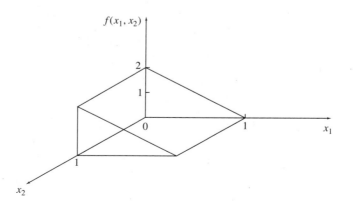

$$= 0.75(0.7 - 0.4)$$
$$= 0.75(0.3)$$
$$= 0.225$$

Thus, the fraction of such samples having less than 50% impurities and a relative proportion of type I impurities of between 40% and 70% is 0.225. ∎

Just as the univariate (or marginal) probabilities were computed by summing over rows or columns in the discrete case, the univariate density function for X_1 in the continuous case can be found by integrating ("summing") over values of X_2. Thus, the marginal density function of X_1, $f_1(x_1)$, is given by

$$f_1(x_1) = \int_{-\infty}^{\infty} f(x_1, x_2) dx_2$$

Similarly, the marginal density function of X_2, $f_2(x_2)$, is given by

$$f_2(x_2) = \int_{-\infty}^{\infty} f(x_1, x_2) dx_1$$

E X A M P L E   **5.3**   For the case given in Example 5.2, find the marginal probability density functions for X_1 and X_2.

Solution   Let us first try to visualize what the answers should look like, before going through the integration. To find $f_1(x_1)$, we accumulate all the probabilities in the x_2 direction. Look at Figure 5.2 and think of collapsing the wedge-shaped figure back onto the $[x_1, \ f_1(x_1, x_2)]$ plane. Then more probability mass will build up toward the zero point of the x_1-axis than toward the unity point. In other words, the function $f_1(x_1)$ should be high at 0 and low at 1. Formally,

$$f_1(x_1) = \int_{-\infty}^{\infty} f(x_1, x_2) dx_2$$
$$= \int_{0}^{1} 2(1 - x_1) dx_2$$
$$= 2(1 - x_1), \qquad 0 < x_1 \leq 1$$

F I G U R E **5.3**
Probability density function
$f_1(x_1)$ for Example 5.3.

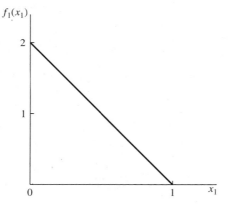

The function graphs as in Figure 5.3. Notice that our conjecture is correct. Thinking of how $f_2(x_2)$ should look geometrically, imagine that the wedge of Figure 5.2 has been forced back onto the $[x_2, f(x_1, x_2)]$ plane. Then the probability mass should accumulate equally all along the $(0, 1)$ interval on the x_2-axis. Mathematically,

$$
\begin{aligned}
f_2(x_2) &= \int_{-\infty}^{\infty} f(x_1, x_2)\mathrm{d}x_1 \\
&= \int_0^1 2(1 - x_1)\mathrm{d}x_1 \\
&= [-(1 - x_1)^2]_0^1, \qquad 0 < x_2 \leq 1 \\
&= 1
\end{aligned}
$$

and again our conjecture is verified, as shown in Figure 5.4.   ∎

F I G U R E **5.4**
Probability density function
$f_2(x_2)$ for Example 5.3.

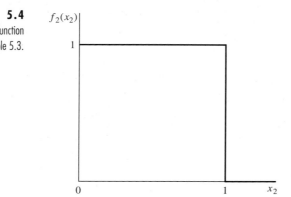

The bivariate continuous case is summarized in Definition 5.2.

DEFINITION **5.2**   Let X_1 and X_2 be continuous random variables. The **joint probability density function** of X_1 and X_2, if it exists, is given by a nonnegative function $f(x_1, x_2)$ such that

$$P(a_1 \leq X_1 \leq a_2, \quad b_1 \leq X_2 \leq b_2) = \int_{b_1}^{b_2} \int_{a_1}^{a_2} f(x_1, x_2)dx_1 dx_2$$

The **marginal probability density functions** of X_1 and X_2, respectively, are given by

$$f_1(x_1) = \int_{-\infty}^{\infty} f(x_1, x_2)dx_2$$

and

$$f_2(x_2) = \int_{-\infty}^{\infty} f(x_1, x_2)dx_1 \quad \blacksquare$$

Following is another (and somewhat more complicated) example.

EXAMPLE **5.4**   Gasoline is to be stocked in a bulk tank once each week and then sold to customers. Let X_1 denote the proportion of the tank that is stocked in a particular week, and let X_2 denote the proportion of the tank that is sold in the same week. Due to limited supplies, X_1 is not fixed in advance but varies from week to week. Suppose that a study of many weeks shows the joint relative frequency behavior of X_1 and X_2 to be such that the following density function provides an adequate model:

$$f(x_1, x_2) = \begin{cases} 3x_1 & \text{for } 0 \leq x_2 \leq x_1 \leq 1 \\ 0 & \text{elsewhere} \end{cases}$$

Notice that X_2 must always be less than or equal to X_1. This density function is graphed in Figure 5.5. Find the probability that X_2 will be between 0.2 and 0.4 for a given week.

Solution   The question refers to the marginal behavior of X_2. Thus, it is necessary to find

$$f_2(x_2) = \int_{-\infty}^{\infty} f(x_1, x_2)dx_1$$

$$= \int_{x_2}^{1} 3x_1 dx_1$$

$$= \frac{3}{2}x_1^2 \big|_{x_2}^{1}$$

$$= \frac{3}{2}(1 - x_2^2), \qquad 0 < x_2 \leq 1$$

FIGURE **5.5**
The joint density function for
Example 5.4.

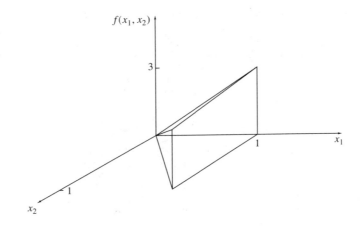

It follows directly that

$$P(0.2 \le X_2 \le 0.4) = \int_{0.2}^{0.4} \frac{3}{2}(1 - x_2^2)dx_2$$

$$= \frac{3}{2}\left(x_2 - \frac{x_2^3}{3}\right)\Bigg|_{0.2}^{0.4}$$

$$= \frac{3}{2}\left\{\left[0.4 - \frac{(0.4)^3}{3}\right] - \left[0.2 - \frac{(0.2)^3}{3}\right]\right\}$$

$$= 0.272$$

Notice that the marginal density of X_2 graphs as a function that is high at $x_2 = 0$ and then tends toward 0 as x_2 tends toward 1. Does this agree with your intuition after looking at Figure 5.5? ∎

5.2 Conditional Probability Distributions

Recall that, in the bivariate discrete case, the conditional probabilities for X_1 for a given X_2 were found by fixing attention on the particular row in which $X_2 = x_2$, and then looking at the relative probabilities within that row; that is, the individual cell probabilities were divided by the marginal total for that row, to obtain conditional probabilities.

In the bivariate continuous case, the form of the probability density function representing the conditional behavior of X_1 for a given value of X_2 is found by slicing through the joint density in the x_1 direction at the particular value of X_2. The function then has to be weighted by the marginal density function for X_2 at that point. Let's look at a specific example before examining the general definition of conditional density functions.

E X A M P L E   **5.5**   Referring to the joint density function of Example 5.4, find the conditional probability that X_2 is less than 0.2, given that X_1 is known to be 0.5.

Solution   Slicing through $f(x_1, x_2)$ in the x_2 direction at $x_1 = 0.5$ yields

$$f(0.5, x_2) = 3(0.5) \doteq 1.5, \qquad 0 \le x_2 \le 0.5$$

Thus, the conditional behavior of X_2 for a given X_1 of 0.5 is constant over the interval $(0, 0.5)$. The marginal value of $f(x_1)$ at $x_1 = 0.5$ is obtained as follows:

$$\begin{aligned} f_1(x_1) &= \int_{-\infty}^{\infty} f(x_1, x_2) dx_2 \\ &= \int_0^{x_1} 3x_1 dx_2 \\ &= 3x_1^2, \qquad 0 \le x_1 \le 1 \\ f_1(0.5) &= 3(0.5)^2 \\ &= 0.75 \end{aligned}$$

Upon dividing, we see that the conditional behavior of X_2 for a given X_1 of 0.5 is represented by the function

$$\begin{aligned} f(x_2 | x_1 = 0.5) &= \frac{f(0.5, x_2)}{f_1(0.5)} \\ &= \frac{1.5}{0.75} \\ &= 2, \qquad 0 < x_2 < 0.5 \end{aligned}$$

This function of x_2 has all the properties of a probability density function and is the conditional density function for X_2 at $X_1 = 0.5$. Then,

$$\begin{aligned} P(X_2 < 0.2 | X_1 = 0.5) &= \int_0^{0.2} f(x_2 | x_1 = 0.5) dx_2 \\ &= \int_0^{0.2} 2 dx_2 \\ &= 0.2(2) \\ &= 0.4 \end{aligned}$$

Thus, among all weeks in which the tank was half full immediately after stocking, sales amounted to less than 20% of the tank 40% of the time.   ∎

Clearly, these manipulations to obtain conditional density functions in the continuous case are analogous to those used to obtain conditional probabilities in the discrete case, except that integrals are used in place of sums. The formal definition of a conditional density function follows.

DEFINITION   **5.3**

Let X_1 and X_2 be jointly continuous random variables with joint probability density function $f(x_1, x_2)$ and marginal densities $f_1(x_1)$ and $f_2(x_2)$, respectively. Then, the **conditional probability density function** of X_1, given $X_2 = x_2$, is defined by

$$f(x_1|x_2) = \begin{cases} \dfrac{f(x_1, x_2)}{f_2(x_2)} & \text{for } f_2(x_2) > 0 \\ 0 & \text{elsewhere} \end{cases}$$

and the conditional probability density function of X_2, given $X_1 = x_1$, is defined by

$$f(x_2|x_1) = \begin{cases} \dfrac{f(x_1, x_2)}{f_1(x_1)} & \text{for } f_1(x_1) \geq 0 \\ 0 & \text{elsewhere} \end{cases}$$

EXAMPLE   **5.6**

A soft-drink machine has a random supply Y_2 at the beginning of a given day and dispenses a random amount Y_1 during the day (with measurements in gallons). It is not resupplied during the day; hence, $Y_1 \leq Y_2$. It has been observed that Y_1 and Y_2 have joint density

$$f(y_1, y_2) = \begin{cases} 1/2 & \text{for } 0 \leq y_1 \leq y_2, \text{ and } 0 \leq y_2 \leq 2 \\ 0 & \text{elsewhere} \end{cases}$$

In other words, the points (y_1, y_2) are uniformly distributed over the triangle with the given boundaries. Find the conditional density of Y_1, given $Y_2 = y_2$. Evaluate the probability that less than 1/2 gallon is sold, given that the machine contains 1 gallon at the start of the day.

Solution   The marginal density of Y_2 is given by

$$f_2(y_2) = \int_{-\infty}^{\infty} f(y_1, y_2) dy_1$$

$$= \begin{cases} \int_0^{y_2} (1/2) dy_1 = (1/2) y_2 & \text{for } 0 \leq y_2 \leq 2 \\ 0 & \text{elsewhere} \end{cases}$$

By Definition 5.3,

$$f_1(y_1|y_2) = \frac{f(y_1, y_2)}{f_2(y_2)}$$

$$= \begin{cases} \dfrac{1/2}{(1/2) y_2} = \dfrac{1}{y_2} & \text{for } 0 \leq y_1 \leq y_2 \leq 2 \\ 0 & \text{elsewhere} \end{cases}$$

The probability of interest is

$$P(Y_1 < 1/2 | Y_2 = 1) = \int_{-\infty}^{1/2} f(y_1 | y_2 = 1) dy_1$$

$$= \int_{0}^{1/2} (1) dy_1$$

$$= \frac{1}{2}$$

Notice that, if the machine had contained 2 gallons at the start of the day, then

$$P(Y_1 \leq 1/2 | Y_2 = 2) = \int_{0}^{1/2} (1/2) dy_1$$

$$= \frac{1}{4}$$

Thus, the amount sold is highly dependent on the amount in supply.   ■

5.3 Independent Random Variables

Before defining independent random variables, let us recall once again that two events A and B are independent if $P(AB) = P(A)P(B)$. Somewhat analogously, two discrete random variables are independent if

$$P(X_1 = x_1,\ X_2 = x_2) = P(X_1 = x_1)P(X_2 = x_2)$$

for all real numbers x_1 and x_2. A similar idea carries over to the continuous case.

DEFINITION   **5.4**

Discrete random variables X_1 and X_2 are said to be **independent** if

$$P(X_1 = x_1,\ X_2 = x_2) = P(X_1 = x_1)P(X_2 = x_2)$$

for all real numbers x_1 and x_2.
    Continuous random variables X_1 and X_2 are said to be **independent** if

$$f(x_1, x_2) = f_1(x_1)f_2(x_2)$$

for all real numbers x_1 and x_2.   ■

The concepts of joint probability density functions and independence extend immediately to n random variables, where n is any finite positive integer. The n random variables $X_1, X_2, \ldots, X_n$ are said to be independent if their joint density function, $f(x_1, \ldots, x_n)$, is given by

$$f(x_1, \ldots, x_n) = f_1(x_1)f_2(x_2) \cdots f_n(x_n)$$

for all real numbers $x_1, x_2, \ldots, x_n$.

E X A M P L E   **5.7**   Show that the random variables that have the joint distribution identified in Table 5.2 are not independent.

Solution   It is necessary to check only one entry in the table. We see that $P(X_1 = 0, \ X_2 = 0) = 0.38$, whereas $P(X_1 = 0) = 0.76$ and $P(X_2 = 0) = 0.55$. Because

$$P(X_1 = 0, \ X_2 = 0) \neq P(X_1 = 0)P(X_2 = 0)$$

the random variables cannot be independent.   ▪

E X A M P L E   **5.8**   Show that the random variables in Example 5.2 are independent.

Solution   Here,

$$f(x_1, x_2) = \begin{cases} 2(1 - x_1) & \text{for } 0 \leq x_1 \leq 1 \quad \text{and} \quad 0 \leq x_2 \leq 1 \\ 0 & \text{elsewhere} \end{cases}$$

We saw in Example 5.3 that

$$f_1(x_1) = 2(1 - x_1), \qquad 0 \leq x_1 \leq 1$$

and

$$f_2(x_2) = 1, \qquad 0 \leq x_2 \leq 1$$

Thus, $f(x_1, x_2) = f_1(x_1)f_2(x_2)$, for all real numbers x_1 and x_2; and X_1 and X_2 are independent random variables.   ▪

Exercises

5.1   Two construction contracts are to be randomly assigned to one or more of three firms. Numbering the firms I, II, and III, let X_1 be the number of contracts assigned to firm I, and let X_2 be the number assigned to firm II. (A firm may receive more than one contract.)

  a   Find the joint probability distribution for X_1 and X_2.
  b   Find the marginal probability distribution for X_1.
  c   Find $P(X_1 = 1 | X_2 = 1)$.

5.2   A radioactive particle is randomly located in a square area with sides that are 1 unit in length. Let X_1 and X_2 denote the coordinates of the particle. Since the particle is equally likely to fall in any subarea of fixed size, a reasonable model for (X_1, X_2) is given by

$$f(x_1, x_2) = \begin{cases} 1 & \text{for } 0 \leq x_1 \leq 1, \quad \text{and} \quad 0 \leq x_2 \leq 1 \\ 0 & \text{elsewhere} \end{cases}$$

  a   Sketch the probability density surface.
  b   Find $P(X_1 \leq 0.2, \ X_2 \leq 0.4)$.

c   Find $P(0.1 \leq X_1 \leq 0.3, X_2 > 0.4)$.

5.3   The same study that produced the seat-belt safety data of Table 5.1 also took into account the ages of the children involved in fatal accidents. The results for children wearing *no* seat belts are shown in the accompanying table. Here, $X_1 = 1$ if the child did not survive, and X_2 indicates the child's age in years. (An age of 0 implies that the child was less than 1 year old; an age of 1 implies that the child was more than 1 year old but not yet 2; and so on.)

Children Without Seat Belts in Serious Car Accidents

Age	Survivors	Fatalities
0	104	127
1	165	91
2	267	107
3	277	90
4	316	94

a   Construct an approximate joint probability distribution for X_1 and X_2.

b   Construct the conditional distributions of X_1 for fixed values of X_2. Discuss the implications of these results.

c   Construct the conditional distributions of X_2 for fixed values of X_1. Are the implications the same here as in part (b)?

5.4   An environmental engineer measures the amount (by weight) of particulate pollution in air samples (of a certain volume) collected over the smokestack of a coal-fueled power plant. Let X_1 denote the amount of pollutant per sample when a certain cleaning device on the stack is not operating, and let X_2 denote the amount of pollutant per sample when the cleaning device is operating, under similar environmental conditions. It is observed that X_1 is always greater than $2X_2$; and the relative frequency of (X_1, X_2) can be modeled by

$$f(x_1, x_2) = \begin{cases} k & \text{for } 0 \leq x_1 \leq 2, \quad 0 \leq x_2 \leq 1, \quad \text{and} \quad 2x_2 \leq x_1 \\ 0 & \text{elsewhere} \end{cases}$$

(In other words, X_1 and X_2 are randomly distributed over the region inside the triangle bounded by $x_1 = 2$, $x_2 = 0$, and $2x_2 = x_1$.)

a   Find the value of k that makes this a probability density function.

b   Find $P(X_1 \geq 3X_2)$. (In other words, find the probability that the cleaning device will reduce the amount of pollutant by one-third or more.)

5.5   Refer to the situation described in Exercise 5.2.

a   Find the marginal density function for X_1.

b   Find $P(X_1 \leq 0.5)$.

c   Are X_1 and X_2 independent?

5.6   Refer to the situation described in Exercise 5.4.

a   Find the marginal density function of X_2.

b   Find $P(X_2 \leq 0.4)$.

c   Are X_1 and X_2 independent?

d   Find $P(X_2 \leq 1/4 | X_1 = 1)$.

5.7   Let X_1 and X_2 denote the proportions of two different chemicals found in a sample mixture of chemicals used as an insecticide. Suppose that X_1 and X_2 have a joint probability density given by

$$f(x_1, x_2) = \begin{cases} 2 & \text{for } 0 \le x_1 \le 1, \quad 0 \le x_2 \le 1, \quad \text{and} \quad 0 \le x_1 + x_2 \le 1 \\ 0 & \text{elsewhere} \end{cases}$$

(Notice that $X_1 + X_2$ must be equal to unity at most, because the random variables denote proportions within the same sample.)

a   Find $P(X_1 \le 3/4, \ X_2 \le 3/4)$.

b   Find $P(X_1 \le 1/4, \ X_2 \le 1/2)$.

c   Find $P(X_1 \le 1/2 | X_2 \le 1/2)$.

5.8   Refer to the situation described in Exercise 5.7.

a   Find the marginal density functions for X_1 and X_2.

b   Are X_1 and X_2 independent?

c   Find $P(X_1 > 1/2 | X_2 = 1/4)$.

5.9   Let X_1 and X_2 denote the proportions of time, out of one workweek, that employees I and II, respectively, actually spend performing their assigned tasks. The joint relative frequency behavior of X_1 and X_2 is modeled by the probability density function

$$f(x_1, x_2) = \begin{cases} x_1 + x_2 & \text{for } 0 \le x_1 \le 1, \quad \text{and} \quad 0 \le x_2 \le 1 \\ 0 & \text{elsewhere} \end{cases}$$

a   Find $P(X_1 < 1/2, \ X_2 > 1/4)$.

b   Find $P(X_1 + X_2 \le 1)$.

c   Are X_1 and X_2 independent?

5.10   Referring to Exercise 5.9, find the probability that employee I spends more than 75% of the workweek on his assigned task, given that employee II spends exactly 50% of the workweek on her assigned task.

5.11   An electronic surveillance system has one of each of two different types of components in joint operations. If X_1 and X_2 denote the random life lengths of the components of type I and type II, respectively, the joint probability density function is given by

$$f(x_1, x_2) = \begin{cases} (1/8)x_1 e^{-(x_1+x_2)/2} & \text{for } x_1 > 0, \quad \text{and} \quad x_2 > 0 \\ 0 & \text{elsewhere} \end{cases}$$

(measurements are in hundreds of hours).

a   Are X_1 and X_2 independent?

b   Find $P(X_1 > 1, \ X_2 > 1)$.

5.12   A bus arrives at a bus stop at a randomly selected time within a 1-hour period. A passenger arrives at the bus stop at a randomly selected time within the same hour. The passenger is willing to wait for the bus for up to 1/4 of an hour. What is the probability that the passenger will catch the bus? [*Hint:* Let X_1 denote the bus arrival time, and let X_2 denote the passenger arrival time. If these arrivals are independent, then

$$f(x_1, x_2) = \begin{cases} 1 & \text{for } 0 \le x_1 \le 1, \quad \text{and} \quad 0 \le x_2 \le 1 \\ 0 & \text{elsewhere} \end{cases}$$

Now find $P(X_2 \le X_1 \le X_2 + 1/4)$.]

5.13   Two friends are to meet at a library. Each arrives at an independently selected time within a fixed 1-hour period. Each agrees to wait no more than 10 minutes for the other. Find the probability that they will meet.

5.14   Two quality control inspectors each interrupt a production line at randomly (and independently) selected times within a given day (of 8 hours). Find the probability that the two interruptions will be more than 4 hours apart.

5.15   Two telephone calls come into a switchboard at random times within a fixed 1-hour period. Assume that the calls are independent of each other.

   a   Find the probability that both are made in the first half hour.

   b   Find the probability that the two calls are within 5 minutes of each other.

5.16   A bombing target is in the center of a circle with radius 1 mile. A bomb falls at a randomly selected point inside that circle. If the bomb destroys everything within $\frac{1}{2}$ mile of its landing point, what is the probability that it will destroy the target?

5.4   Expected Values of Functions of Random Variables

When we encounter problems that involve more than one random variable, we often combine the variables into a single function. We may be interested in the life lengths of five different electronic components within the same system or the difference between two strength test measurements on the same section of cable. We now discuss how to find expected values of functions of more than one random variable. Definition 5.5 gives the basic result for finding expected values in the bivariate case. The definition, of course, can be generalized to more variables.

DEFINITION   **5.5**

Suppose that the discrete random variables (X_1, X_2) have a joint probability function given by $p(x_1, x_2)$. If $g(X_1, X_2)$ is any real-valued function of (X_1, X_2), then

$$E[g(X_1, X_2)] = \sum_{x_1} \sum_{x_2} g(x_1, x_2) p(x_1, x_2)$$

The sum is over all values of (x_1, x_2) for which $p(x_1, x_2) > 0$. If (X_1, X_2) are continuous random variables, with probability density function $f(x_1, x_2)$, then

$$E[g(X_1, X_2)] = \int_{-\infty}^{\infty} \int_{-\infty}^{\infty} g(x_1, x_2) f(x_1, x_2) dx_1 dx_2. \quad \blacksquare$$

Notice that, if X_1 and X_2 are independent, then

$$E[g(X_1)h(X_2)] = E[g(X_1)]E[h(X_2)]$$

A function of two variables that is commonly of interest in probabilistic and statistical problems is the *covariance*.

DEFINITION   **5.6**

The **covariance** between two random variables X_1 and X_2 is given by

$$\text{cov}(X_1, X_2) = E[(X_1 - \mu_1)(X_2 - \mu_2)]$$

where

$$\mu_1 = E(X_1) \text{ and } \mu_2 = E(X_2) \quad \blacksquare$$

The covariance helps us assess the relationship between two variables in the following sense. If X_2 tends to be large when X_1 is large, and if X_2 tends to be small when X_1 is small, then X_1 and X_2 have a positive covariance. If, on the other hand, X_2 tends to be small when X_1 is large and large when X_1 is small, then the two variables have a negative covariance.

While covariance measures the direction of the association between two random variables, *correlation* measures the strength of the association.

DEFINITION   **5.7**

The **correlation** coefficient between two random variables X_1 and X_2 is given by

$$\rho = \frac{\text{cov}(X_1, X_2)}{\sqrt{V(X_1) \times V(X_2)}} \quad \blacksquare$$

The correlation coefficient is a unitless quantity that takes on values between -1 and $+1$. If $\rho = +1$ or $\rho = -1$, then X_2 must be a linear function of X_1. The next theorem gives an easier computational form for the covariance.

THEOREM   **5.1**

If X_1 has a mean of μ_1 and X_2 has a mean of μ_2, then

$$\text{cov}(X_1, X_2) = E(X_1 X_2) - \mu_1 \mu_2$$

The proof (not included here) is an obvious application of the properties of expected value. $\blacksquare$

If X_1 and X_2 are independent random variables, then $E(X_1 X_2) = E(X_1)E(X_2)$. Using Theorem 5.1, it is then clear that independence between X_1 and X_2 implies that $\text{cov}(X_1, X_2) = 0$. The converse is not necessarily true; that is, zero covariance does not imply that the variables are independent.

To see how this theorem works, look at the joint distribution in Table 5.4. Clearly,

TABLE **5.4**
Joint Distribution to Illustrate
Theorem 5.1

X_1

X_2		-1	0	1	
	-1	1/8	1/8	1/8	3/8
	0	1/8	0	1/8	2/8
	1	1/8	1/8	1/8	3/8
		3/8	2/8	3/8	1

$E(X_1) = E(X_2) = 0$, and $E(X_1 X_2) = 0$ as well. Therefore $\text{cov}(X_1, X_2) = 0$. On the other hand,

$$P(X_1 = -1, \ X_2 = -1) = 1/8 \neq P(X_1 = -1) P(X_2 = -1)$$

so the random variables are dependent.

We shall now calculate the covariance in two examples.

E X A M P L E   **5.9**   A firm that sells word-processing systems keeps track of the number of customers who call on any one day and the number of orders placed on any one day. Let X_1 denote the number of calls, let X_2 denote the number of orders placed, and let $p(x_1, x_2)$ denote the joint probability function for (X_1, X_2). Records indicate that

$p(0, 0) = 0.04$	$p(2, 0) = 0.20$
$p(1, 0) = 0.16$	$p(2, 1) = 0.30$
$p(1, 1) = 0.10$	$p(2, 2) = 0.20$

Thus, for any given day the probability of, say, two calls and one order is 0.30. Find $\text{cov}(X_1, X_2)$ and the correlation between X_1 and X_2.

Solution   Theorem 5.1 suggests that we first find $E(X_1 X_2)$, which is

$$
\begin{aligned}
E(X_1 X_2) &= \sum_{x_1} \sum_{x_2} x_1 x_2 p(x_1, x_2) \\
&= (0 \times 0) p(0, 0) + (1 \times 0) p(1, 0) + (1 \times 1) p(1, 1) \\
&\quad + (2 \times 0) p(2, 0) + (2 \times 1) p(2, 1) + (2 \times 2) p(2, 2) \\
&= 0(0.4) + 0(0.16) + 1(0.10) + 0(0.20) + 2(0.30) + 4(0.20) \\
&= 1.50
\end{aligned}
$$

Now we must find $E(X_1) = \mu_1$ and $E(X_2) = \mu_2$. The marginal distributions of X_1 and X_2 are given on the following charts:

x_1	$p(x_1)$
0	0.04
1	0.26
2	0.70

x_2	$p(x_2)$
0	0.40
1	0.40
2	0.20

It follows that

$$\mu_1 = 1(0.26) + 2(0.70) = 1.66$$

and

$$\mu_2 = 1(0.40) + 2(0.20) = 0.80$$

Thus,

$$\text{cov}(X_1, X_2) = E(X_1 X_2) - \mu_1 \mu_2$$

$$= 1.50 - 1.66(0.80)$$
$$= 1.50 - 1.328$$
$$= 0.172$$

From the marginal distributions of X_1 and X_2, it follows that

$$V(X_1) = E(X_1^2) - \mu_1$$
$$= 3.06 - (1.66)^2$$
$$= 0.30$$

and

$$V(X_2) = E(X_2^2) - \mu_1^2$$
$$= 1.2 - (0.8)^2$$
$$= 0.56$$

Hence,

$$\rho = \frac{\text{cov}(X_1, X_2)}{\sqrt{V(X_1) \times V(X_2)}}$$
$$= \frac{0.172}{\sqrt{(0.30)(0.56)}}$$
$$= 0.42$$

A moderate positive association exists between the number of calls and the number of orders placed. Do the positive covariance the strength of the correlation agree with your intuition? ▪

When a problem involves n random variables $X_1, X_2, \ldots, X_n$, we are often interested in studying linear combinations of those variables. For example, if the random variables measure the quarterly incomes for n plants in a corporation, we may want to look at their sum or average. If X_1 represents the monthly cost of servicing defective plants before a new quality control system was installed, and X_2 denotes that cost after the system went into operation, then we may want to study $X_1 - X_2$. Theorem 5.2 gives a general result for the mean and the variance of a linear combination of random variables.

THEOREM   **5.2**

Let $Y_1, \ldots, Y_n$ and $X_1, \ldots, X_m$ be random variables, with $E(Y_i) = \mu_i$ and $E(X_i) = \xi_i$. Define

$$U_1 = \sum_{i=1}^{n} a_i Y_i, \qquad U_2 = \sum_{j=1}^{m} b_j X_j$$

for constants $a_1, \ldots, a_n,$ and $b_1, \ldots, b_m.$ Then the following relationships hold:

(a) $E(U_1) = \sum\limits_{i=1}^{n} a_i \mu_i$

(b) $V(U_1) = \sum\limits_{i=1}^{n} a_i^2 V(Y_i) + 2 \sum\sum\limits_{i<j} a_i a_j \text{cov}(Y_i, Y_j)$

where the double sum is over all pairs (i, j) with $i < j$

(c) $\text{cov}(U_1, U_2) = \sum\limits_{i=1}^{n} \sum\limits_{j=1}^{m} a_i b_j \text{cov}(Y_i, X_j)$

Proof

(a) The proof for part (a) follows directly from the definition of expected value and from properties of sums or integrals.

(b) To prove this part, we resort to the definition of *variance* and write

$$V(U_1) = E[U_1 - E(U_1)]^2$$

$$= E\left[\sum_{i=1}^{n} a_i Y_i - \sum_{i=1}^{n} a_i \mu_i\right]^2$$

$$= E\left[\sum_{i=1}^{n} a_i (Y_i - \mu_i)\right]^2$$

$$= E\left[\sum_{i=1}^{n} a_i^2 (Y_i - \mu_i)^2 + \sum\sum_{i \neq j} a_i a_j (Y_i - \mu_i)(Y_j - \mu_j)\right]$$

$$= \sum_{i=1}^{n} a_i^2 E(Y_i - \mu_i)^2 + \sum\sum_{i \neq j} a_i a_j E[(Y_i - \mu_i)(Y_j - \mu_j)]$$

By the definition of *variance* and *covariance*, we then have

$$V(U_1) = \sum_{i=1}^{n} a_i^2 V(Y_i) + \sum\sum_{i \neq j} a_i a_j \text{cov}(Y_i, Y_j)$$

Notice that $\text{cov}(Y_i, Y_j) = \text{cov}(Y_j, Y_i)$; hence, we can write

$$V(U_1) = \sum_{i=1}^{n} a_i^2 V(Y_i) + 2 \sum\sum_{i<j} a_i a_j \text{cov}(Y_i, Y_j)$$

(c) The proof of this part is obtained by similar steps. We have

$$\text{cov}(U_1, U_2) = E\{[U_1 - E(U_1)][U_2 - E(U_2)]\}$$

$$= E\left[\left(\sum_{i=1}^{n} a_i Y_i - \sum_{i=1}^{n} a_i \mu_i\right)\left(\sum_{j=1}^{m} b_j X_j - \sum_{j=1}^{m} b_j \xi_j\right)\right]$$

$$= E\left\{\left[\sum_{i=1}^{n} a_i (Y_i - \mu_i)\right]\left[\sum_{j=1}^{m} b_j (X_j - \xi_j)\right]\right\}$$

$$= E\left[\sum_{i=1}^{n}\sum_{j=1}^{m} a_i b_j (Y_i - \mu_i)(X_j - \xi_j)\right]$$

$$= \sum_{i=1}^{n}\sum_{j=1}^{m} a_i b_j E[(Y_i - \mu_i)(X_j - \xi_j)]$$

$$= \sum_{i=1}^{n}\sum_{j=1}^{m} a_i b_j \text{cov}(Y_i, X_j)$$

On observing that $\text{cov}(Y_i, Y_i) = V(Y_i)$, we see that part (b) is a special case of part (c).   ∎

EXAMPLE **5.10**   Let $Y_1, Y_2, \ldots, Y_n$ be independent random variables, with $E(Y_i) = \mu$ and $V(Y_i) = \sigma^2$. (These variables may denote the outcomes of n independent trials of an experiment.) Defining

$$\bar{Y} = \frac{1}{n}\sum_{i=1}^{n} Y_i$$

show that $E(\bar{Y}) = \mu$ and $V(\bar{Y}) = \sigma^2/n$.

Solution   Notice that $\bar{Y}$ is a linear function with all constants a_i equal to $1/n$; that is,

$$\bar{Y} = \left(\frac{1}{n}\right)Y_1 + \cdots + \left(\frac{1}{n}\right)Y_n$$

By Theorem 5.2 part (a),

$$E(\bar{Y}) = \sum_{i=1}^{n} a_i \mu$$

$$= \mu \sum_{i=1}^{n} a_i$$

$$= \mu \sum_{i=1}^{n} \frac{1}{n}$$

$$= \frac{n\mu}{n}$$

$$= \mu$$

By Theorem 5.2 part (b),

$$V(\bar{Y}) = \sum_{i=1}^{n} a_i^2 V(Y_i) + 2\sum\sum_{i<j} a_i a_j a_{ij} \text{cov}(Y_i, Y_j)$$

but the covariance terms are all zero, since the random variables are independent. Thus,

$$V(\bar{Y}) = \sum_{i=1}^{n} \left(\frac{1}{n}\right)^2 \sigma^2 = \frac{n\sigma^2}{n^2} = \frac{\sigma^2}{n} \quad \blacksquare$$

E X A M P L E   **5.11**   Let X_1 denote the amount of gasoline stocked in a bulk tank at the beginning of a week, and let X_2 denote the amount sold during the week; then $Y = X_1 - X_2$ represents the amount left over at the end of the week. Find the mean and the variance of Y, assuming that the joint density function of (X_1, X_2) is given by

$$f(x_1, x_2) = \begin{cases} 3x_1 & \text{for } 0 \le x_2 \le x_1 \le 1 \\ 0 & \text{elsewhere} \end{cases}$$

Solution   We must first find the means and the variances of X_1 and X_2. The marginal density of X_1 is found to be

$$f_1(x_1) = \begin{cases} 3x_1^2 & \text{for } 0 \le x_1 \le 1 \\ 0 & \text{elsewhere} \end{cases}$$

Thus,

$$E(X_1) = \int_0^1 x_1(3x_1^2)dx_1$$

$$= 3\left[\frac{x_1^4}{4}\right]_0^1$$

$$= \frac{3}{4}$$

The marginal density of X_2 is found to be

$$f_2(x_2) = \begin{cases} \frac{3}{2}(1 - x_2^2) & \text{for } 0 \le x_2 \le 1 \\ 0 & \text{elsewhere} \end{cases}$$

Thus,

$$E(X_2) = \int_0^1 (x_2)\frac{3}{2}(1 - x_2^2)dx_2$$

$$= \frac{3}{2}\int_0^1 (x_2 - x_2^3)dx_2$$

$$= \frac{3}{2}\left\{\left[\frac{x_2^2}{2}\right]_0^1 - \left[\frac{x_2^4}{4}\right]_0^1\right\}$$

$$= \frac{3}{2}\left\{\frac{1}{2} - \frac{1}{4}\right\}$$

$$= \frac{3}{8}$$

By similar arguments, it follows that

$$E(X_1^2) = \frac{3}{5}$$

$$V(X_1) = \frac{3}{5} - \left(\frac{3}{4}\right)^2$$

$$= 0.0375$$

$$E(X_2^2) = \frac{1}{5}$$

and

$$V(X_2) = \frac{1}{5} - \left(\frac{3}{8}\right)^2 = 0.0594$$

The next step is to find $\text{cov}(X_1, X_2)$. Now

$$E(X_1 X_2) = \int_0^1 \int_0^{x_1} (x_1 x_2) 3x_1 \, dx_2 \, dx_1$$

$$= 3 \int_0^1 \int_0^{x_1} x_1^2 x_2 \, dx_2 \, dx_1$$

$$= 3 \int_0^1 x_1^2 \left[\frac{x_2^2}{2}\right]_0^{x_1} dx_1$$

$$= \frac{3}{2} \int_0^1 x_1^4 \, dx_1$$

$$= \frac{3}{2} \left[\frac{x_1^5}{5}\right]_0^1$$

$$= \frac{3}{10}$$

and

$$\text{cov}(X_1, X_2) = E(X_1 X_2) - \mu_1 \mu_2$$

$$= \frac{3}{10} - \left(\frac{3}{4}\right)\left(\frac{3}{8}\right)$$

$$= 0.0188$$

From Theorem 5.2,

$$E(Y) = E(X_1) - E(X_2)$$

$$= \frac{3}{4} - \frac{3}{8}$$

$$= \frac{3}{8}$$

$$= 0.375$$

and

$$V(Y) = V(X_1) + V(X_2) + 2(1)(-1)\text{cov}(X_1, X_2)$$
$$= 0.0375 + 0.0594 - 2(0.0188)$$
$$= 0.0593 \quad \blacksquare$$

When the random variables in use are independent, calculating the variance of a linear function simplifies, since the covariances are zero.

E X A M P L E   **5.12**   A firm purchases two types of industrial chemicals. The amount of type I chemical purchased per week, X_1, has $E(X_1) = 40$ gallons, with $V(X_1) = 4$. The amount of type II chemical purchased, X_2, has $E(X_2) = 65$ gallons, with $V(X_2) = 8$. Type I costs $3 per gallon, whereas type II costs $5 per gallon. Find the mean and the variance of the total weekly amount spent on these types of chemicals, assuming that X_1 and X_2 are independent.

Solution   The dollar amount spent per week is given by

$$Y = 3X_1 + 5X_2$$

From Theorem 5.2,

$$E(Y) = 3E(X_1) + 5E(X_2)$$
$$= 3(40) + 5(65)$$
$$= 445$$

and

$$V(Y) = (3)^2 V(X_1) + (5)^2 V(X_2)$$
$$= 9(4) + 25(8)$$
$$= 236$$

The firm can expect to spend $445 per week on chemicals. $\blacksquare$

E X A M P L E   **5.13**   Sampling problems involving finite populations can be modeled by selecting balls from urns. Suppose that an urn contains r white balls and $(N - r)$ black balls. A random sample of n balls is drawn without replacement; and Y, the number of white balls in the sample, is observed. From Chapter 3, we know that Y has a hypergeometric probability distribution. Find the mean and the variance of Y.

Solution   We first observe some characteristics of sampling without replacement. Suppose that the sampling is done sequentially, and we observe outcomes for $X_1, X_2, \ldots, X_n$, where

$$X_i = \begin{cases} 1 & \text{if the } i\text{th draw results in a white ball} \\ 0 & \text{otherwise} \end{cases}$$

Unquestionably, $P(X_1 = 1) = r/N$. But it is also true that $P(X_2 = 1) = r/N$, because

$$\begin{aligned} P(X_2 = 1) &= P(X_1 = 1, \ X_2 = 1) + P(X_1 = 0, \ X_2 = 1) \\ &= P(X_1 = 1)P(X_2 = 1|X_1 = 1) + P(X_1 = 0)P(X_2 = 1|X_1 = 0) \\ &= \frac{r}{N}\left(\frac{r-1}{N-1}\right) + \frac{N-r}{N}\left(\frac{r}{N-1}\right) \\ &= \frac{r(N-1)}{N(N-1)} \\ &= \frac{r}{N} \end{aligned}$$

The same is true for X_k; that is,

$$P(X_k = 1) = \frac{r}{N}, \qquad k = 1, \ldots, n$$

Thus, the probability of drawing a white ball on any draw, given no knowledge of the outcomes of previous draws, is r/N.

In a similar way, it can be shown that

$$P(X_j = 1, \ X_k = 1) = \frac{r(r-1)}{N(N-1)}, \qquad j \neq k$$

Now observe that $Y = \sum_{i=1}^{n} X_i$; hence,

$$E(Y) = \sum_{i=1}^{n} E(X_i) = n\left(\frac{r}{N}\right)$$

To find $V(Y)$, we need $V(X_i)$ and $\text{cov}(X_i, X_j)$. Because X_i is 1 with probability r/N and 0 with probability $1 - (r/N)$, it follows that

$$V(X_i) = \frac{r}{N}\left(1 - \frac{r}{N}\right)$$

Also,

$$\begin{aligned} \text{cov}(X_i, X_j) &= E(X_i X_j) - E(X_i)E(X_j) \\ &= \frac{r(r-1)}{N(N-1)} - \left(\frac{r}{N}\right)^2 \\ &= -\frac{r}{N}\left(1 - \frac{r}{N}\right)\frac{1}{N-1} \end{aligned}$$

because $X_i X_j = 1$ if and only if $X_i = 1$ and $X_j = 1$. From Theorem 5.2, we know that

$$V(Y) = \sum_{i=1}^{n} V(X_i) + 2 \sum_{i<j} \sum cov(X_i, X_j)$$

$$= n\frac{r}{N}\left(1 - \frac{r}{N}\right) + 2 \sum_{i<j} \sum \left[-\frac{r}{N}\left(1 - \frac{r}{N}\right)\frac{1}{N-1}\right]$$

$$= n\frac{r}{N}\left(1 - \frac{r}{N}\right) - n(n-1)\frac{r}{N}\left(1 - \frac{r}{N}\right)\frac{1}{N-1}$$

because there are $n(n-1)/2$ terms in the double summation. A little algebra yields

$$V(Y) = n\frac{r}{N}\left(1 - \frac{r}{N}\right)\frac{N-n}{N-1} \quad \blacksquare$$

The usefulness of Theorem 5.2 becomes especially clear when one tries to find the expected value and the variance for the hypergeometric random variable without it—by proceeding directly from the definition of an expectation. The necessary summations are exceedingly difficult to obtain.

Exercises

5.17 Table 5.1 shows the joint probability distribution of fatalities (X_1) and number of seat belts used (X_2) for children in accidents.
a Find $E(X_1)$, $V(X_1)$, $E(X_2)$, and $V(X_2)$.
b Find $cov(X_1, X_2)$.

5.18 In a study of particulate pollution in air samples over a smokestack, X_1 represents the amount of pollutant per sample when a cleaning device is not operating, and X_2 represents the amount per sample when the cleaning device is operating. Assume that (X_1, X_2) has a joint probability density function

$$f(x_1, x_2) = \begin{cases} 1 & \text{for } 0 \le x_1 \le 2, \quad 0 \le x_2 \le 1, \quad \text{and } 2x_2 \le x_1 \\ 0 & \text{elsewhere} \end{cases}$$

The random variable $Y = X_1 - X_2$ represents the amount by which the weight of emitted pollutant can be reduced by using the cleaning device.
a Find $E(Y)$ and $V(Y)$.
b Find an interval within which values of Y should lie at least 75% of the time.

5.19 The proportions X_1 and X_2 of two chemicals found in samples of an insecticide have the joint probability density function

$$f(x_1, x_2) = \begin{cases} 2 & \text{for } 0 \le x_1 \le 1, \quad 0 \le x_2 \le 1, \quad \text{and } 0 \le x_1 + x_2 \le 1 \\ 0 & \text{elsewhere} \end{cases}$$

The random variable $Y = X_1 + X_2$ denotes the proportion of the insecticides due to the combination of both chemicals.

a   Find $E(Y)$ and $V(Y)$.

b   Find an interval within which values of Y should lie for at least 50% of the samples of insecticide.

5.20   For a sheet-metal stamping machine in a certain factory, the time between failures, X_1, has a mean time between failures (MTBF) of 56 hours and a variance of 16 hours. The repair time, X_2, has a mean time to repair (MTTR) of 5 hours and a variance of 4 hours.

a   If X_1 and X_2 are independent, find the expected value and the variance of $Y = X_1 + X_2$, which represents one operation-repair cycle.

b   Would you expect an operation-repair cycle to last more than 75 hours? Why?

5.21   A particular fast-food outlet is interested in the joint behavior of the random variables Y_1, defined as the total time between a customer's arrival at the store and his or her leaving the service window, and Y_2, the time that the customer waits in line before reaching the service window. Because Y_1 includes the time a customer waits in line, we must have $Y_1 \geq Y_2$. The relative frequency distribution of observed values of Y_1 and Y_2 can be modeled by the probability density function

$$f(y_1, y_2) = \begin{cases} e^{-y_1} & \text{for } 0 \leq y_2 \leq y_1 < \infty \\ 0 & \text{elsewhere} \end{cases}$$

a   Find $P(Y_1 < 2, \ Y_2 > 1)$.

b   Find $P(Y_1 \geq 2Y_2)$.

c   Find $P(Y_1 - Y_2 \geq 1)$. [*Note:* $Y_1 - Y_2$ denotes the time spent at the service window.]

d   Find the marginal density functions for Y_1 and Y_2.

5.22   Referring to Exercise 5.21, if a randomly observed customer's total waiting time for service is known to have been more than 2 minutes, find the probability that the customer waited less than 1 minute to be served.

5.23   For the situation described in Exercise 5.21, the random variable $Y_1 - Y_2$ represents the time spent at the service window.

a   Find $E(Y_1 - Y_2)$.

b   Find $V(Y_1 - Y_2)$.

c   Is it highly likely that a customer will spend more than 2 minutes at the service window?

5.24   Referring to Exercise 5.21, suppose that a customer spends a length of time y_1 at the store. Find the probability that this customer spends less than half of that time at the service window.

5.25   Prove Theorem 5.1.

5.5   The Multinomial Distribution

Suppose that an experiment consists of n independent trials, much like the binomial case, except that each trial can result in any one of k possible outcomes. For example, a customer checking out of a grocery store may choose any one of k checkout counters. Now suppose that the probability that a particular trial results in outcome i is denoted by p_i, where $i = 1, \ldots, k$, and where p_i remains constant from trial to trial. Let Y_i, with $i = 1, \ldots, k$, denote the number of the n trials that result in outcome i. In developing a formula for $P(Y_1 = y_1, \ \ldots, \ Y_k = y_k)$, we first observe that, because of independence of trials, the probability of having y_1 outcomes of type 1 through y_k outcomes of type k in a particular order is

$$p_1^{y_1} p_2^{y_2} \cdots p_k^{y_k}$$

The number of such orderings equals the number of ways to partition the n trials into y_1 type 1 outcomes, y_2 type 2 outcomes, and so on through y_k type k outcomes, or

$$\frac{n!}{y_1! \, y_2! \cdots y_k!}$$

where

$$\sum_{i=1}^{k} y_i = n$$

Hence,

$$P(Y_1 = y_1, \ \ldots, \ Y_k = y_k) = \frac{n!}{y_1! \, y_2! \cdots y_k!} \, p_1^{y_1} p_2^{y_2} \cdots p_k^{y_k}$$

This is called the *multinomial probability distribution*. Notice that, for $k = 2$, we are back into the binomial case, because Y_2 is then equal to $n - Y_1$. The following example illustrates the computations involved.

EXAMPLE **5.14**   Items under inspection are subject to two types of defects. About 70% of the items in a large lot are judged to be free of defects; another 20% have only a type A defect; and the final 10% have only a type B defect (none has both types of defect). If six of these items are randomly selected from the lot, find the probability that three have no defects, one has a type A defect, and two have type B defects.

Solution   If we can legitimately assume that the outcomes are independent from trial to trial (that is, from item to item in our sample), which they usually would be in a large lot, then the multinomial distribution provides a useful model. Letting Y_1, Y_2, and Y_3 denote the number of trials resulting in zero, type A, and type B defectives, respectively, we have $p_1 = 0.7$, $p_2 = 0.2$, and $p_3 = 0.1$. It follows that

$$P(Y_1 = 3, \ Y_2 = 1, \ Y_3 = 2) = \frac{6!}{3! \, 1! \, 2!} (0.7)^3 (0.2)(0.1)^2$$
$$= 0.041 \quad \blacksquare$$

EXAMPLE **5.15**   Find $E(Y_i)$ and $V(Y_i)$ for the multinomial probability distribution.

Solution   We are concerned with the marginal distribution of Y_i, the number of trials that fall in cell i. Imagine that all the cells, excluding cell i, are combined into a single large cell. Hence, every trial will result either in cell i or not in cell i, with probabilities p_i and $1 - p_i$, respectively; and Y_i possesses a binomial marginal probability distribution. Consequently,

$$E(Y_i) = np_i$$
$$V(Y_i) = np_i q_i$$

where

$$q_i = 1 - p_i$$

The same results can be obtained by setting up the expectations and evaluating. For example,

$$E(Y_1) = \sum_{y_1} \sum_{y_2} \cdots \sum_{y_k} y_1 \frac{n!}{y_1! \, y_2! \cdots y_k!} \, p_1^{y_1} p_2^{y_2} \cdots p_k^{y_k}$$

Because we have already derived the expected value and the variance of Y_i, the tedious summation of this expectation is left to the interested reader. ∎

EXAMPLE 5.16 If $Y_1, \ldots, Y_k$ have the multinomial distribution given in Example 5.14, find $\text{cov}(Y_s, Y_t)$, with $s \neq t$.

Solution Thinking of the multinomial experiment as a sequence of n independent trials, we define

$$U_i = \begin{cases} 1 & \text{if trial } i \text{ results in class } s \\ 0 & \text{otherwise} \end{cases}$$

and

$$W_i = \begin{cases} 1 & \text{if trial } i \text{ results in class } t \\ 0 & \text{otherwise} \end{cases}$$

Then,

$$Y_s = \sum_{i=1}^{n} U_i$$

and

$$Y_t = \sum_{j=1}^{n} W_j$$

To evaluate $\text{cov}(Y_s, Y_t)$, we need the following results:

$$E(U_i) = p_s$$
$$E(W_j) = p_t$$
$$\text{cov}(U_i, W_j) = 0$$

if $i \neq j$, since the trials are independent; and

$$\text{cov}(U_i, W_i) = E(U_i W_i) - E(U_i) E(W_i)$$
$$= 0 - p_s p_t$$

because $U_i W_i$ always equals zero. From Theorem 5.2, we then have

$$\text{cov}(Y_s, Y_t) = \sum_{i=1}^{n} \sum_{j=1}^{n} \text{cov}(U_i, W_j)$$

$$= \sum_{i=1}^{n} \text{cov}(U_i, W_i) + \sum \sum_{i \neq j} \text{cov}(U_i, W_j)$$

$$= \sum_{i=1}^{n} (-p_s p_t) + 0$$

$$= -n p_s p_t$$

The covariance is negative, which is to be expected because a large number of outcomes in cell s would force the number in cell t to be small. ■

Multinomial Experiment

1   The experiment consists of n identical trials.

2   The outcome of each trial falls into one of k classes or cells.

3   The probability that the outcome of a single trial will fall in a particular cell—say, cell i—is $p_i (i = 1, 2, \ldots, k)$, and remains the same from trial to trial. Notice that $p_1 + p_2 + p_3 + \cdots + p_k = 1$.

4   The trials are independent.

5   The random variables of interest are $Y_1, Y_2, \ldots, Y_k$, where $Y_i (i = 1, 2, \ldots, k)$ is equal to the number of trials in which the outcome falls in cell i. Notice that $Y_1 + Y_2 + Y_3 + \cdots + Y_k = n$.

The Multinomial Distribution

$$P(Y_1 = y_1, \ldots, Y_K = y_k) = \frac{n!}{y_1! y_2! \cdots y_k!} p_1^{y_1} p_2^{y_2} \cdots p_k^{y_k}$$

where $\sum_{i=1}^{k} y_i = n$ and $\sum_{i=1}^{k} p_i = 1$

$E(Y_i) = n p_i$ $\qquad V(Y_i) = n p_i (1 - p_i),$ $\qquad i = 1, \ldots, k$

$\text{cov}(Y_i, Y_j) = -n p_i p_j,$ $\qquad i \neq j$

Exercises

5.26   The National Fire Incident Reporting Service says that, among residential fires, approximately 73% are in family homes, 20% are in apartments, and the other 7% are in other types of dwellings. If four fires are reported independently in one day, find the probability that two are in family homes, one is in an apartment, and one is in another type of dwelling.

5.27 The typical cost of damages for a fire in a family home is $20,000, whereas the typical cost of damages in an apartment fire is $10,000, and in other dwellings only $2000. Using the information given in Exercise 5.26, find the expected total damage cost for the four independently reported fires.

5.28 In inspections of commercial aircraft, wing cracks are reported as nonexistent, detectable, or critical. The history of a certain fleet reveals that 70% of the planes inspected have no wing cracks, 25% have detectable wing cracks, and 5% have critical wing cracks. For the next five planes inspected, find the probabilities of the following events.

a One has a critical crack, two have detectable cracks, and two have no cracks.

b At least one critical crack is observed.

5.29 With the recent emphasis on solar energy, solar radiation has been carefully monitored at various sites in Florida. Among typical July days in Tampa, 30% have total radiation of at most 5 calories, 60% have total radiation of at most 6 calories, and 10% have total radiation of at most 8 calories. A solar collector for a hot water system is to be run for 6 days. Find the probability that 3 days will produce no more than 5 calories each, 1 day will produce between 5 and 6 calories, and 2 days will produce between 6 and 8 calories each. What assumptions must be true for your answer to be correct?

5.30 The U.S. Bureau of Labor Statistics reported that, as of 1981, approximately 21% of the adult population under age 65 was between 18 and 24 years of age, 28% was between 25 and 34, 19% was between 35 and 44, and 32% was between 45 and 64. In 1982, an automobile manufacturer wanted to obtain opinions on a new design from five randomly chosen adults from the under-65 group of adults. Of the five people so selected, find the approximate probability that two were between 18 and 24, two were between 25 and 44, and one was between 45 and 64.

5.31 Customers leaving a subway station can exit through any one of three gates. Assuming that any particular customer is equally likely to select any one of the three gates, find the probabilities of the following events among a sample of four customers.

a Two select gate A, one selects gate B, and one selects gate C.

b All four select the same gate.

c All three gates are used.

5.32 Among a large number of applicants for a certain position, 60% have only a high-school education, 30% have some college training, and 10% have completed a college degree. If five applicants are selected to be interviewed, find the probability that at least one will have completed a college degree. What assumptions must be true for your answer to be valid?

5.33 In a large lot of manufactured items, 10% contain exactly one defect, and 5% contain more than one defect. Ten items are randomly selected from this lot for sale, and the repair costs total

$$Y_1 + 3Y_2$$

where Y_1 denotes the number among the ten having exactly one defect, and Y_2 denotes the number among the ten having two or more defects. Find the expected value of the repair costs. Find the variance of the repair costs.

5.34 Referring to Exercise 5.33, if Y denotes the number of items that contain at least one defect, find the probabilities of the following events among the ten sampled items.

a Y is exactly 2.

b Y is at least 1.

5.35 Vehicles arriving at an intersection can turn right or left or can continue straight ahead. In a study of traffic patterns at this intersection over a long period of time, engineers have noted that 40% of the vehicles turn left, 25% turn right, and the remainder continue straight ahead.

a For the next five cars entering this intersection, find the probability that one turns left, one turns right, and three continue straight ahead.

b For the next five cars entering the intersection, find the probability that at least one turns right.

c If 100 cars enter the intersection in a day, find the expected values and the variance of the number that turn left. What assumptions must be true for your answer to be valid?

5.6   More on the Moment-generating Function

The use of moment-generating functions to identify distributions of random variables is particularly useful in work with sums of independent random variables. For example, suppose that X_1 and X_2 are independent, exponential random variables, each with mean θ, and $Y = X_1 + X_2$. Now, the moment-generating function of Y is given by

$$
\begin{aligned}
M_Y(t) = E(e^{tY}) &= E(e^{t(X_1+X_2)}) \\
&= E(e^{tX_1} \times e^{tX_2}) \\
&= E(e^{tX_1})E(e^{tX_2}) \\
&= M_{X_1}(t)M_{X_2}(t)
\end{aligned}
$$

because X_1 and X_2 are independent. From Chapter 4,

$$
M_{X_1}(t) = M_{X_2}(t) = (1 - \theta t)^{-1}
$$

and so

$$
M_Y(t) = (1 - \theta t)^{-2}
$$

Upon recognizing the form of this moment-generating function, we can immediately conclude that Y has a gamma distribution, with $\alpha = 2$ and $\beta = \theta$. This result can be generalized to the sum of independent gamma random variables with common scale parameter β.

EXAMPLE   **5.17**   Let X_1 denote the number of vehicles passing a particular point on the eastbound lane of a highway in 1 hour. Suppose that the Poisson distribution with mean λ_1 is a reasonable model for X_1. Now, let X_2 denote the number of vehicles passing a point on the westbound lane of the same highway in 1 hour. Suppose that X_2 has a Poisson distribution with mean λ_2. Of interest is $Y = X_1 + X_2$, the total traffic count in both lanes in one hour. Find the probability distribution for Y if X_1 and X_2 are assumed to be independent.

Solution   It is known from Chapter 4 that

$$
M_{X_1}(t) = e^{\lambda_1(e^t - 1)}
$$

and

$$M_{X_2}(t) = e^{\lambda_2(e^t - 1)}$$

By the property of moment-generating functions described earlier,

$$
\begin{aligned}
M_Y(t) &= M_{X_1}(t) M_{X_2}(t) \\
&= e^{\lambda_1(e^t - 1)} e^{\lambda_2(e^t - 1)} \\
&= e^{(\lambda_1 + \lambda_2)(e^t - 1)}
\end{aligned}
$$

Now the moment-generating function for Y has the form of a Poisson moment-generating function, with mean $\lambda_1 + \lambda_2$. Thus, by the uniqueness property, Y must have a Poisson distribution with mean $\lambda_1 + \lambda_2$. Being able to add independent Poisson random variables and still retain the Poisson properties is important in many applications. ■

Exercises

5.36　Find the moment-generating function for the negative binomial random variable. Use it to derive the mean and the variance of the distribution.

5.37　A parking lot has two entrances. Cars arrive at entrance I according to a Poisson distribution, with an average of three per hour; and they arrive at entrance II according to a Poisson distribution, with an average of four per hour. Find the probability that exactly three cars will arrive at the parking lot in a given hour.

5.38　Let X_1 and X_2 denote independent, normally distributed random variables, not necessarily having the same mean or variance. Show that, for any constants a and b, $Y = aX_1 + bX_2$ is normally distributed.

5.39　Resistors of a certain type have resistances that are normally distributed, with a mean of 100 ohms and a standard deviation of 10 ohms. Two such resistors are connected in series, which causes the total resistance in the circuit to be the sum of the individual resistances. Find the probabilities of the following events.

　a　The total resistance exceeds 220 ohms.

　b　The total resistance is less than 190 ohms.

5.40　A certain type of elevator has a maximum weight capacity X_1, which is normally distributed with a mean and a standard deviation of 5000 and 300 pounds, respectively. For a certain building equipped with this type of elevator, the elevator loading X_2 is a normally distributed random variable with a mean and a standard deviation of 4000 and 400 pounds, respectively. For any given time that the elevator is in use, find the probability that it will be overloaded, assuming that X_1 and X_2 are independent.

5.7 Conditional Expectations

Section 5.2 contains a discussion of conditional probability functions and conditional density functions, which we shall now relate to conditional expectations. Conditional expectations are defined in the same manner as univariate expectations are, except that the conditional density is used in place of the marginal density function.

DEFINITION 5.8

If X_1 and X_2 are any two random variables, the **conditional expectation** of X_1 given that $X_2 = x_2$ is defined to be

$$E(X_1|X_2 = x_2) = \int_{-\infty}^{\infty} x_1 f(x_1|x_2)dx_1$$

if X_1 and X_2 are jointly continuous, and

$$E(X_1|X_2 = x_2) = \sum_{y_1} x_1 p(x_1|x_2)$$

if X_1 and X_2 are jointly discrete. ▪

EXAMPLE 5.18

Refer to Example 5.6, with X_1 denoting the amount of drink sold and X_2 denoting the amount in supply, and

$$f(x_1, x_2) = \begin{cases} 1/2 & \text{for } 0 \leq x_1 \leq x_2, \text{ and } 0 \leq x_2 \leq 2 \\ 0 & \text{elsewhere} \end{cases}$$

Find the conditional expectation of amount of sales X_1, given that $X_2 = 1$.

Solution   In Example 5.6, we found that

$$f(x_1|x_2) = \begin{cases} 1/x_2 & \text{for } 0 \leq x_1 \leq x_2 \leq 2 \\ 0 & \text{elsewhere} \end{cases}$$

Thus, from Definition 5.7,

$$E(X_1|X_2 = 1) = \int_{-\infty}^{\infty} x_1 f(x_1|x_2)dx_1$$

$$= \int_0^1 x_1(1)dx_1$$

$$= \left. \frac{x_1^2}{2} \right|_0^1$$

$$= \frac{1}{2}$$

It follows that, if the soft-drink machine contains 1 gallon at the start of the day, the expected amount that will be sold that day is 1/2 gallon. ▪

The conditional expectation of X_1, given $X_2 = x_2$, is a function of x_2. If we now let X_2 range over all its possible values, we can think of the conditional expectation as being a function of the random variable X_2; hence, we can find the expected value of the conditional expectation. The result of this type of iterated expectation is given in Theorem 5.3.

THEOREM **5.3**

Let X_1 and X_2 denote random variables. Then

$$E(X_1) = E[E(X_1|X_2)]$$

where, on the right-hand side, the inside expectation is stated with respect to the conditional distribution of X_1 given X_2, and the outside expectation is stated with respect to the distribution of X_2.

Proof

Let X_1 and X_2 have the joint density function $f(x_1, x_2)$ and marginal densities $f_1(x_1)$ and $f_2(x_2)$, respectively. Then

$$
\begin{aligned}
E(X_1) &= \int_{-\infty}^{\infty} x_1 f_1(x_1)dx_1 \\
&= \int_{-\infty}^{\infty} \int_{-\infty}^{\infty} x_1 f(x_1, x_2)dx_1, dx_2 \\
&= \int_{-\infty}^{\infty} \int_{-\infty}^{\infty} x_1 f_1(x_1|x_2) f_2(x_2)dx_1 dx_2 \\
&= \int_{-\infty}^{\infty} \left[\int_{-\infty}^{\infty} x_1 f_1(x_1|x_2)dx_1 \right] f_2(x_2)dx_2 \\
&= \int_{-\infty}^{\infty} E(X_1|X_2 = x_2) f_x(x_2)dx_2 \\
&= E[E(X_1|X_2)]
\end{aligned}
$$

The proof is similar for the discrete case. ∎

Conditional expectations can also be used to find the variance of a random variable, in the spirit of Theorem 5.3. The main result is given in Theorem 5.4.

THEOREM **5.4**

For random variables X_1 and X_2,

$$V(X_1) = E[V(X_1|X_2)] + V[E(X_1|X_2)]$$

Proof

We begin with $E[V(X_1|X_2)]$ and write

$$
\begin{aligned}
E[V(X_1|X_2)] &= E[E(X_1^2|X_2) - [E(X_1|X_2)]^2] \\
&= E[E(X_1^2|X_2)] - E[E(X_1|X_2)]^2 \\
&= E[X_1^2] - E[E(X_1|X_2)]^2 \\
&= E[X_1^2] - [E(X_1)]^2 - E[E(X_1|X_2)]^2 + [E(X_1)]^2 \\
&= V(X_1) - E[[E(X_1|X_2)]^2] + [E[E(X_1|X_2)]^2]
\end{aligned}
$$

$$= V(X_1) - V[E(X_1|X_2)]$$

Rearranging terms, we have

$$V(X_1) = E[V(X_1|X_2)] + V[E(X_1|X_2)] \quad \blacksquare$$

E X A M P L E **5.19** A quality control plan for an assembly line involves sampling $n = 10$ finished items per day and counting Y, the number of defective items. If p denotes the probability of observing a defective item, then Y has a binomial distribution, when the number of items produced by the line is large. However, p varies from day to day and is assumed to have a uniform distribution on the interval from 0 to 1/4.

1 Find the expected value of Y for any given day.

2 Find the standard deviation of Y.

Solution **1** From Theorem 5.3, we know that

$$E(Y) = E[E(Y|p)]$$

For a given p, Y has binomial distribution; hence,

$$E(Y|p) = np$$

Thus,

$$\begin{aligned} E(Y) &= E(np) \\ &= nE(p) \\ &= n \int_0^{1/4} 4p\,dp \\ &= n\left(\frac{1}{8}\right) \end{aligned}$$

and for $n = 10$,

$$E(Y) = \frac{10}{8} = \frac{5}{4}$$

2 From Theorem 5.4,

$$\begin{aligned} V(Y) &= E[V(Y|p)] + V[E(Y|p)] \\ &= E[np(1-p)] + V[np] \\ &= nE(p) - nE(p^2) + n^2V(p) \\ &= nE(p) - nV(p) + n[E(p)]^2 + n^2V(p) \\ &= nE(p)[1 + E(p)] + n(n-1)V(p) \end{aligned}$$

For the specific case under consideration,

$$V(Y) = 10 \left(\frac{1}{8}\right)\left(\frac{7}{8}\right) + 10(9)\frac{1}{(4)^2(12)}$$
$$= 1.56$$

and

$$\sqrt{VY} = 1.25$$

In the long run, this inspection policy should discover, on average, 5/4 defective items per day, with a standard deviation of 1.25. The calculations can be checked by actually finding the unconditional distribution of Y and computing $E(Y)$ and $V(Y)$ directly. ∎

Exercises

5.41   Suppose that Y denotes the number of bacteria per cubic centimeter in water samples and that, for a given location, Y has a Poisson distribution with mean λ. But λ varies from location to location and has a gamma distribution with parameters σ and β. Find expressions for $E(Y)$ and $V(Y)$.

5.42   The number of eggs N found in nests of a certain species of turtle has a Poisson distribution with mean λ. Each egg has probability p of being viable, and this event is independent from egg to egg. Find the mean and the variance of the number of viable eggs per nest.

5.43   Let Y have a geometric distribution; that is, let

$$p(y) = p(1-p)^{y-1}, \qquad y = 1, 2, \ldots$$

The goal is to find $V(Y)$ through a conditional argument.

a   Let $X = 1$ if the first trial is a success, and let $X = 0$ if the first trial is a failure. Argue that

$$E(Y^2|X = 1) = 1$$
$$E(Y^2|X = 0) = E(1 + Y)^2$$

b   Write $E(Y^2)$ in terms of $E(Y^2|X)$.

c   Simplify the expression in part (b) to show that

$$E(Y^2) = \frac{2-p}{p^2}$$

d   Show that

$$V(Y) = \frac{1-p}{p^2}$$

5.8 Compounding and Its Applications

The univariate probability distributions of Chapters 3 and 4 depend on one or more parameters; once the parameters are known, the distributions are specified completely. These parameters frequently are unknown, however, and (as in Example 5.18) sometimes may be regarded as random quantities. Assigning distributions to these parameters and then finding the marginal distribution of the original random variable is known as *compounding*. This process has theoretical as well as practical uses, as we see next.

E X A M P L E **5.20**   Suppose that Y denotes the number of bacteria per cubic centimeter in a certain liquid and that, for a given location, Y has a Poisson distribution with mean λ. Suppose, too, that λ varies from location to location and that, for a location chosen at random, λ has a gamma distribution with parameters α and β, where α is a positive integer. Find the probability distribution for the bacteria count Y at a randomly selected location.

Solution   Because λ is random, the Poisson assumption applies to the conditional distribution of Y for fixed λ. Thus,

$$p(y|\lambda) = \frac{\lambda^y e^{-\lambda}}{y!}, \qquad y = 0, 1, 2, \ldots.$$

Also,

$$f(\lambda) = \begin{cases} \dfrac{1}{\Gamma(\alpha)\beta^\alpha} \lambda^{\alpha-1} e^{-\lambda/\beta} & \text{for } \lambda > 0 \\ 0 & \text{elsewhere} \end{cases}$$

Then the joint distribution of λ and Y is given by

$$\begin{aligned} g(y, \lambda) &= p(y|\lambda) f(\lambda) \\ &= \frac{1}{y!\,\Gamma(\alpha)\beta^\alpha} \lambda^{y+\alpha-1} e^{-\lambda[1+(1/\beta)]} \end{aligned}$$

The marginal distribution of Y is found by integrating over λ; it yields

$$\begin{aligned} p(y) &= \frac{1}{y!\Gamma(\alpha)\beta^\alpha} \int_0^\infty \lambda^{y+\alpha-1} e^{-\lambda[1+(1/\beta)]} dx \\ &= \frac{1}{y!\Gamma(\alpha)\beta^\alpha} \Gamma(y+\alpha) \left(1 + \frac{1}{\beta}\right)^{-(y+\alpha)} \end{aligned}$$

Since α is an integer,

$$\begin{aligned} p(y) &= \frac{(y+\alpha-1)!}{(\alpha-1)!y!} \left(\frac{1}{\beta}\right)^\alpha \left(\frac{\beta}{1+\beta}\right)^{y+\alpha} \\ &= \binom{y+\alpha-1}{\alpha-1} \left(\frac{1}{1+\beta}\right)^\alpha \left(\frac{\beta}{1+\beta}\right)^y \end{aligned}$$

If we let $y + \alpha = n$ and $1/(1 + \beta) = p$, then $p(y)$ has the form of a negative binomial distribution. Hence, the negative binomial distribution is a reasonable model for counts in which the mean count may be random.   ∎

E X A M P L E   **5.21**   Suppose that a customer arrives at a checkout counter in a store just as the counter is opening. A random number of customers N will be ahead of him, since some customers may arrive early. Suppose that this number has the probability distribution

$$p(n) = P(N = n) = pq^n, \qquad n = 0, 1, 2, \ldots$$

where $0 < p < 1$ and $q = 1 - p$ (this is a form of the geometric distribution).

Customer service times are assumed to be independent and identically distributed exponential random variables, with a mean of θ. Find the expected waiting time for this customer to complete his checkout.

Solution   For a given value of n, the waiting time W is the sum of $n + 1$ independent exponential random variables and thus has a gamma distribution, with $\alpha = n + 1$ and $\beta = \theta$; that is,

$$f(w|n) = \frac{1}{\Gamma(n + 1)\theta^{n+1}} w^n e^{-w/\theta}$$

Hence,

$$f(w, n) = \frac{p}{\Gamma(n + 1)\theta^{n+1}} (qw)^n e^{-w/\theta}$$

and

$$
\begin{aligned}
f(w) &= \frac{p}{\theta} e^{-w/\theta} \sum_{n=0}^{\infty} \left(\frac{qw}{\theta}\right)^n \left(\frac{1}{n!}\right) \\
&= \frac{p}{\theta} e^{-w/\theta} e^{qw/\theta} \\
&= \frac{p}{\theta} e^{-(w/\theta)(1-q)} \\
&= \frac{p}{\theta} e^{-w(p/\theta)}
\end{aligned}
$$

The waiting time W remains exponential, but its expected value is (θ/p).   ∎

5.9   Summary

Most phenomena that one studies, such as the effect of a drug on the body or the life length of a computer system, are the result of many variables acting jointly. In such studies, it is important to look not only at the **marginal distributions** of random variables individually, but also at the **joint distribution** of the variables acting together and at the **conditional distributions** of one variable for fixed values of another. Expected values of joint distributions lead to **covariance** and **correlation**, which help assess the direction and strength of association between two random variables. **Moment-generating functions** help identify distributions as being of a certain type (more on this in the next chapter) and help in the calculation of expected values.

As in the univariate case, multivariate probability distributions are models of reality and do not precisely fit real situations. Nevertheless, they serve as very useful approximations that help build understanding of the world around us.

Supplementary Exercises

5.44 Let X_1 and X_2 have the joint probability density function given by

$$f(x_1, x_2) = \begin{cases} Kx_1, x_2 & \text{for } 0 \leq x_1 \leq 1, \quad \text{and} \quad 0 \leq x_2 \leq 1 \\ 0 & \text{elsewhere} \end{cases}$$

a Find the value of K that makes this a probability density function.

b Find the marginal densities of X_1 and X_2.

c Find the joint distribution function for X_1 and X_2.

d Find the probability $P(X_1 < 1/2, \ X_2 < 3/4)$.

e Find the probability $P(X_1 \leq 1/2 | X_2 > 3/4)$.

5.45 Let X_1 and X_2 have the joint density function given by

$$f(x_1, x_2) = \begin{cases} 3x_1 & \text{for } 0 \leq x_2 \leq x_1 \leq 1 \\ 0 & \text{elsewhere} \end{cases}$$

a Find the marginal density functions of X_1 and X_2.

b Find $P(X_1 \leq 3/4, \ X_2 \leq 1/2)$.

c Find $P(X_1 \leq 1/2 | X_2 \geq 3/4)$.

5.46 A committee of three persons is to be randomly selected from a group consisting of four Republicans, three Democrats, and two independents. Let X_1 denote the number of Republicans on the committee, and let X_2 denote the number of Democrats on the committee.

a Find the joint probability distribution of X_1 and X_2.

b Find the marginal distributions of X_1 and X_2.

c Find the probability $P(X_1 = 1 | X_2 \geq 1)$.

5.47 For Exercise 5.44, find the conditional density of X_1, given $X_2 = x_2$. Are X_1 and X_2 independent?

5.48 Refer to the situation described in Exercise 5.45.

a Find the conditional density of X_1, given $X_2 = x_2$.

b Find the conditional density of X_2, given $X_1 = x_1$.

c Show that X_1 and X_2 are dependent.

d Find the probability $P(X_1 \leq 3/4 | X_2 = 1/2)$.

5.49 Let x_1 denote the amount of a certain bulk item stocked by a supplier at the beginning of a week, and suppose that X_1 has a uniform distribution over the interval $0 \leq x_1 \leq 1$. Let X_2 denote the amount of this item sold by the supplier during the week, and suppose that X_2 has a uniform distribution over the interval $0 \leq x_2 \leq x_1$, where x_1 is a specific value of X_1.

a Find the joint density function of x_1 and x_2.

b If the supplier stocks an amount of the item of 1/2, what is the probability that she sells an amount greater than 1/4?

c If it is known that the supplier sold an amount equal to 1/4, what is the probability that she had stocked an amount greater than 1/2?

5.50 Let (X_1, X_2) denote the coordinates of a point at random inside a unit circle, with center at the origin; that is, X_1 and X_2 have a joint density function given by

$$f(x_1, x_2) = \begin{cases} 1/\pi & \text{for } x_1^2 + x_2^2 \le 1 \\ 0 & \text{elsewhere} \end{cases}$$

a Find the marginal density function of X_1.

b Find $P(X_1 \le X_2)$.

5.51 Let X_1 and X_2 have the joint density function given by

$$f(x_1, x_2) = \begin{cases} x_1 + x_2 & \text{for } 0 \le x_1 \le 1, \text{ and } 0 \le x_2 \le 1 \\ 0 & \text{elsewhere} \end{cases}$$

a Find the marginal density functions of X_1 and X_2.

b Are X_1 and X_2 independent?

c Find the conditional density of X_1, given $X_2 = x_2$.

5.52 Let X_1 and X_2 have the joint density function given by

$$f(x_1, x_2) = \begin{cases} K & \text{for } 0 \le x_1 \le 2, \ 0 \le x_2 \le 1, \text{ and } 2x_2 \le x_1 \\ 0 & \text{elsewhere} \end{cases}$$

a Find the value of K that makes the function a probability density.

b Find the marginal densities of X_1 and X_2.

c Find the conditional density of X_1, given $X_2 = x_2$.

d Find the conditional density of X_2, given $X_1 = x_1$.

e Find $P(X_1 \le 1.5, \ X_2 \le 0.5)$.

f Find $P(X_2 \le 0.5 | X_1 \le 1.5)$.

5.53 Let X_1 and X_2 have a joint distribution that is uniform over the region shaded in the accompanying diagram.

a Find the marginal density for X_2.

b Find the marginal density for X_1.

c Find $P[(X_1 - X_2) \ge 0]$.

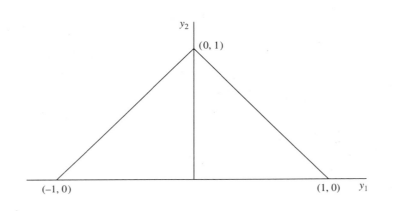

5.54   Refer to the situation described in Exercise 5.44.

   a   Find $E(X_1)$.

   b   Find $V(X_1)$.

   c   Find $\text{cov}(X_1, X_2)$.

5.55   Referring to Exercise 5.45, find $\text{cov}(X_1, X_2)$.

5.56   Refer to the situation described in Exercise 5.46.

   a   Find $\text{cov}(X_1, X_2)$.

   b   Find $E(X_1 + X_2)$ and $V(X_1 + X_2)$ by finding the probability distribution of $X_1 + X_2$.

   c   Find $E(X_1 + X_2)$ and $V(X_1 + X_2)$ by using Theorem 5.2.

5.57   Refer to the situation described in Exercise 5.51.

   a   Find $\text{cov}(X_1, X_2)$.

   b   Find $E(3X_1 - 2X_2)$.

   c   Find $V(3X_1 - 2X_2)$.

5.58   Refer to the situation described in Exercise 5.52.

   a   Find $E(X_1 + 2X_2)$.

   b   Find $V(X_1 + 2X_2)$.

5.59   A quality control plan calls for randomly selecting three items from the daily production (assumed to be large) of a certain machine and observing the number of defectives. The proportion p of defectives produced by the machine varies from day to day and is assumed to have a uniform distribution on the interval (0, 1). For a randomly chosen day, find the unconditional probability that exactly two defectives are observed in the sample.

5.60   The number of defects per yard, denoted by X, for a certain fabric is known to have a Poisson distribution with parameter λ. However, λ is not known and is assumed to be random, with its probability density function given by

$$f(\lambda) = \begin{cases} e^{-\lambda} & \text{for } \lambda \geq 0 \\ 0 & \text{elsewhere} \end{cases}$$

Find the unconditional probability function for X.

5.61   The length of life X of a fuse has the probability density

$$f(x) = \begin{cases} e^{-\lambda/\theta}/\theta & \text{for } x > 0 \quad \text{and} \quad \theta > 0 \\ 0 & \text{elsewhere} \end{cases}$$

Three such fuses operate independently. Find the joint density of their lengths of life, X_1, X_2, and X_3.

5.62   A retail grocer figures that his daily gain X from sales is a normally distributed random variable, with $\mu = 50$ and $\sigma^2 = 10$ (measurements in dollars). X could be negative if he is forced to dispose of perishable goods. He also figures daily overhead costs Y to have a gamma distribution, with $\alpha = 4$ and $\beta = 2$. If X and Y are independent, find the expected value and the variance of his net daily gain. Would you expect his net gain for tomorrow to go above $70?

5.63   A coin has probability p of coming up heads when tossed. In n independent tosses of the coin, let $X_i = 1$ if the ith toss results in heads and $X_i = 0$ if the ith toss results in tails. Then Y, the number of heads in the n tosses, has a binomial distribution and can be represented as

$$Y = \sum_{i=1}^{n} X_i$$

Find $E(Y)$ and $V(Y)$, using Theorem 5.2.

5.64   Refer to the situation and results of Exercises 5.45 and 5.48.

a Find $E(X_2|X_1 = x_1)$.

b Use Theorem 5.3 to find $E(X_2)$.

c Find $E(X_2)$ directly from the marginal density of X_2.

5.65 Refer to the situation described in Exercise 5.60.

a Find $E(X)$ by first finding the conditional expectation of X for given λ and then using Theorem 5.3.

b Find $E(X)$ directly from the probability distribution of X.

5.66 Referring to Exercise 5.49, if the supplier stocks an amount equal to 3/4, what is the expected amount that she will sell during the week?

5.67 Let X be a continuous random variable with distribution function $F(x)$ and density function $f(x)$. We can then write, for $x_1 \leq x_2$,

$$P(X \leq x_2 | X \geq x_1) = \frac{F(x_2) - F(x_1)}{1 - F(x_1)}$$

As a function of x_2 for fixed x_1, the right-hand side of this expression is called the *conditional distribution function* of X, given that $X \geq x_1$. On taking the derivative with respect to x_2, we see that the corresponding conditional density function is given by

$$\frac{f(x_2)}{1 - F(x_1)}, \qquad x_2 \geq x_1$$

Suppose that a certain type of electronic component has life length X, with the density function (life length measured in hours)

$$f(x) = \begin{cases} (1/200)e^{-x/200} & \text{for } x \geq 0 \\ 0 & \text{elsewhere} \end{cases}$$

Find the expected length of life for a component of this type that has already been in use for 100 hours.

5.68 Let X_1, X_2, and X_3 be random variables, either continuous or discrete. The joint moment-generating function of X_1, X_2, and X_3 is defined by

$$M(t_1, t_2, t_3) = E(e^{t_1 X_1 + t_2 X_2 + t_3 X_3})$$

a Show that $M(t, t, t)$ gives the moment-generating function of $X_1 + X_2 + X_3$.

b Show that $M(t, t, 0)$ gives the moment-generating function of $X_1 + X_2$.

c Show that

$$\frac{\partial^{k_1 + k_2 + k_3} M(t_1, t_2, t_3)}{\partial t_1^{k_1} \partial t_2^{k_2} \partial t_3^{k_3}} \bigg|_{t_1 = t_2 = t_3 = 0} = E(X_1^{k_1} X_2^{k_2} X_3^{k_3})$$

5.69 Let X_1, X_2, and X_3 have a multinomial distribution, with probability function

$$p(x_1, x_2, x_3) = \frac{n!}{x_1! \, x_2! \, x_3!} p_1^{x_1} p_2^{x_2} p_3^{x_3}, \qquad \sum_{i=1}^{n} x_i = n$$

Employ the results of Exercise 5.65 to perform the following tasks.

a Find the joint moment-generating function of X_1, X_2, and X_3.

b Use the joint moment-generating function to find $\text{cov}(X_1, X_2)$.

5.70 The negative binomial variable X is defined as the number of the trial on which the rth success occurs in a sequence of independent trials with constant probability p of success on each trial.

Let X_i denote a geometric random variable, defined as the number of the trial on which the first success occurs. Then, we can write

$$X = \sum_{i=1}^{n} X_i$$

for independent random variables $X_1, \ldots, X_r$. Use Theorem 5.2 to show that $E(X) = r/p$ and $V(X) = r(1-p)/p^2$.

5.71   A box contains four balls, numbered 1 through 4. One ball is selected at random from this box. Let

$$X_1 = 1 \text{ if ball 1 or ball 2 is drawn}$$
$$X_2 = 1 \text{ if ball 1 or ball 3 is drawn}$$
$$X_3 = 1 \text{ if ball 1 or ball 4 is drawn}$$

and let the values of X_i be zero otherwise. Show that any two of the random variables X_1, X_2, and X_3 are independent, but that the three together are not.

5.72   Let X_1 and X_2 be jointly distributed random variables with finite variances.

a   Show that $[E(X_1 X_2)]^2 \leq E(X_1^2)E(X_2^2)$. (*Hint:* Observe that, for any real number t, $E[(tX_1 - X_2)^2] \geq 0$; or equivalently,

$$t^2 E(X_1^2) - 2E(X_1 X_2) + E(X_2^2) \geq 0$$

This is a quadratic expression of the form $At^2 + Bt + C$; and because it is not negative, we must have $B^2 - 4AC \leq 0$. The preceding inequality follows directly.)

b   Let ρ denote the correlation coefficient of X_1 and X_2; that is,

$$\rho = \frac{\text{cov}(X_1, X_2)}{\sqrt{V(X_1)V(X_2)}}$$

Using the inequality of part (a), show that $\rho^2 \leq 1$.

5.73   A box contains N_1 white balls, N_2 black balls, and N_3 red balls ($N_1 + N_2 + N_3 = N$). A random sample of n balls is selected from the box, without replacement. Let X_1, X_2, and X_3 denote the number of white, black, and red balls, respectively, observed in the sample. Find the correlation coefficient for X_1 and X_2. (Let $p_i = N_i/N$, for $i = 1, 2, 3$.)

5.74   Let $X_1, X_2, \ldots, X_n$ be independent random variables with $E(X_i) = \mu$ and $V(X_i) = \sigma^2$, for $i = 1, \ldots, n$. Let

$$U_1 = \sum_{i=1}^{n} a_i X_i$$

and

$$U_2 = \sum_{i=1}^{n} b_i X_i$$

where $a_1, \ldots, a_n$, and $b_1, \ldots, b_n$ are constants. U_1 and U_2 are said to be orthogonal if $\text{cov}(U_1, U_2) = 0$. Show that U_1 and U_2 are orthogonal if and only if $\sum_{i=1}^{n} a_i b_i = 0$.

5.75   The life length X of fuses of a certain type is modeled by the exponential distribution with

$$f(x) = \begin{cases} (1/3)e^{-x/3} & \text{for } x > 0 \\ 0 & \text{elsewhere} \end{cases}$$

The measurements are in hundreds of hours.

a   If two such fuses have independent life lengths X_1 and X_2, find their joint probability density function.

b   One fuse in part (a) is in a primary system, and the other is in a backup system that comes into use only if the primary system fails. The total effective life length of the two fuses is then $X_1 + X_2$. Find $P(X_1 + X_2 \leq 1)$.

5.76   Referring to Exercise 5.75, suppose that three such fuses are operating independently in a system.

a   Find the probability that exactly two of the three last longer than 500 hours.

b   Find the probability that at least one of the three fails before 500 hours.

Functions of Random Variables

6.1 Introduction

As we observed in Chapter 5, many situations we wish to study produce a set of random variables, $X_1, \ldots, X_n$, instead of a single random variable. Questions about the average life of components, the maximum price of a stock during a quarter, the time between two incoming telephone calls, or the total production costs across the plants in a manufacturing firm all involve the study of functions of random variables. This chapter considers the problem of finding the probability density function for a function of random variables with known probability distributions.

We have already seen some results along these lines in Chapter 5. Moment-generating functions were used to show that sums of exponential random variables have gamma distributions, and that linear functions of independent normal random variables are again normal. Those were special cases, however, and now more general methods for finding distributions of functions of random variables will be introduced.

If X_1 and X_2 represent two different features of the phenomenon under study, it is sometimes convenient (and even necessary) to look at one random variable as a function of the other. For example, if the probability distribution for the velocity of a molecule in a uniform gas is known, the distribution of the kinetic energy can be found (see Exercise 6.47). And if the distribution of the radius of a sphere is known, the distribution of the volume can be found (see Exercise 6.46). Similarly, knowledge of the probability distribution for points in a plane allows us to find, in some cases, the distribution of the distance from a selected point to its nearest neighbor. Knowledge of the distribution of life lengths of components in a complex system allows us to find the probability distribution for the life length of the system as a whole. Examples are endless, but those mentioned here should adequately illustrate the point that we are embarking on the study of a very important area of applied probability.

6.2 Method of Distribution Functions

If X has a probability density function $f(x)$, and if U is some function of X, then we can find $F_U(u) = P(U \leq u)$ directly by integrating $f(x)$ over the region for which $U \leq u$. We can find the probability density function for U by differentiating $F_U(u)$. The method is illustrated in Example 6.1.

E X A M P L E   **6.1**   The percentage of time X that a lathe is in use during a typical 40-hour workweek is a random variable whose probability density function is given by

$$f(x) = \begin{cases} 3x^2 & \text{for } 0 \leq x \leq 1 \\ 0 & \text{elsewhere} \end{cases}$$

The actual number of *hours*, out of a 40-hour week, that the lathe is *not* in use, then, is

$$U = 40(1 - X)$$

Find the probability density function for U.

Solution   Because the distribution function for U looks at a region of the form $U \leq u$, we must first find that region on the x scale. Now

$$U \leq u \Rightarrow 40(1 - X) \leq u$$
$$\Rightarrow X > 1 - \frac{u}{40}$$

so

$$F_U(u) = P(U \leq u)$$
$$= P[40(1 - X) \leq u]$$
$$= P\left(X > 1 - \frac{u}{40}\right)$$
$$= \int_{1-u/40}^{1} f(x)dx$$
$$= \int_{1-u/40}^{1} 3x^2$$
$$= \left[x^3\right]_{1-u/40}^{1}$$

or

$$F_U(u) = 1 - \left(1 - \frac{u}{40}\right)^3, \qquad 0 \leq u \leq 40.$$

Now, the probability density function is found by differentiating the distribution function, so

$$f_U(u) = \frac{dF_U(u)}{du}$$

$$= \begin{cases} \frac{3}{40}\left(1 - \frac{u}{40}\right)^2 & \text{for } 0 \le u \le 40 \\ 0 & \text{elsewhere} \end{cases}$$

We could now use $f_U(u)$ to evaluate probabilities or to find expected values related to the number of hours that the lathe is not in use. The reader should verify that $f_U(u)$ has all the properties of a probability density function.   ∎

The bivariate case is handled similarly, although it is often more difficult to transform the bivariate regions from statements about U—a function of (X_1, X_2)—to statements about X_1 and X_2. Example 6.2 illustrates the point.

E X A M P L E   **6.2**   Two friends plan to meet at the library during a given 1-hour period. Their arrival times are independent and randomly distributed across the 1-hour period. Each agrees to wait for 15 minutes, or until the end of the hour. If the friend doesn't appear during that time, she will leave. What is the probability that the two friends will meet?

Solution   If X_1 denotes one person's arrival time in $(0, 1)$, the 1-hour period, and if X_2 denotes the second person's arrival time, then (X_1, X_2) can be modeled as having a two-dimensional uniform distribution over the unit square; that is,

$$f(x_1, x_2) = \begin{cases} 1 & \text{for } 0 \le x_1 \le 1 \quad \text{and} \quad 0 \le x_2 \le 1 \\ 0 & \text{elsewhere} \end{cases}$$

The event that the two friends will meet depends on the time U between their arrivals, where

$$U = |X_1 - X_2|$$

We shall solve the specific problem by finding the probability density for U. Now,

$$U \le u \Rightarrow |X_1 - X_2| \le u$$
$$\Rightarrow -u \le X_1 - X_2 \le u$$

Figure 6.1 shows the square region over which (X_1, X_2) has positive probability and the region defined by $U \le u$. The probability of $U \le u$ can be found by integrating the joint density function of (X_1, X_2) over the six-sided region shown in the center of Figure 6.1. This can be simplified by integrating over the triangles (A_1 and A_2) and subtracting from 1, as we now see. We have

$$F_U(u) = P(U \le u) = \iint\limits_{|x_1 - x_2| \le u} f(x_1, x_2) dx_1 dx_2$$

$$= 1 - \int\int_{A_1} f(x_1, x_2)dx_1 dx_2 - \int\int_{A_2} f(x_1, x_2)dx_1 dx_2$$

$$= 1 - \int_u^1 \int_0^{x_2-u} (1)dx_1 dx_2 - \int_0^{1-u} \int_{x_2+u}^1 (1)dx_1 dx_2$$

$$= 1 - \int_u^1 (x_2 - u)dx_2 - \int_0^{1-u} (1 - u - x_2)dx_2$$

$$= 1 - \frac{1}{2}[(x_2 - u)^2]_u^1 - \frac{1}{2}[-(1 - u - x_2)^2]_0^{1-u}$$

$$= 1 - \frac{1}{2}(1 - u)^2 - \frac{1}{2}(1 - u)^2$$

$$= 1 - (1 - u)^2, \qquad 0 \le u \le 1$$

(Notice that the two double integrals really evaluate the volumes of prisms with triangular bases.) Because the probability density function is found by differentiating the distribution function, we have

$$f_U(u) = \frac{dF_U(u)}{du} = \begin{cases} 2(1 - u), & \text{for } 0 \le u \le 1 \\ 0 & \text{elsewhere} \end{cases}$$

FIGURE 6.1
Region $U \le u$ for Example 6.2.

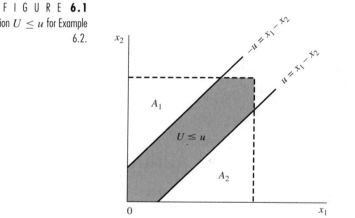

The problem asks for the probability of a meeting if each friend waits for up to 15 minutes. Because 15 minutes is 1/4 hour, this can be found by evaluating

$$P\left(U \le \frac{1}{4}\right) = \int_0^{1/4} f_U(u)\, du$$

$$= F_u\left(\frac{1}{4}\right)$$

$$= 1 - \left(1 - \frac{1}{4}\right)^2$$

$$= 1 - \left(\frac{3}{4}\right)^2$$

$$= \frac{7}{16}$$

$$= 0.4375$$

There is less than a 50–50 chance that the two friends will meet under this rule.   ■

Applications of probability often call for the use of sums of random variables. A study of downtimes of computer systems might require knowledge of the sum of the downtimes over a day or a week. The total cost of a building project can be studied as the sum of the costs for the major components of the project. The size of an animal population can be modeled as the sum of the sizes of the colonies within the population. The list of examples is endless. We now present an example that shows how to use distribution functions to find the probability distribution of a sum of independent random variables. This complements the work in Chapter 5 on using moment-generating functions to find distributions of sums.

EXAMPLE 6.3   Suppose that X_1 and X_2 are independent, exponentially distributed random variables, each with a mean of one. Find the probability density function for $U = X_1 + X_2$.

Solution   The joint density function of X_1 and X_2 is given by

$$f(x_1, x_2) = f_1(x_1) f_2(x_2)$$
$$= \begin{cases} e^{-x_1} e^{-x_2} & \text{for } x_1 \geq 0 \quad \text{and} \quad x_2 \geq 0 \\ 0 & \text{elsewhere} \end{cases}$$

and the distribution function of either X_1 or X_2 is given by

$$F(x) = 1 - e^{-x}, \qquad x \geq 0$$

To find $P(U \leq u)$, we must integrate over the region shown in Figure 6.2. Thus,

$$F_U(u) = P(U \leq u) = \int_0^u \int_0^{u-x_2} f_1(x_1) f_2(x_2) dx_1 dx_2$$
$$= \int_0^u F(u - x_2) f_2(x_2) dx_2$$
$$= \int_0^u (1 - e^{-(u-x_2)}) e^{-x_2} dx_2$$
$$= [1 - e^{-x_2}]_0^u - [x_2 e^{-u}]_0^u$$
$$= 1 - e^{-u} - u e^{-u} \qquad u \geq 0$$

and

$$f_U(u) = \frac{dF_U(u)}{du}$$

$$= \begin{cases} ue^{-u} & \text{for } u \geq 0 \\ 0 & \text{elsewhere} \end{cases}$$

Notice that $f_U(u)$ is a gamma function, which is consistent with results for the distribution of sums of independent exponential random variables found by using moment-generating functions. ∎

F I G U R E **6.2**
Region $U \leq u$ for Example
6.3.

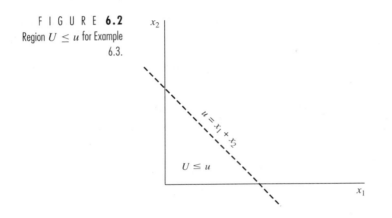

Summary of the Distribution Function Method

Let U be a function of the continuous random variables $X_1, X_2, \ldots, X_n$. Then,

1 Find the region $U = u$ in the $(x_1, x_2, \ldots, x_n)$ space.

2 Find the region $U \leq u$.

3 Find $F_U(u) = P(U \leq u)$ by integrating $f(x_1, x_2, \ldots, x_n)$ over the region $U \leq u$.

4 Find the density function $f_U(u)$ by differentiating $F_U(u)$. Thus, $f_U(u) = dF_U(u)/du$.

Exercises

6.1 Let X be a random variable with a probability density function given by

$$f(x) = \begin{cases} 2(1-x) & \text{for } 0 \leq x \leq 1 \\ 0 & \text{elsewhere} \end{cases}$$

Find the density functions of the following variables.

a $U_1 = 2X - 1$

b $U_2 = 1 - 2X$

c $U_3 = X^2$

6.2 Let X be a random variable whose density function is given by

$$f(x) = \begin{cases} (3/2)x^2 & \text{for } -1 \leq x \leq 1 \\ 0 & \text{elsewhere} \end{cases}$$

Find the density functions of the following variables.

a $U_1 = 3X$

b $U_2 - 3 - X$

c $U_3 = X^2$

6.3 A supplier of kerosene has a weekly demand X possessing a probability density function given by

$$f(x) = \begin{cases} x & \text{for } 0 \leq x \leq 1 \\ 1 & \text{for } 1 < x \leq 1.5 \\ 0 & \text{elsewhere} \end{cases}$$

with measurements in hundreds of gallons. The supplier's profit is given by $U = 10X - 4$.

a Find the probability density function for U.

b Use the answer in part (a) to find $E(U)$.

c Find $E(U)$ by the methods of Chapter 4.

6.4 The waiting time X for delivery of a new component for an industrial operation is uniformly distributed over the interval from 1 to 5 days. The cost of this delay is given by $U = 2X^2 + 3$. Find the probability density function for U.

6.5 The joint distribution of amount of pollutant emitted from a smokestack without a cleaning device (X_1) and with a cleaning device (X_2) is given by

$$f(x_1, x_2) = \begin{cases} k & \text{for } 0 \leq x_1 \leq 2, \quad 0 \leq x_2 \leq 1, \quad \text{and} \quad 2x_2 \leq x_1 \\ 0 & \text{elsewhere} \end{cases}$$

The reduction in amount of pollutant emitted due to the cleaning device is given by $U = X_1 - X_2$.

a Find the probability density function for U.

b Use the answer in part (a) to find $E(U)$. (Compare this with the result of Exercise 5.4.)

6.6 The total time X_1 from arrival to completion of service at a fast-food outlet and the time spent waiting in line X_2 before arriving at the service window have a joint density function given by

$$f(x_1, x_2) = \begin{cases} e^{-x_1} & \text{for } 0 \leq x_2 \leq x_1 < \infty \\ 0 & \text{elsewhere} \end{cases}$$

where $U = X_1 - X_2$ represents the time spent at the service window.

a Find the probability density function for U.

b Find $E(U)$ and $V(U)$, using the answer to part (a). (Compare this with the result of Exercise 5.21.)

6.7 Suppose that a unit of mineral ore contains a proportion X_1 of metal A and a proportion X_2 of metal B. Experience has shown that the joint probability density function of (X_1, X_2) is uniform over the region $0 \le x_1 \le 1, \quad 0 \le x_2 \le 1, \quad 0 \le x_1 + x_2 \le 1$. Let $U = X_1 + X_2$, the proportion of metals A and B per unit.

a Find the probability density function for U.

b Find $E(U)$ by using that answer to part (a).

c Find $E(U)$ by using only the marginal densities of X_1 and X_2.

6.8 Suppose that a continuous random variable X has the distribution function $F(x)$. Show that $F(X)$ is uniformly distributed over the interval $(0, 1)$.

6.3 Method of Transformations

The transformation method for finding the probability distribution of a function of random variables is simply a generalization of the distribution function method (Section 6.2). Through the distribution function approach, we can arrive at a single method of writing down the density function of $U = h(X)$, provided that $h(x)$ is either decreasing or increasing. [By $h(x)$ increasing, we mean that, if $x_1 < x_2$, then $h(x_1) < h(x_2)$ for any real numbers x_1 and x_2.]

Suppose that $h(x)$ is an increasing function of x and that $U = h(X)$, where X has density function $f_X(x)$. The symbol $h^{-1}(u)$ denotes the inverse function; that is, if $u = h(x)$, we can solve for x, obtaining $x = h^{-1}(u)$. The graph of an increasing function $h(x)$ appears in Figure 6.3, where we see that the set of points x such that $h(x) \le u_1$ is precisely the same as the set of points x such that $x \le h^{-1}(u_1)$. To find the density of $U = h(X)$ by the distribution function method, we write

$$\begin{aligned}
F_U(u) &= P(U \le u) \\
&= P[h(X) \le u] \\
&= P[X \le h^{-1}(u)] \\
&= F_X[h^{-1}(u)]
\end{aligned}$$

where $F_X(x)$ is the distribution function of X.

F I G U R E **6.3**
Increasing function.

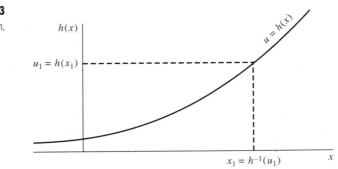

To find the density function of U—namely $f_U(u)$—we must differentiate $F_U(u)$. Because $x = h^{-1}(u)$,

$$F_U(u) = F_X[h^{-1}(u)] = F_X(x)$$

Then,

$$
\begin{aligned}
f_U(u) &= \frac{dF_U(u)}{du} \\
&= \frac{dF_X(x)}{dx}\left(\frac{dx}{du}\right) \\
&= f_X(x)\frac{dx}{du} \\
&= f_X[h^{-1}(u)]\frac{dx}{du}
\end{aligned}
$$

Notice that $dx/du = 1/(du/dx)$.

E X A M P L E   **6.4**   Let X have the probability density function given by

$$f_X(x) = \begin{cases} 2x & \text{for } 0 < x < 1 \\ 0 & \text{elsewhere} \end{cases}$$

Find the density function of $U = 3X - 1$.

Solution   The function of interest here is $h(x) = 3x - 1$, which is increasing in x. If $u = 3x - 1$, then

$$x = h^{-1}(u) = \frac{u + 1}{3}$$

and

$$\frac{dx}{du} = \frac{1}{3}$$

Thus,

$$
\begin{aligned}
f_U(u) &= f_X[h^{-1}(u)]\frac{dx}{du} \\
&= 2x\frac{dx}{du} \\
&= 2\left(\frac{u + 1}{3}\right)\left(\frac{1}{3}\right) \\
&= \begin{cases} \dfrac{2(u + 1)}{9}, & \text{for } -1 < u < 2 \\ 0 & \text{elsewhere} \end{cases}
\end{aligned}
$$

The range over which $f_U(u)$ is positive is simply the interval $0 < x < 1$ transformed to the u-axis by the function $u = 3x - 1$.   ▪

If $h(x)$ is a decreasing function, as in Figure 6.4, then the set of points x such that $h(x) \leq u_1$ is the same as the set of points x such that $x \geq h^{-1}(u_1)$. It follows that, for $U = h(X)$,

$$
\begin{aligned}
F_U(u) &= P(U \leq u) \\
&= P[h(X) \leq u] \\
&= P[X \geq h^{-1}(u)] \\
&= 1 - F_X[h^{-1}(u)]
\end{aligned}
$$

On taking derivatives with respect to u, we have

$$
f_U(u) = -f_X[h^{-1}(u)]\frac{dx}{du}
$$

Since dx/du is negative for a decreasing function, this equation is equivalent to

$$
f_U(u) = f_X[h^{-1}(u)]\left|\frac{dx}{du}\right|
$$

The results for h increasing and for h decreasing can now be combined into the following statement.

FIGURE 6.4
Decreasing function.

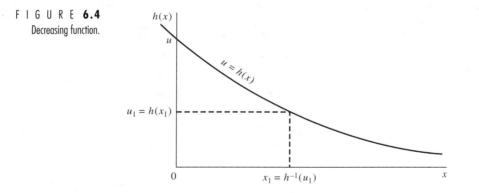

Let X have the probability density function $f_x(x)$. If $h(x)$ is either increasing or decreasing in x, then $U = h(X)$ has the density function given by

$$
f_U(u) = f_X[h^{-1}(u)]\left|\frac{dx}{du}\right|
$$

EXAMPLE   **6.5**   Let X have the probability density function given by

$$
f_X(x) = \begin{cases} 2x & \text{for } 0 < x < 1 \\ 0 & \text{elsewhere} \end{cases}
$$

Find the density function of $U = -4X + 3$.

Solution   The function of interest here, $h(x) = -4x + 3$, is decreasing in x. If $u = -4x + 3$, then

$$x = h^{-1}(u) = \frac{3 - u}{4}$$

and

$$\frac{dx}{du} = -\frac{1}{4}$$

Thus,

$$f_U(u) = f_X[h^{-1}(u)]\left|\frac{dx}{du}\right|$$

$$= 2x\left|\frac{dx}{du}\right|$$

$$= 2\left(\frac{3 - u}{4}\right)\left(\frac{1}{4}\right)$$

$$= \begin{cases} \dfrac{3 - u}{8} & \text{for } -1 < u < 3 \\ 0 & \text{elsewhere} \end{cases}$$ ∎

The transformation method, as outlined here, can be applied readily to some functions that are neither increasing nor decreasing. To illustrate, consider the case $U = h(X) = X^2$, where X is still continuous, with distribution function $F_X(x)$ and density function $f_X(x)$. We then have

$$\begin{aligned} F_U(u) &= P(U \le u) \\ &= P(X^2 \le u) \\ &= P(-\sqrt{u} \le X \le \sqrt{u}) \\ &= F_X(\sqrt{u}) - F_X(-\sqrt{u}) \end{aligned}$$

On differentiating with respect to u, we see that

$$\begin{aligned} f_U(u) &= f_X(\sqrt{u})\frac{1}{2\sqrt{u}} + f_X(-\sqrt{u})\frac{1}{2\sqrt{u}} \\ &= \frac{1}{2\sqrt{u}}[f_X(\sqrt{u}) + f_X(-\sqrt{u})] \end{aligned}$$

E X A M P L E   **6.6**   Let X have the probability density function given by

$$f_X(x) = \begin{cases} \dfrac{x + 1}{2} & \text{for } -1 \le x \le 1 \\ 0 & \text{elsewhere} \end{cases}$$

Find the density function for $U = X^2$.

Solution   We know that

$$f_U(u) = \frac{1}{2\sqrt{u}}[f_X(\sqrt{u}) + f_X(-\sqrt{u})]$$

and on substituting into this equation, we obtain

$$f_U(u) = \frac{1}{2\sqrt{u}}\left(\frac{\sqrt{u}+1}{2} + \frac{-\sqrt{u}+1}{2}\right)$$

$$= \begin{cases} \dfrac{1}{2\sqrt{u}} & \text{for } 0 \le u \le 1 \\ 0 & \text{elsewhere} \end{cases}$$

Notice that, because X has positive density over the interval $-1 \le x \le 1$, it follows that $U = X^2$ has positive density over the interval $0 \le u \le 1$. ∎

The transformation method can also be used in multivariate situations. The following example illustrates the bivariate case.

EXAMPLE 6.7   Let X_1 and X_2 have joint density function given by

$$f(x_1, x_2) = \begin{cases} e^{-(x_1+x_2)} & \text{for } 0 \le x_1, \text{ and } 0 \le x_2 \\ 0 & \text{elsewhere} \end{cases}$$

Find the density function for $U = X_1 + X_2$.

Solution   This problem must be solved in two stages: first, we find the joint density of X_1 and U; and second, we find the marginal density of U. The approach is to let one of the original variables—say, X_1—be fixed at a value x_1. Then $U = x_1 + X_2$, and we can consider the one-dimensional transformation problem in which $U = h(X_2) = x_1 + X_2$. Letting $g(x_1, u)$ denote the joint density of X_1 and U, we have

$$g(x_1, u) = f[x_1, h^{-1}(u)]\left|\frac{dx_2}{du}\right|$$

$$= \begin{cases} e^{-u}(1) & \text{for } 0 \le u, \text{ and } 0 \le x_1 \le u \\ 0 & \text{otherwise} \end{cases}$$

(Notice that $X_1 \le U$.) The marginal density of U is then given by

$$f_U(u) = \int_{-\infty}^{\infty} g(x_1, u)dx_1$$

$$= \int_0^u e^{-u} dx_1$$

$$= \begin{cases} ue^{-u} & \text{for } 0 \le u \\ 0 & \text{otherwise} \end{cases}$$ ∎

Summary of the Transformation Method

Let U be an increasing or decreasing function of the random variable X—say, $U = h(X)$.

1   Find the inverse function, $X = h^{-1}(U)$.
2   Evaluate dx/du.
3   Find $f_U(u)$ by calculating

$$f_U(u) = f_X[h^{-1}(u)]\left|\frac{dx}{du}\right|$$

6.4 Method of Conditioning

Conditional density functions frequently provide a convenient path for finding distributions of random variables. Suppose that X_1 and X_2 have a joint density function, and that we want to find the density function for $U = h(X_1, X_2)$. We observe that the density of U—namely, $f_U(u)$—can be written

$$f_U(u) = \int_{-\infty}^{\infty} f(u, x_2)dx_2$$

Since the conditional density of U gives X_2 is equal to

$$f(u|x_2) = \frac{f(u, x_2)}{f_2(x_2)}$$

it follows that

$$f_U(u) = \int_{-\infty}^{\infty} f(u|x_2)f_2(x_2)dx_2$$

E X A M P L E   **6.8**   Let X_1 and X_2 be independent random variables, each with the density function

$$f_X(x) = \begin{cases} e^{-x} & \text{for } x > 0 \\ 0 & \text{elsewhere} \end{cases}$$

Find the density function for $U = X_1/X_2$.

Solution   We want to find the conditional density of U given $X_2 = x_2$, which is obtained by the transformation method. When X_2 is held fixed at the constant value x_2, U is simply X_1/x_2, where X_1 has the exponential density function. Thus, we want the density function of $U = X_1/x_2$. The inverse function is $X_1 = Ux_2$. By the method of Section 6.3,

$$f(u \mid x_2) = f_{X_1}(ux_2)\left|\frac{dx_1}{du}\right|$$

$$= \begin{cases} e^{-ux_2}x_2 & \text{for } u > 0 \\ 0 & \text{elsewhere} \end{cases}$$

Now,

$$\begin{aligned}
f_U(u) &= \int_{-\infty}^{\infty} f(u|x_2) f(x_2) dx_2 \\
&= \int_0^{\infty} e^{-ux_2} x_2 (e^{-x_2}) dx_2 \\
&= \int_0^{\infty} x_2 e^{-x_2(u+1)} dx_2 \\
&= \begin{cases} (u+1)^{-2} & \text{for } u > 0 \\ 0 & \text{elsewhere} \end{cases}
\end{aligned}$$

Evaluating this integral is made easier by observing that the integrand is a gamma function. ∎

Summary of the Conditioning Method

Let U be a function of the random variables X_1 and X_2.

1 Find the conditional density of U given $X_2 = x_2$. (This is usually found by means of transformations.)

2 Find $f_U(u)$ from the relation

$$f_U(u) = \int_{-\infty}^{\infty} f(u \mid x_2) f(x_2) dx_2$$

Exercises

6.9 Referring to Exercise 6.1, find the answers by using the method of transformations.

6.10 Referring to Exercise 6.2, find the answers by using the method of transformations.

6.11 In a process that involves sintering two types of copper powder, the density function for X_1, the volume proportion of solid copper in a sample, is given by

$$f_1(x_1) = \begin{cases} 6x_1(1 - x_1) & \text{for } 0 \le x_1 \le 1 \\ 0 & \text{elsewhere} \end{cases}$$

The density function for X_2, the proportion of type A crystals among the solid copper, is given by

$$f_2(x_2) = \begin{cases} 3x_2^2 & \text{for } 0 \le x_2 \le 1 \\ 0 & \text{elsewhere} \end{cases}$$

The variable $U = X_1 X_2$ represents the proportion of the sample volume due to type A crystals. Find the probability density function for U, assuming that X_1 and X_2 are independent.

6.12 A density function sometimes used to model lengths of life of electronic components is the Rayleigh density, which is given by

$$f(x) = \begin{cases} \left(\dfrac{2x}{\theta}\right) e^{-x^2/\theta} & \text{for } x > 1 \\ 0 & \text{elsewhere} \end{cases}$$

a If X has a Rayleigh density, find the probability density function for $U = X^2$.

b Use the result of part (a) to find $E(X)$ and $V(X)$.

6.13 The Weibull density function is given by

$$f(x) = \begin{cases} \dfrac{1}{\alpha} m x^{m-1} e^{-x^m/\alpha} & \text{for } x > 0 \\ 0 & \text{elsewhere} \end{cases}$$

where α and m are positive constants. If X has the Weibull density,

a Find the density function of $U = X^m$.

b Find $E(X^k)$ for any positive integer k.

6.5 Method of Moment-generating Functions

The definitions and uses of moment-generating functions have been introduced in Chapters 3, 4, and 5. Now, we will pull those ideas together and show how they form a powerful technique for finding distributions of functions of random variables in some common situations.

To review, the moment-generating function of a random variable Y is given by

$$M(t) = E(e^{tY})$$

when such an expected value exists for values of t in a neighborhood of zero.

Key to our use of moment-generating functions is the *uniqueness* theorem.

THEOREM 6.1

Let X and Y be two random variables with distribution functions $F_X(x)$ and $F_Y(y)$, respectively. Suppose that the respective moment-generating functions, $M_X(t)$ and $M_Y(t)$, exist and are equal for $|t| < h$, $h > 0$. Then the two functions $F_X(x)$ and $F_Y(y)$ are equal.
    The proof requires theory beyond the level of this book. ∎

Theorem 6.1 tells us that, if we can identify the moment-generating function of Y as belonging to some particular distribution, then Y *must* have that particular distribution.

Before looking at special cases, let's review two other basic properties of moment-generating functions. If

$$Y = aX + b$$

for constants a and b, then

$$M_Y(t) = e^{bt} M_X(at)$$

If X and Y are independent random variables, then

$$M_{X+Y}(t) = M_X(t) M_Y(t)$$

6.5.1   Gamma Case

The moment-generating function is quite useful in dealing with certain functions of random variables that come from the gamma family, in which the probability density function is given by

$$f(x) = \begin{cases} \dfrac{1}{\Gamma(\alpha)\beta^\alpha} x^{\alpha-1} e^{-x/\beta} & \text{for } x \geq 0 \\ 0 & \text{elsewhere} \end{cases}$$

For this distribution,

$$\begin{aligned}
M_x(t) &= \frac{1}{\Gamma(\alpha)\beta^\alpha} \int_0^\infty e^{tx} x^{\alpha-1} e^{-x/\beta} dx \\
&= \frac{1}{\Gamma(\alpha)\beta^\alpha} \int_0^\infty x^{\alpha-1} \exp[-x(\frac{1}{\beta} - t)] dx \\
&= \frac{1}{\Gamma(\alpha)\beta^\alpha} \Gamma(\alpha)(\frac{1}{\beta} - t)^{-\alpha} \\
&= (1 - \beta t)^{-\alpha}
\end{aligned}$$

As we have seen, sums of random variables occur frequently in applications. If $X_1, \ldots, X_n$ are independent gamma random variables, each having the preceding density function, then

$$Y = \sum_{i=1}^n X_i$$

has the moment-generating function

$$\begin{aligned}
M_y(t) &= M_{x_1}(t) M_{x_2}(t) \cdots M_{x_n}(t) \\
&= [(1 - \beta t)^{-\alpha}]^n \\
&= (1 - \beta t)^{-n\alpha}
\end{aligned}$$

Thus, Y has a gamma distribution with parameters $n\alpha$ and β. What happens to Y if the α-parameter is allowed to change from X_i to X_j?

6.5.2   Normal Case

Moment-generating functions are extremely useful in finding distributions of functions of normal random variables. If X has a normal distribution with mean μ and variance σ^2, then we know from Chapters 4 and 5 that

$$M_x(t) = \exp\left[\mu t + \frac{t^2\sigma^2}{2}\right]$$

Suppose that $X_1, \ldots, X_n$ are independent normal random variables with

$$E(X_i) = \mu_i \quad \text{and} \quad V(X_i) = \sigma_i^2$$

Let

$$Y = \sum_{i=1}^{n} a_i X_i$$

for constants $a_1, \ldots, a_n$. The moment-generating function for Y is, then,

$$
\begin{aligned}
M_y(t) &= M_{a_1 X_1}(t) M_{a_2 X_2}(t) \cdots M_{a_n X_n}(t) \\
&= \exp\left[a_1\mu_1 t + \frac{t^2 a_1^2 \sigma_1^2}{2}\right] \cdots \exp\left[a_n\mu_n t + \frac{t^2 a_n^2 \sigma_n^2}{2}\right] \\
&= \exp\left[t \sum_{i=1}^{n} a_i\mu_i + \frac{t^2}{2}\sum_{i=1}^{n} a_i^2\sigma_i^2\right]
\end{aligned}
$$

Thus, Y has a normal distribution with mean $\sum a_i\mu_i$ and variance $\sum a_i^2\sigma_i^2$. This distribution-preserving property of linear functions of normal random variables has numerous theoretical and applied uses.

6.5.3   Normal and Gamma Relationships

The two cases considered so far—gamma and normal—have connections to each other. Understanding these connections begins with looking at squares of normally distributed random variables.

Suppose that Z has a standard normal distribution. What can we say about the distribution of Z^2? Moment-generating functions might help. We have

$$
\begin{aligned}
M_{Z^2}(t) &= \int_{-\infty}^{\infty} e^{tz^2} \frac{1}{\sqrt{2\pi}} e^{-z^2/2} dz \\
&= \frac{1}{\sqrt{2\pi}} \int_{-\infty}^{\infty} \exp\left[-\frac{1}{2}z^2(1-2t)\right] dz \\
&= (1-2t)^{-1/2} \frac{1}{\sqrt{2\pi}(1-2t)^{-1/2}} \int_{-\infty}^{\infty} \exp\left[-\frac{1}{2}z^2(1-2t)\right] dz \\
&= (1-2t)^{-1/2}(1) \\
&= (1-2t)^{-1/2}
\end{aligned}
$$

Now we begin to see the connection. Z^2 has a gamma distribution, with parameters $\frac{1}{2}$ and 2.

What about the sum of squares of independent standard normal random variables? Suppose that $Z_1, Z_2, \ldots, Z_n$ are independent standard normals; and let

$$Y = \sum_{i-1}^{n} Z_i^2$$

Then,

$$
\begin{aligned}
M_Y(t) &= M_{Z_1^2}(t) M_{Z_2^2}(t) \cdots M_{Z_n^2}(t) \\
&= [(1 - 2t)^{-1/2}]^n \\
&= (1 - 2t)^{-n/2}
\end{aligned}
$$

and Y has a gamma distribution with parameters $n/2$ and 2. This particular type of gamma function comes up so often that it is given a special name—the *chi-square distribution*. In summary, a gamma distribution with parameters $n/2$ and 2 is called a *chi-square distribution with n degrees of freedom*.

In order to pursue more of the connections between normal and chi-square distributions, we must review some basic properties of joint moment-generating functions. If X_1 and X_2 have the joint probability density function $f(x_1, x_2)$, then their joint moment-generating function is given by

$$
\begin{aligned}
M_{X_1, X_2}(t_1, t_2) &= E\left[e^{t_1 X_1 + t_2 X_2}\right] \\
&= \int_{-\infty}^{\infty} \int_{-\infty}^{\infty} e^{t_1 x_1 + t_2 x_2} f(x_1, x_2) dx_1 dx_2
\end{aligned}
$$

If X_1 and X_2 are independent,

$$M_{X_1, X_2}(t_1, t_2) = M_{X_1}(t_1) M_{X_2}(t_2)$$

We make use of these results in the ensuing discussion.

If $X_1, \ldots, X_n$ represent a set of independent random variables from a common normal (μ, σ^2) distribution, then the sample mean

$$\overline{X} = \frac{1}{n} \sum_{i=1}^{n} X_i$$

is taken to be the best estimator of an unknown μ. The sample variance

$$S^2 = \frac{1}{n-1} \sum_{i=1}^{n} (X_i - \overline{X})^2$$

is taken to be the best estimator of σ^2. From our earlier work on linear functions, it is clear that $\overline{X}$ has another normal distribution. But what about the distribution of S^2?

The first task is to show that $\overline{X}$ and S^2 are independent random variables. It is possible to write S^2 as

$$S^2 = \frac{1}{2n(n-1)} \sum_{i=1}^{n} \sum_{j=1}^{n} (X_i - X_j)^2$$

so that we only need show that differences $(X_i - X_j)$ are independent of $X_1 + X_2 + \dots + X_n$, the random part of $\overline{X}$. Let's see how this can be shown for the case $n = 2$; the general case follows similarly.

Consider, for two independent normal (μ, σ^2) random variables X_1 and X_2

$$T = X_1 + X_2 \qquad \text{and} \qquad D = X_2 - X_1$$

Now,

$$M_T(t) = \exp[t(2\mu) + t^2\sigma^2]$$

and

$$M_D(t) = \exp[t^2\sigma^2]$$

Also,

$$
\begin{aligned}
M_{T,D}(t_1, t_2) &= E\{\exp[t_1(X_1 + X_2) + t_2(X_2 - X_1)]\} \\
&= E\{\exp[(t_1 - t_2)X_1 + (t_1 + t_2)X_2]\} \\
&= \exp\left[(t_1 - t_2)\mu + \frac{(t_1 + t_2)^2}{2}\sigma^2 + (t_1 + t_2)\mu + \frac{(t_1 + t_2)^2}{2}\sigma^2 \right] \\
&= \exp[t_1(2\mu) + t_1^2\sigma^2 + t_2^2\sigma^2] \\
&= M_T(t_1)M_D(t_2)
\end{aligned}
$$

Thus, T and D are independent (and incidently, each has a normal distribution).

The second task is to show how S^2 relates to quantities whose distribution we know. We begin with $\sum(X_i - \mu)^2$, which can be written

$$
\begin{aligned}
\sum_{i=1}^{n} (X_i - \mu)^2 &= \sum_{i=1}^{n} [(X_i - \overline{X}) + (\overline{X} - \mu)]^2 \\
&= \sum_{i=1}^{n} (X_i - \overline{X})^2 + n(\overline{X} - \mu)^2
\end{aligned}
$$

(Why does the cross-product term vanish?) Now,

$$\frac{1}{\sigma^2} \sum_{i=1}^{n} (X_i - \mu)^2 = \frac{1}{\sigma^2} \sum_{i=1}^{n} (X_i - \overline{X})^2 + \frac{n}{\sigma^2} (\overline{X} - \mu)^2$$

or

$$\sum_{i=1}^{n} \left(\frac{X_i - \mu}{\sigma} \right)^2 = \frac{1}{\sigma^2} \sum (X_i - \overline{X})^2 + \left(\frac{\overline{X} - \mu}{\sigma/\sqrt{n}} \right)^2$$

or

$$\sum_{i=1}^{n} Z_i^2 = \frac{(n-1)S^2}{\sigma^2} + Z^2$$

where Z and all values of Z_i have a standard normal distribution. Relating these functions to known distributions, the quantity on the left has a chi-square distribution with n degrees of freedom, and the one on the right has a chi-square distribution with 1 degree of freedom. Relating these distributions to moment-generating functions (and using the independence of S^2 and $\overline{X}$), we have

$$(1 - 2t)^{-n/2} = M_{(n-1)S^2/\sigma^2}(t)(1 - 2t)^{-1/2}$$

Thus,

$$M_{(n-1)S^2/\sigma^2}(t) = (1 - 2t)^{\frac{-n}{2} + \frac{1}{2}}$$
$$= (1 - 2t)^{-\frac{(n-1)}{2}}$$

and therefore, $(n-1)S^2/\sigma^2$ has a chi-square distribution with $(n-1)$ degrees of freedom. This is a fundamental result in classical statistical inference. Its derivation shows something of the power of the moment-generating function approach.

Exercises

6.14   Suppose that $X_1, \ldots, X_k$ are independent random variables, each having a binomial distribution with parameters n and p.

a   If n and p are the same for all distributions, find the distribution of $\sum_{i=1}^{k} X_i$.

b   What happens to the distribution of $\sum_{i=1}^{k} X_i$ if the n's change from X_i to X_j?

c   What happens to the distribution of $\sum_{i=1}^{k} X_i$ if the p's change from X_i to X_j?

6.15   Suppose that $X_1, \ldots, X_k$ are independent Poisson random variables, each with mean λ.

a   Find the distribution of $\sum_{i=1}^{k} X_i$.

b   What happens to the distribution of $\sum_{i=1}^{k} X_i$ if the λ's change from X_i to X_j?

c   For constants $a_1, \ldots, a_x$, does $\sum_{i=1}^{k} a_i X_i$ have a Poisson distribution?

6.16   Suppose that $X_1, \ldots, X_n$ are independent random variables, each having a gamma distribution, but with X_i having parameters α_i and β.

a   What is the distribution of $\sum_{i=1}^{k} X_i$?

b   Suppose that the β parameter changes from X_i to X_j. Does the sum still have a gamma distribution?

6.17   For $X_1, \ldots, X_n$ independent observations from a normal (μ, σ^2) distribution, find the distribution of $\overline{X}$.

6.18   Show that

$$S^2 = \frac{1}{n-1} \sum_{i=1}^{n} (X_i - \overline{X})^2$$

can be written as

$$S^2 = \frac{1}{2n(n-1)} \sum_{i=1}^{n} \sum_{j=1}^{n} (X_i - X_j)^2$$

[*Hint:* Add and subtract an $\overline{X}$ inside $(X_i - X_j)^2$.]

6.19 Suppose that $X_1, \ldots, X_n$ is a set of independent observations from a normal (μ, σ^2) distribution.

a Find $E(S^2)$.

b Find $V(S^2)$.

c Find $E(S)$.

6.20 Projectiles fired at a target land at coordinates X_1, X_2), where X_1 and X_2 are independent standard normal random variables. Suppose that two such projectiles land independently at (X_1, X_2) and at (X_3, X_4).

a Find the probability distribution for D^2, the square of the distance between the two points.

b Find the probability distribution for D.

6.6 Order Statistics

Many functions of random variables of interest in practice depend on the relative magnitudes of the observed variables. For instance, we may be interested in the fastest time in an automobile race or the heaviest mouse among those fed a certain diet. Thus, we often order observed random variables according to their magnitudes. The resulting ordered variables are called *order statistics*.

Formally, let $X_1, X_2, \ldots, X_n$ denote independent continuous random variables with distribution function $F(x)$ and density function $f(x)$. We shall denote the ordered random variables X_i by $X_{(1)}, X_{(2)}, \ldots, X_{(n)}$, where $X_{(1)} \le X_{(2)} \le \cdots \le X_{(n)}$. (Because the random variables are continuous, the equality signs can be ignored.) Thus,

$$X_{(1)} = \min(X_1, \ldots, X_n),$$

the minimum value of X_i, and

$$X_{(n)} = \max(X_i, \ldots, X_n).$$

the maximum value of X_i.

The probability density functions for $X_{(1)}$ and $X_{(n)}$ are found easily. Looking at $X_{(n)}$ first, we see that

$$P[X_{(n)} \le x] = P(X_1 \le x, \quad X_2 \le x, \quad \ldots, \quad X_n \le x)$$

because $[X_{(n)} \le x]$ implies that all the values of X_i must be less than or equal to x, and vice versa. However, $X_1, \ldots, X_n$ are independent; hence,

$$P[X_{(n)} \le x] = [F(x)]^n$$

Letting $g_n(x)$ denote the density function of $X_{(n)}$, we see, on taking derivatives on both sides, that

$$g_n(x) = n[F(x)]^{n-1} f(x)$$

The density function of $X_{(1)}$, denoted by $g_1(x)$, can be found by a similar device. We have

$$
\begin{aligned}
P[X_{(1)} \leq x] &= 1 - P[X_{(1)} > x] \\
&= 1 - P(X_1 > x, \quad X_2 > x, \quad \ldots, \quad X_n > x) \\
&= 1 - [1 - F(x)]^n
\end{aligned}
$$

Hence,

$$g_1(x) = n[1 - F(x)]^{n-1} f(x)$$

Similar reasoning can be used to derive the probability density function of $X_{(j)}$ for any j from 1 to n, but the details are a bit tedious. Nevertheless, we present one way of doing this, beginning with constructing the distribution function for $X_{(j)}$. Now,

$$
\begin{aligned}
P(X_{(j)} \leq x) &= P(\text{At least } j \text{ of the values of } X_j \text{ are less than or equal to } x) \\
&= \sum_{k=j}^{n} P(\text{Exactly } k \text{ of the values of } X_j \text{ are less than or equal to } x) \\
&= \sum_{k=j}^{n} \binom{n}{k} [F(x)]^k [1 - F(x)]^{n-k}
\end{aligned}
$$

The last expression comes about because the event "exactly k of the values of X_i are less than or equal to x" has a binomial construction with $p = F(x) = P(X_i \leq x)$.

The probability density function of $X_{(j)}$ is the derivative of the distribution function for $X_{(j)}$, as given in the preceding equation. This derivative, with respect to x, is

$$\sum_{k=j}^{n} \binom{n}{k} \left\{ k[F(x)]^{k-1} [1 - F(x)]^{n-k} f(x) - (n-k)[F(x)]^k [1 - F(x)]^{n-k-1} f(x) \right\}$$

We now separate the term for $k = j$ from the first half of the expression, obtaining

$$\binom{n}{j} j [F(x)]^{j-1} [1 - F(x)]^{n-j} f(x)$$

$$+ \sum_{k=j+1}^{n} \binom{n}{k} k [F(x)]^{k-1} [1 - F(x)]^{n-k} f(x)$$

$$- \sum_{k=j}^{n-1} \binom{n}{k} (n-k)[F(x)]^k [1 - F(x)]^{n-k-1} f(x)$$

The second summation can end at $(n-1)$, since the $k = n$ term is zero. To make these two summations extend over the same range of values, we change the index of summation in the first one from k to $k + 1$ (that is, we replace k by $k + 1$ in the first summation). Then, the sums are

$$\sum_{k=j}^{n-1} \binom{n}{k+1}(k+1)[F(x)]^k[1 - F(x)]^{n-k-1} f(x)$$

$$- \sum_{k=j}^{n-1} \binom{n}{k}(n-k)[F(x)]^k[1 - F(x)]^{n-k-1} f(x)$$

By writing out the full factorial versions, we can see that

$$\binom{n}{k+1}(k+1) = \binom{n}{k}(n-k)$$

and therefore, the two summations are identical, canceling each other out. Thus, we are left with the probability density function for $X_{(j)}$:

$$g_j(x) = \binom{n}{j} j [F(x)]^{j-1}[1 - F(x)]^{n-j} f(x)$$

$$= \frac{n!}{(j-1)!(n-j)!}[F(x)]^{j-1}[1 - F(x)]^{n-j} f(x)$$

Careful inspection of this form will reveal a pattern that can be generalized to joint density functions of two order statistics. Think of $f(x)$ as being proportional to the probability that one of the values of X_i falls into a small interval about the value x. The density function of $X_{(j)}$ requires $(j - 1)$ of the values of X_i to fall below x, one of them to fall on (or at least close to) x and $(n - j)$ of the m to exceed x. Thus, the probability that $X_{(j)}$ falls close to x is proportional to a multinomial probability with three cells, as follows:

$$\frac{(j-1)X_i's \overset{\text{one}}{\underset{X_i}{}} (n-j)X_i's}{x}$$

The probability that a value of X_i will fall in the first cell is $F(x)$; the probability for the second cell is proportional to $f(x)$; and the probability for the third cell is $[1 - F(x)]$. Thus,

$$g_j(x) = \frac{n!}{(j-1)!1!(n-j)!}[F(x)]^{j-1} f(x)[1 - F(x)]^{n-j}$$

If we generalize to the joint probability density function for $X_{(i)}$ and $X_{(j)}$, where $i < j$ the diagram is

$$\frac{(i-1)X_i's \overset{\text{one}}{\underset{X_i}{}} (j-i-1)X_i's \overset{\text{one}}{\underset{X_i}{}} (n-j)X_j's}{x_i \qquad\qquad x_j}$$

and the joint density (as proportional to a multinomial probability) becomes

$$g_{ij}(x_i, x_j) = \frac{n!}{(i-1)!(j-i-1)!(n-j)!}[F(x_i)]^{i-1}[F(x_j)$$
$$- F(x_i)]^{j-i-1}[1 - F(x_j)]^{n-j} f(x_i) f(x_j)$$

In particular, the joint density function of $X_{(1)}$ and $X_{(n)}$ becomes

$$g_{1n}(x_1, x_n) = \frac{n!}{(n-2)!}[F(x_n) - F(x_1)]^{n-2} f(x_1)f(x_n)$$

The same method can be used to find the joint density of $X_{(1)}, \ldots, X_{(n)}$, which turns out to be

$$g_{12\cdots n}(x_1, \ldots, x_n) = \begin{cases} n!f(x_1), & \ldots, & f(x_n) & \text{for } x_1 \leq x_2 \leq \cdots \leq x_n \\ 0 & & & \text{elsewhere} \end{cases}$$

The marginal density function for any of the order statistics can be found from this joint density function, but we shall not pursue the matter in this text.

E X A M P L E **6.9** Electronic components of a certain type have a life length X, with a probability density given by

$$f(x) = \begin{cases} (1/100)e^{-x/100} & \text{for } x > 0 \\ 0 & \text{elsewhere} \end{cases}$$

(Life length is measured in hours.) Suppose that two such components operate independently and in series in a certain system; that is, the system fails when either component fails. Find the density function for U, the life length of the system.

Solution Because the system fails at the failure of the first component, $U = \min(X_1, X_2)$, where X_1 and X_2 are independent random variables with the given density. Then, because $F(x) = 1 - e^{-x/100}$ for $x \geq 0$,

$$\begin{aligned} f_U(u) &= g_1(x) \\ &= n[1 - F(x)]^{n-1} f(x) \\ &= 2e^{-x/100}\left(\frac{1}{100}\right)e^{-x/100} \\ &= \begin{cases} \left(\frac{1}{50}\right)e^{-x/50} & \text{for } x > 0 \\ 0 & \text{elsewhere} \end{cases} \end{aligned}$$

Thus, we see that the minimum of two exponentially distributed random variables has an exponential distribution. ∎

E X A M P L E **6.10** Suppose that, in the situation described in Example 6.9, the components operate in parallel; that is, the system does not fail until both components fail. Find the density functions for U, the life length of the system.

Solution Now, $U = \max(X_1, X_2)$ and

$$\begin{aligned} f_U(u) &= g_2(x) \\ &= n[F(x)]^{n-1} f(x) \\ &= \begin{cases} 2(1 - e^{-x/100})\left(\frac{1}{100}\right)e^{-x/100} & \text{for } x > 0 \\ 0 & \text{elsewhere} \end{cases} \end{aligned}$$

$$= \begin{cases} \frac{1}{50}(e^{-x/100} - e^{-x/50}) & \text{for } x > 0 \\ 0 & \text{elsewhere} \end{cases}$$

We see that the maximum of two exponential random variables is not an exponential random variable. ∎

E X A M P L E   **6.11**   Suppose that $X_1, \ldots, X_n$ are independent random variables, each with a uniform distribution on the interval $(0, 1)$.

1   Find the probability density function of the range $R = X_{(n)} - X_{(1)}$

2   Find the mean and the variance of R.

Solution   1   For the uniform case,

$$f(x) = \begin{cases} 1 & \text{for } 0 < x < 1 \\ 0 & \text{elsewhere} \end{cases}$$

and

$$F(x) = \begin{cases} 0 & \text{for } x \leq 0 \\ x & \text{for } 0 < x < 1 \\ 1 & \text{for } x \geq 1 \end{cases}$$

Thus, the joint density function of $X_{(1)}, X_{(n)}$ is

$$g_{1n}(x_1, x_n) = \begin{cases} n(n-1)[x_n - x_1]^{n-2} & \text{for } 0 < x_1 < x_n < 1 \\ 0 & \text{elsewhere} \end{cases}$$

To find the density function for $R = X_{(n)} - X_{(1)}$, fix the value x_1 and let $r = x_n - x_1$. Thus, $r = h(x_n) = x_n - x_1$, and the inverse function is $x_n = h^{-1}(r) = r + x_1$. From the transformation method,

$$g_{1r}(x_1, r) = n(n-1)(r)^{n-2}(1)$$

Now, this is still a function of x_1 and r, and x_1 cannot be larger than $1 - r$ (why?). By integrating out x_1 we obtain

$$g_R(r) = \int_0^{1-r} n(n-1)r^{n-2}dx_1$$
$$= n(n-1)r^{n-2}(1-r), \qquad 0 < r < 1$$

2   Notice that this can be written

$$g_R(r) = n(n-1)r^{(n-1)-1}(1-r)^{2-1}, \qquad 0 < r < 1$$

which is a beta density function with parameters $\alpha = n - 1$ and $\beta = 2$. By inspection,

$$E(R) = \frac{\alpha}{\alpha + \beta} = \frac{n-1}{n+1}$$

and

$$V(R) = \left(\frac{\alpha}{\alpha+\beta}\right)\left(\frac{\beta}{\alpha+\beta}\right)\left(\frac{1}{\alpha+\beta+1}\right)$$

$$= \left(\frac{n-1}{n+1}\right)\left(\frac{2}{n+1}\right)\left(\frac{1}{n+2}\right)$$

$$= \frac{2(n-1)}{(n+1)^2(n+2)} \quad \blacksquare$$

Exercises

6.21 Let X_1 and X_2 be independent and uniformly distributed over the interval $(0, 1)$. Find the probability density functions of the following variables.

a $U_1 = \min(X_1, X_2)$

b $U_2 = \max(X_1, X_2)$

6.22 The opening prices per share of two similar stocks, X_1 and X_2, are independent random variables, each with density function

$$f(x) = \begin{cases} (1/2)e^{-(1/2)(x-4)} & \text{for } x \geq 4 \\ 0 & \text{elsewhere} \end{cases}$$

On a given morning, Mr. A is going to buy shares of whichever stock is less expensive. Find the probability density function for the price per share that Mr. A will have to pay.

6.23 Suppose that the length of time it takes a worker to complete a certain task X has the probability density function

$$f(x) = \begin{cases} e^{-(x-\theta)} & \text{for } x > \theta \\ 0 & \text{elsewhere} \end{cases}$$

where θ is a positive constant that represents the minimum time from start to completion of the task. Let $X_1, \ldots, X_n$ denote independent random variables from this distribution.

a Find the probability density function for $X_{(1)} = \min(X_1, \ldots, X_n)$.

b Find $E(X_{(1)})$.

6.24 Suppose that $X_1, \ldots, X_n$ constitute a random sample from a uniform distribution over the interval $(0, \theta)$ for some constant $\theta > 0$.

a Find the probability density function for $X_{(n)}$.

b Find $E(X_{(n)})$.

c Find $V(X_{(n)})$.

d Find a constant b such that $E(bX_{(n)}) = \theta$.

e Find the probability density function for the range $R = X_{(n)} - X_{(1)}$.

f Find $E(R)$.

6.25 Suppose that $X_1, \ldots, X_n$ represent a random sample from the uniform distribution over the interval (θ_1, θ_2).

a Find $E(X_{(1)})$.

b Find $E(X_{(n)})$.

c Find a function of $X_{(1)}$ and $X_{(n)}$ such that the expected value of that function is $(\theta_2 - \theta_1)$.

6.26 Suppose that Y_1 and Y_2 are independent standard normal random variables. Find the probability density function for $R = Y_{(2)} - Y_{(1)}$.

6.7 Probability-generating Functions: Applications to Random Sums of Random Variables

Probability-generating functions (see Section 3.10) also are useful tools for determining the probability distribution of a function of random variables. We shall examine their applicability by developing some properties of random sums of random variables.

Let $X_1, X_2, \ldots$ be a sequence of independent and identically distributed random variables, each with the probability-generating function $R(s)$. Let N be an integer-valued random variable with the probability-generating function $Q(s)$. Define S_N as

$$S_N = X_1 + X_2 + \cdots + X_N$$

The following are illustrative of the many applications of such random variables. N may represent the number of customers per day at a checkout counter, and X_i may represent the number of items purchased by the ith customer; then, S_N represents the total number of items sold per day. Or N may represent the number of seeds produced by a plant, and X_i may be defined by

$$X_i = \begin{cases} 1 & \text{if the } i\text{th seed germinates} \\ 0 & \text{otherwise} \end{cases}$$

Then S_N is the total number of germinating seeds produced by the plant.

We now assume that S_N has the probability-generating function $P(s)$, and we investigate its construction. By definition,

$$P(s) = E(s^{S_N})$$

which can be written

$$P(s) = E[E(s^{S_n}|N = n)]$$

$$= \sum_{n=0}^{\infty} E(s^{S_n}|N = n)P(N = n)$$

Notice that

$$E(s^{S_n}) = E(s^{X_1 + X_2 + \cdots + X_n})$$

$$= E(s^{X_1}s^{X_2} \cdots s^{X_n})$$

$$= E(s^{X_1})E(s^{X_2}) \cdots (S^{X_n})$$

$$= [R(s)]^n$$

Also,

$$Q(s) = E(s^N) = \sum_{n=0}^{\infty} s^n P(N = n)$$

If N is independent of the values of X_i, then

$$P(s) = \sum_{n=0}^{\infty} E(s^{S_n}) P(N = n)$$

$$= \sum_{n=0}^{\infty} [R(s)]^n P(N = n)$$

or

$$P(s) = Q[R(s)]$$

The expression $Q[R(s)]$ is the probability-generating function $Q(s)$ evaluated at $R(s)$ instead of at s.

Suppose that N has a Poisson distribution with mean λ. Then,

$$Q(s) = e^{\lambda(s-1)} = e^{-\lambda+\lambda s}$$

It follows that

$$P(s) = Q[R(s)]$$
$$= e^{-\lambda+\lambda R(s)}$$

This is often referred to as the probability-generating function of the compound Poisson distribution.

E X A M P L E **6.12**   Suppose that N, the number of animals caught in a trap per day, has a Poisson distribution with mean λ. The probability of any one animal's being male is p. Find the probability-generating function and the expected value of the number of males caught per day.

Solution   Let $X_i = 1$ if the ith animal caught is male, and let $X_i = 0$ otherwise. Then $S_N = X_1 + \cdots + X_N$ denotes the total number of males caught per day. Now,

$$R(s) = E(s^X) = q + ps$$

where $q = 1 - p$; and $P(s)$, the probability-generating function of S_N, is given by

$$P(s) = e^{-\lambda+\lambda R(s)}$$
$$= e^{-\lambda+\lambda(q+ps)}$$
$$= e^{-\lambda(1-q)+\lambda ps}$$
$$= e^{-\lambda p+\lambda ps}$$

Thus, S_N has a Poisson distribution with mean λp.   ∎

Exercises

6.27 In the compound Poisson distribution of Section 6.6, let

$$P(X_i = n) = \frac{1}{n}\alpha q^n, \qquad n = 1, 2, \ldots$$

for some constants α and q, where $0 < q < 1$. (X_i is said to have a *logarithmic series distribution*, which is frequently used in ecological models.) Show that the resulting compound Poisson distribution is related to the negative binomial distribution.

6.28 Show that the random variable S_N defined in Section 6.6 has mean $E(N)E(X)$ and variance $E(N)V(X) + V(N)E^2(X)$. Use probability-generating functions.

6.29 Let $X_1, X_2, \ldots$ denote a sequence of independent and identically distributed random variables, with $P(X_i = 1) = p$ and $P(X_i = -1) = q = 1 - p$. If a value of 1 signifies success, then

$$S_n = X_1 + \cdots + X_n$$

is the excess of successes over failures for n trials. A success could represent, for example, the completion of a sale or a win at a gambling device. Let $p_n^{(y)}$ equal the probability that S_n reaches the value y for the first time on the nth trial. We outline the steps for obtaining the probability-generating function $P(s)$ for $p_n^{(1)}$:

$$P(s) = \sum_{n=1}^{\infty} s^n p_n^{(1)}$$

a   Let $P^{(2)}(s)$ denote the probability-generating function for $p_n^{(2)}$; that is,

$$P^{(2)}(s) = \sum_{n=1}^{\infty} s^n p_n^{(2)}.$$

Since a first accumulation of S_n to 2 represents two successive first accumulations to 1, reason that

$$P^{(2)}(s) = P^2(s).$$

b   For $n > 1$, if $S_n = 1$ for the first time on trial n, this implies that S_n must have been at 2 on trial $n - 1$, and that a failure must have occurred on trial n. Since $p_n^{(1)} = q p_{n-1}^{(2)}$, show that

$$P(s) = ps + qs P^2(s)$$

c   Solve the quadratic equation in part (b) to show that

$$P(s) = \frac{1}{2qs}[1 - (1 - 4pqs^2)^{1/2}]$$

d   Show that

$$P(1) = \frac{1}{2q}(1 - |p - q|)$$

so the probability that S_n ever becomes positive equals 1 or p/q, whichever is smaller.

e   Show that, for the case $p = q = 1/2$, the expected number of trials until the first passage of S_n through 1 is infinite.

6.8 Summary

Many interesting applications of probability involve one or more functions of random variables, such as the sum, the average, the maximum value, and the range. Four methods for finding distributions of functions of random variables are outlined in this chapter. The method of distribution functions works well when the cumulative probability of the transformed variable can be calculated in closed form. If the transformed variable is a monotonic function of a univariate random variable, the method of transformations is appropriate and easy to apply. If the transformed variable is a function of more than one random variable, the method of conditioning (an extension of the method of transformations) might be the method of choice. The method of moment-generating functions is particularly useful for finding the distributions of sums of independent random variables.

Order statistics for independent random variables are of great practical utility, since it is often important to consider extreme values, medians, and ranges, among other statistical entities. A general form for the joint or marginal distribution of any set of order statistics is developed and used as a model in practical applications.

In finding the distribution of a function of random variables, one should bear in mind that there is no single best method for solving any one type of problem. In light of the preceding guidelines, one might try a number of approaches to see which one seems to be the most efficient.

Supplementary Exercises

6.30 Let X_1 and X_2 be independent and uniformly distributed over the interval $(0, 1)$. Find the probability density functions of the following variables.

a $U_1 = X_1^2$

b $U_2 = X_1/X_2$

c $U_3 = -\ln(X_1 X_2)$

d $U_4 = X_1 X_2$

6.31 Let X_1 and X_2 be independent Poisson random variables, with mean λ_1 and λ_2, respectively.

a Find the probability function of $X_1 + X_2$.

b Find the conditional probability function of X_1, given $X_1 + X_2 = m$.

6.32 Refer to the situation described in Exercise 6.1.

a Find the expected values of U_1, U_2, and U_3 directly (without using the density functions of U_1, U_2, and U_3).

b Find the expected values of U_1, U_2, and U_3 by using the derived density functions for these random variables.

6.33 A parachutist wants to land at a target T, but finds that she is equally likely to land at any point on a straight line (A, B) of which T is the midpoint. Find the probability density function of the distance between her landing point and the target. (*Hint:* Denote A by -1, B by $+1$, and T by 0. Then, the parachutist's landing point has a coordinate X that is uniformly distributed between -1 and $+1$. The distance between X and T is $|X|$.)

6.34 Let X_1 denote the amount of a bulk item stocked by a supplier at the beginning of a day, and let X_2 denote the amount of that item sold during the day. Suppose that X_1 and X_2 have the joint density function

$$f(x_1, x_2) = \begin{cases} 2 & \text{for } 0 \leq x_2 \leq x_1 \leq 1 \\ 0 & \text{elsewhere} \end{cases}$$

Of interest to this supplier is the random variable $U = X_1 - X_2$, which denotes the amount of the item left at the end of the day.

a Find the probability density function for U.

b Find $E(U)$.

c Find $V(U)$.

6.35 An efficiency expert takes two independent measurements, X_1 and X_2, on the length of time required for workers to complete a certain task. Each measurement is assumed to have the density function given by

$$f(x) = \begin{cases} (1/4)xe^{-x/2} & \text{for } x > 0 \\ 0 & \text{elsewhere} \end{cases}$$

Find the density function for the average

$$U = (1/2)(X_1 + X_2)$$

6.36 The length of time that a certain machine operates without failure is denoted by X_1, and the length of time needed to repair a failure is denoted by X_2. After repair, the machine is assumed to operate as if it were a new machine. X_1 and X_2 are independent, and each has the density function

$$f(x) = \begin{cases} e^{-x} & \text{for } x > 0 \\ 0 & \text{elsewhere} \end{cases}$$

Find the probability density function for

$$U = \frac{X_1}{X_1 + X_2}$$

which represents the proportion of time that the machine is in operation during any one operation-repair cycle.

6.37 Two sentries are sent to patrol a road that is 1 mile long. The sentries are sent to points chosen independently and at random along the road. Find the probability that the sentries will be less than 1/2 mile apart when they reach their assigned posts.

6.38 Let X_1 and X_2 be independent standard normal random variables. Find the probability density function of $U = X_1/X_2$.

6.39 Let X be uniformly distributed over the interval $(-1, 3)$. Find the probability density function of $U = X^2$.

6.40 If X denotes the life length of a component, and if $F(x)$ is the distribution function of X, then $P(X > x) = 1 - F(x)$ is called the *reliability* of the component. Suppose that a system consists of four components with identical reliability functions, $1 - F(x)$, operating as indicated in the following figure.

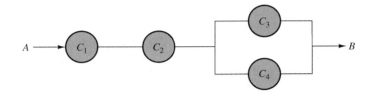

The system operates correctly if an unbroken chain of components is in operation between A and B. Assuming that the four components operate independently, find the reliability of the system, in terms of $F(x)$.

6.41 Let $X_1, X_2, \ldots, X_n$ denote a random sample from the uniform distribution, $f(x) = 1$, where $0 \le x \le 1$. Find the probability density function for the range $R = X_{(n)} - X_{(1)}$.

6.42 Suppose that U and V are independent random variables, with U having a standard normal distribution and V having a chi-square distribution with n degrees of freedom. Define T by

$$T = \frac{U}{\sqrt{V/n}}$$

Then T has a t (or Student's t) distribution, and the density can be obtained as follows.

a If V is fixed at v, then T is given by U/c, where $c = \sqrt{v/n}$. Use this idea to find the conditional density of T for fixed $V = v$.

b Find the joint density of T and V—namely, $f(t, v)$—by using

$$f(t, v) = f(t|v) f(v)$$

c Integrate over v to show that

$$f(t) = \frac{\Gamma[(n+1)/2]}{\sqrt{\pi n}\,\Gamma(n/2)}(1 + t^2/n)^{-(n+1)/2}, \qquad -\infty < t < \infty$$

6.43 Suppose that V and W are independent chi-square random variables with n_1 and n_2 degrees of freedom, respectively. Then F, defined by

$$F = \frac{V/n_1}{W/n_2}$$

is said to have an F distribution with n_1 and n_2 degrees of freedom, and the density function can be obtained as follows.

a If W is fixed at w, then $F = V/c$, where $c = wn_1/n_2$. Find the conditional density of F for fixed $W = w$.

b Find the joint density of F and W.

c Integrate over w to show that the probability density function of F—namely, $g(f)$—is given by

$$g(f) = \frac{\Gamma[(n_1 + n_2)/2](n_1/n_2)^{n_1/2}}{\Gamma(n_1/2)\Gamma(n_2/2)}(f)^{(n_1/2)-1}\left(1 + \frac{n_1 f}{n_2}\right)^{-(n_1+n_2)/2}, \qquad 0 < f < \infty$$

6.44 An object is to be dropped at point 0 in the plane, but lands at point (X, Y) instead, where X and Y denote horizontal and vertical distances, respectively, on a coordinate system centered at 0. If X and Y are independent, normally distributed random variables, each with a mean of zero and a variance of σ^2, find the distribution of the distances between the landing point (X, Y) and 0. (The resulting distribution is called the *Rayleigh distribution*.)

6.45 Suppose that n electronic components, each having an exponentially distributed life length with a mean of θ, are put into operation at the same time. The components operate independently and are observed until r of them have failed ($r \leq n$). Let W_j denote the length of time until the jth failure ($W_1 \leq W_2 \leq \cdots \leq W_r$). Let $T_j = W_j - W_{j-1}$ for $j \geq 2$ and $T_1 = W_1$.

a Show that T_j, for $j = 1, \ldots, r$, has an exponential distribution, with mean $\theta/(n - j + 1)$.

b Show that $U_r = \sum_{j=1}^{r} W_j + (n - r)W_r = \sum_{j=1}^{r}(n - j + 1)T_j$; and, hence, that $E(U_r) = r\theta$. [This suggests that $(1/r)U_r$ can be used as an estimator of θ.]

6.46 A machine produces spherical containers whose radii vary according to the probability density function

$$f(r) = \begin{cases} 2r & \text{for } 0 \leq r \leq 1 \\ 0 & \text{elsewhere} \end{cases}$$

Find the probability density function for the volume of the containers.

6.47 Let V denote the velocity of a molecule having mass m in a uniform gas at equilibrium. The probability density function of V is known to be

$$f(v) = \frac{4}{\sqrt{\pi}}b^{3/2}v^2e^{-bv^2}, \qquad v > 0$$

where $b = m/2KT$; in the latter equation, T denotes the absolute temperature of the gas, and K represents Boltzmann's constant. Find the probability density function for the kinetic energy E, given by $E = (1/2)mV^2$.

6.48 Suppose that members of a certain animal species are randomly distributed over a planar area, so that the number found in a randomly selected quadrant of unit area has a Poisson distribution with mean λ. Over a specified period of time, each animal has a probability of θ of dying; and deaths are assumed to be independent from animal to animal. Ignoring births and other population changes, what is the distribution of animals at the end of the time period?

Some Approximations to Probability Distributions: Limit Theorems

7.1 Introduction

As was noted in Chapter 6, we frequently are interested in functions of random variables, such as their average or their sum. Unfortunately, the application of the methods in Chapter 6 for finding probability distributions for functions of random variables may lead to intractable mathematical problems. Hence, we need some simple methods for approximating the probability distributions of functions of random variables.

In Chapter 7, we discuss properties of functions of random variables when the number of variables n gets large (approaches infinity). We shall see, for example, that the distribution for certain functions of random variables can easily be approximated for large n, even though the exact distribution for fixed n may be difficult to obtain. Even more important, the approximations are sometimes good for samples of modest size and, in some instances, for samples as small as $n = 5$ or 6. In Section 7.2, we present some extremely useful theorems, called *limit theorems*, that give properties of random variables as n tends toward infinity.

7.2 Convergence in Probability

Suppose that a coin has probability p, with $0 \leq p \leq 1$, of coming up heads on a single toss; and suppose that we toss the coin n times. What can be said about the fraction of heads observed in the n tosses? Intuition tells us that the sampled fraction of heads provides an estimate of p, and we would expect the estimate to fall closer to p for larger sample sizes—that is, as the quantity of information in the sample increases. Although our supposition and intuition may be correct for many problems of estimation, it is *not* always true that larger samples sizes lead to better estimates. Hence, this example gives rise to a question that occurs in all estimation problems: what can be said about the random distance between an estimate and its target parameter?

Notationally, let X denote the number of heads observed in the n tosses. Then $E(X) = np$ and $V(X) = np(1 - p)$. One way to measure the closeness of X/n to p is to ascertain the probability that the distance $|(X/n) - p|$ will be less than a preassigned real number ϵ. This probability

$$P\left(\left|\frac{X}{n} - p\right| \le \epsilon\right)$$

should be close to unity for larger n, if our intuition is correct. The following definition formalizes this convergence concept.

DEFINITION   **7.1**

> The sequence of random variables, $X_1, X_2, \ldots, X_n$, is said to **converge in probability** to the constant c if, for every positive number ϵ,
>
> $$\lim_{n \to \infty} P(|X_n - c| \le \epsilon) = 1$$ ∎

The following theorem often provides a mechanism for proving convergence in probability.

THEOREM   **7.1**

> Let $X_1, \ldots, X_n$ be independent and identically distributed random variables, with $E(X_i) = \mu$ and $V(X_i) = \sigma^2 < \infty$. Let $\overline{X}_n = (1/n)\sum_{i=1}^{n} X_i$. Then, for any positive real number ϵ,
>
> $$\lim_{n \to \infty} P(|\overline{X}_n - \mu| \ge \epsilon) = 0$$
>
> or
>
> $$\lim_{n \to \infty} P(|\overline{X}_n - \mu| < \epsilon) = 1$$
>
> Thus, $\overline{X}_n$ converges in probability toward μ.

Proof

> Notice that $E(\overline{X}_n) = \mu$ and $V(\overline{X}_n) = \sigma^2/n$. To prove the theorem, we appeal to Tchebysheff's theorem (see Section 3.2, pages 73–87, or Section 4.2, pages 146–150), which states that
>
> $$P(|X - \mu| \ge k\sigma) \le \frac{1}{k^2}$$
>
> where $E(X) = \mu$ and $V(X) = \sigma^2$. In the context of our theorem, X is to be replaced by $\overline{X}_n$, and σ^2 is to be replaced by σ^2/n. It then follows that
>
> $$P\left(|\overline{X}_n - \mu| \ge k\frac{\sigma}{\sqrt{n}}\right) \le \frac{1}{k^2}$$

Notice that k can be any real number, so we shall choose

$$k = \frac{\epsilon}{\sigma}\sqrt{n}$$

Then,

$$P\left[|\overline{X}_n - \mu| \geq \frac{\epsilon\sqrt{n}}{\sigma}\left(\frac{\sigma}{\sqrt{n}}\right)\right] \leq \frac{\sigma^2}{\epsilon^2 n}$$

or

$$P(|\overline{X}_n - \mu| \geq \epsilon) \leq \frac{\sigma^2}{\epsilon^2 n}$$

Now let us take the limit of this expression as n tends toward infinity. Recall that σ^2 is finite and that ϵ is a positive real number. On taking the limit as n tends toward infinity, we have

$$\lim_{n\to\infty} P(|\overline{X}_n - \mu| \geq \epsilon) = 0$$

The conclusion that $\lim_{n\to\infty} P(|\overline{X}_n - \mu| < \epsilon) = 1$ follows directly because

$$P(|\overline{X}_n - \mu| \geq \epsilon) = 1 - P(|\overline{X}_n - \mu| < \epsilon) \quad \blacksquare$$

We now apply Theorem 7.1 to our coin-tossing example.

EXAMPLE   **7.1**   Let X be a binomial random variable with probability of success p and number of trials n. Show that X/n converges in probability toward p.

Solution   We have seen that we can write X as $\sum_{i=1}^{n} X_i$, where $X_i = 1$ if the ith trial results in success, and $X_i = 0$ otherwise. Then

$$\frac{X}{n} = \frac{1}{n}\sum_{i=1}^{n} X_i$$

Also, $E(X_i) = p$ and $V(X_i) = p(1 - p)$. The conditions of Theorem 7.1 are then fulfilled with $\mu = p$ and $\sigma^2 = p(1 - p)$, and we conclude that

$$\lim_{n \to \infty} P\left(\left|\frac{X}{n} - p\right| \geq \epsilon\right) = 0$$

for any positive ϵ.   ∎

Theorem 7.1, sometimes called the *(weak) law of large numbers*, is the theoretical justification for the averaging process employed by many experimenters to obtain precision in measurements. For example, an experimenter may take the average of five measurements of the weight of an animal to obtain a more precise estimate of the animal's weight. His feeling—a feeling borne out by Theorem 7.1—is that the average of a number of independently selected weights has a high probability of being quite close to the true weight.

Like the law of large numbers, the theory of convergence in probability has many applications. Theorem 7.2, which is presented without proof, points out some properties of the concept of convergence in probability.

THEOREM   **7.2**   Suppose that X_n converges in probability toward μ_1 and that Y_n converges in probability toward μ_2. Then, the following statements are also true.

1    $X_n + Y_n$ converges in probability toward $\mu_1 + \mu_2$.
2    $X_n Y_n$ converges in probability toward $\mu_1 \mu_2$.
3   X_n / Y_n converges in probability toward μ_1 / μ_2, provided that $\mu_2 \neq 0$.
4    $\sqrt{X_n}$ converges in probability toward $\sqrt{\mu_1}$, provided that $P(X_n \geq 0) = 1$.   ∎

EXAMPLE   **7.2**   Suppose that $X_1, X_2, \ldots, X_n$ are independent and identically distributed random variables, with $E(X_i) = \mu$, $E(X_i^2) = \mu_2'$, $E(X_i^3) = \mu_3'$, and $E(X_i^4) = \mu_4'$ all assumed finite. Let S'^2 denote the sample variance given by

$$S'^2 = \frac{1}{n} \sum_{i=1}^{n} (X_i - \overline{X})^2$$

Show that S'^2 converges in probability toward $V(X_i)$.

Solution   First, notice that we can write

$$S'^2 = \frac{1}{n} \sum_{i=1}^{n} X_i^2 - \overline{X}^2$$

where

$$\overline{X} = \frac{1}{n} \sum_{i=1}^{n} X_i$$

To show that S'^2 converges in probability to $V(X_i)$, we apply both Theorems 7.1 and 7.2. Look at the terms in S'^2. The quantity $(1/n)\sum_{i=1}^{n} X_i^2$ is the average of n independent and identically distributed variables of the form X_i^2, with $E(X_i^2) = \mu_2'$ and $V(X_i^2) = \mu_4' - (\mu_2')^2$. Because $V(X_i^2)$ is assumed to be finite, Theorem 7.1 tells us that $(1/n)\sum_{i=1}^{n} X_i^2$ converges in probability toward μ_2'. Now consider the limit of $\overline{X}^2$ as n approaches infinity. Theorem 7.1 tells us that $\overline{X}$ converges in probability toward μ; and it follows from Theorem 7.2, part (b), that $\overline{X}^2$ converges in probability toward μ^2. Having shown that $(1/n)\sum_{i=1}^{n} X_i^n$ and $\overline{X}^2$ converge in probability toward μ_2' and μ^2, respectively, we can conclude from Theorem 7.2 that

$$S'^2 = \frac{1}{n} \sum_{i=1}^{n} X_i^2 - \overline{X}^2$$

converges in probability toward $\mu_2' - \mu^2 = V(X_i)$. ∎

This example shows that, for large samples, the sample variance has a high probability of being close to the population variance.

7.3 Convergence in Distribution

In Section 7.2, we dealt only with the convergence of certain random variables toward constants and said nothing about the form of the probability distributions. In this section, we look at what happens to the probability distributions of certain types of random variables as n tends toward infinity. We need the following definition before proceeding.

DEFINITION **7.2**

Let Y_n be a random variable with distribution function $F_n(y)$. Let Y be a random variable with distribution function $F(y)$. If

$$\lim_{n \to \infty} F_n(y) = F(y)$$

at every point y for which $F(y)$ is continuous, then Y_n is said to **converge in distribution** toward Y. $F(y)$ is called the *limiting distribution function* of Y_n. ∎

Definition 7.2 is illustrated by the following example.

EXAMPLE   **7.3**   Let $X_1, \ldots, X_n$ be independent uniform random variables over the interval $(0, \ \theta)$ for a positive constant θ. In addition, let $Y_n = \max(X_1 \ldots, X_n)$. Find the limiting distribution of Y_n.

Solution   The distribution function for the uniform random variable X_i is

$$F_X(y) = P(X_i \leq y) = \begin{cases} 0 & \text{for } y \leq 0 \\ \dfrac{y}{\theta} & \text{for } 0 < y < \theta \\ 1 & \text{for } y \geq \theta \end{cases}$$

In Section 6.5, we found that the distribution function for Y_n is

$$G(y) = P(Y_n \leq y) = [F_X(y)]^n$$

where $F_X(y)$ is the distribution function for each X_i. Then

$$\lim_{n \to \infty} G(y) = \begin{cases} 0 & \text{for } y \leq 0 \\ \lim\limits_{n \to \infty} \left(\frac{y}{\theta}\right)^n = 0 & \text{for } 0 < y < \theta \\ 1 & \text{for } y \geq \theta \end{cases}$$

Thus, Y_n converges in distribution toward a random variable that has a probability of 1 at the point θ and a probability of 0 elsewhere.   ∎

It is often easier to find limiting distributions by working with moment-generating functions. The following theorem gives the relationship between convergence of distribution functions and convergence of moment-generating functions.

THEOREM   **7.3**   Let Y_n and Y be random variables with moment-generating functions $M_n(t)$ and $M(t)$, respectively. If

$$\lim_{n \to \infty} M_n(t) = M(t)$$

for all real t, then Y_n converges in distribution toward Y.
  The proof of Theorem 7.3 is beyond the scope of this text.   ∎

EXAMPLE   **7.4**   Let X_n be a binomial random variable with n trials and probability p of success on each trial. If n tends toward infinity and p tends toward zero, with np remaining fixed, show that X_n converges in distribution toward a Poisson random variable.

Solution   This problem was solved in Chapter 3, when we derived the Poisson probability distribution. We now solve it by using moment-generating functions and Theorem 7.3.

We know that the moment-generating function of X_n—namely, $M_n(t)$—is given by

$$M_n(t) = (q + pe^t)^n$$

where $q = 1 - p$. This can be rewritten as

$$M_n(t) = [1 + p(e^t - 1)]^n$$

Letting $np = \lambda$ and substituting into $M_n(t)$, we obtain

$$M_n(t) = \left[1 + \frac{\lambda}{n}(e^t - 1)\right]^n$$

Now let us take the limit of this expression as n approaches infinity.

From the calculus, you may recall that

$$\lim_{n \to \infty} \left(1 + \frac{k}{n}\right)^n = e^k$$

Letting $k = \lambda(e^t - 1)$, we have

$$\lim_{n \to \infty} M_n(t) = \exp[\lambda(e^t - 1)]$$

We recognize the right-hand expression as the moment-generating function for the Poisson random variable. Hence, it follows from Theorem 7.3 that X_n converges in distribution toward a Poisson random variable. ■

EXAMPLE 7.5   In the interest of pollution control, an experimenter wants to count the number of bacteria per small volume of water. The sample size in this problem is really the volume of water in which the count is made; we do not have a value for n, as in previous problems. For purposes of approximating the probability distribution of counts, we can think of the volume—and hence, of the average count per volume—as the quantity that is getting large.

We let X denote the bacteria count per cubic centimeter of water, and we assume that X has a Poisson probability distribution with mean λ. We want to approximate the probability distribution of X for large values of λ, which we do by showing that

$$Y = \frac{X - \lambda}{\sqrt{\lambda}}$$

converges in distribution toward a standard normal random variable as λ tends toward infinity.

Specifically, if the allowable pollution in a water supply is a count of 110 bacteria per cubic centimeter, approximate the probability that X will be at most 110, assuming that $\lambda = 100$.

Solution   We proceed by taking the limit of the moment-generating function of Y as $\lambda \to \infty$ and then using Theorem 7.3. The moment-generating function of X—namely, $M_X(t)$—is given by

$$M_X(t) = e^{\lambda(e^t - 1)}$$

and hence the moment-generating function for Y—namely, $M_Y(t)$—is

$$M_Y(t) = e^{-\sqrt{\lambda}t} M_X\left(\frac{t}{\sqrt{\lambda}}\right)$$

$$= e^{-\sqrt{\lambda}t} \exp[\lambda(e^{t/\sqrt{\lambda}} - 1)]$$

The term $e^{t/\sqrt{\lambda}} - 1$ can be written as

$$e^{t/\sqrt{\lambda}} - 1 = \frac{t}{\sqrt{\lambda}} + \frac{t^2}{2\lambda} + \frac{t^3}{6\lambda^{3/2}} + \cdots$$

Thus, on adding exponents, we have

$$M_Y(t) = \exp\left[-\sqrt{\lambda}t + \lambda\left(\frac{t}{\sqrt{\lambda}} + \frac{t^2}{2\lambda} + \frac{t^3}{6\lambda^{3/2}} + \cdots\right)\right]$$

$$= \exp\left(\frac{t^2}{2} + \frac{t^3}{6\sqrt{\lambda}} + \cdots\right)$$

In the exponent of $M_Y(t)$, the first term $(t^2/2)$ is free of λ, and the remaining terms all have a λ to some positive power in the denominator. Therefore, as $\lambda \to \infty$, all the terms after the first will tend toward zero sufficiently quickly to allow

$$\lim_{\lambda \to \infty} M_Y(t) = e^{t^2/2}$$

and the right-hand expression is the moment-generating function for a standard normal random variable. We now want to approximate $P(X \le 110)$. Notice that

$$P(X \le 110) = P\left(\frac{X - \lambda}{\sqrt{\lambda}} \le \frac{110 - \lambda}{\sqrt{\lambda}}\right)$$

We have shown that $Y = (X - \lambda)/\sqrt{\lambda}$ is approximately a standard normal random variable for large λ. Hence, for $\lambda = 100$, we have

$$P\left(Y \le \frac{110 - 100}{10}\right) = P(Y \le 1)$$

$$= 0.8413$$

from Table 4 in the Appendix.

The normal approximation to Poisson probabilities works reasonably well for $\lambda \ge 25$.   ∎

Exercises

7.1   Let $Y_1, \ldots, Y_n$ be independent random variables, each with the probability density function

$$f(y) = \begin{cases} 3y^2 & \text{for } 0 \le y \le 1 \\ 0 & \text{elsewhere} \end{cases}$$

Show that

$$\overline{Y} = \frac{1}{n} \sum_{i=1}^{n} Y_i$$

converges in probability toward a constant as $n \to \infty$, and find the constant.

7.2   Let $Y_1, \ldots, Y_n$ be independent gamma-type random variables whose density function is given by

$$f(y) = \begin{cases} \dfrac{1}{b^a \Gamma(a)} y^{a-1} e^{-y/b} & \text{for } y \ge 0 \\ 0 & \text{elsewhere} \end{cases}$$

Show that the mean $\overline{Y}$ converges in probability toward a constant, and find the constant.

7.3   Let $Y_1, \ldots, Y_n$ be independent random variables, each uniformly distributed over the interval $(0, \theta)$.

a   Show that the mean $\overline{Y}$ converges in probability toward a constant as $n \to \infty$, and find the constant.

b   Show that $\max(Y_1, \ldots, Y_n)$ converges in probability toward θ as $n \to \infty$.

7.4   Let $Y_1, \ldots, Y_n$ be independent random variables, each possessing the density function

$$f(y) = \begin{cases} \dfrac{2}{y^2} & \text{for } y \ge 2 \\ 0 & \text{elsewhere} \end{cases}$$

Does the law of large numbers apply to $\overline{Y}$ in this case? If so, find the limit in probability of $\overline{Y}$.

7.5   If the probability that a person will suffer a bad reaction from an injection of a certain serum is 0.001, use the Poisson distribution to approximate the probability that, out of 1000 persons, 2 or more will suffer a bad reaction.

7.6   The number of accidents per year Y at a given intersection is assumed to have a Poisson distribution. Over the past few years, an average of 36 accidents per year have occurred at this intersection. If the number of accidents per year is at least 45, an intersection can qualify to be rebuilt under an emergency program set up by the state. Approximate the probability that the intersection in question will qualify under the emergency program at the end of the next year.

7.4 The Central Limit Theorem

Example 7.5 gives a random variable that converges in distribution toward the standard normal random variable. That this phenomenon is shared by a large class of random variables is shown by the following theorem.

THEOREM **7.4**

The Central Limit Theorem   Let $X_1, \ldots, X_n$ be independent and identically distributed random variables, with $E(X_i) = \mu$ and $V(x_i) = \sigma^2 < \infty$. Define Y_n as

$$Y_n = \sqrt{n}\left(\frac{\overline{X} - \mu}{\sigma}\right)$$

where

$$\overline{X} = \frac{1}{n}\sum_{i=1}^{n} X_i$$

Then Y_n converges in distribution toward a standard normal random variable.

Proof

We sketch a proof for the case in which the moment-generating function for X_i exists. (This is not the most general proof, because moment-generating functions do not always exist.)

Define a random variable Z_i by

$$Z_i = \frac{X_i - \mu}{\sigma}$$

Notice that $E(Z_i) = 0$ and $V(Z_i) = 1$. The moment-generating function of Z_i—namely, $M_Z(t)$—can be written as

$$M_Z(t) = 1 + \frac{t^2}{2} + \frac{t^3}{3!}E(Z_i^3) + \cdots$$

Now,

$$Y_n = \sqrt{n}\left(\frac{\overline{X} - \mu}{\sigma}\right)$$

$$= \frac{1}{\sqrt{n}}\left(\frac{\sum\limits_{i=1}^{n} X_i - n\mu}{\sigma}\right)$$

$$= \frac{1}{\sqrt{n}}\sum_{i=1}^{n} Z_i$$

and the moment-generating function of Y_n—namely, $M_n(t)$—can be written as

$$M_n(t) = \left[M_Z \left(\frac{t}{\sqrt{n}} \right) \right]^n$$

Recall that the moment-generating function of the sum of independent random variables is the product of their individual moment-generating functions. Hence,

$$M_n(t) = \left[M_Z \left(\frac{t}{\sqrt{n}} \right) \right]^n$$

$$= \left(1 + \frac{t^2}{2n} + \frac{t^3}{3!n^{3/2}} k + \cdots \right)^n$$

where $k = E(Z_i^3)$.

Now take the limit of $M_n(t)$ as $n \to \infty$. One way to evaluate the limit is to consider $\log M_n(t)$, where

$$\log M_n(t) = n \log \left[1 + \left(\frac{t^2}{2n} + \frac{t^3 k}{6n^{3/2}} + \cdots \right) \right]$$

A standard series expansion for $\log(1 + x)$ is

$$\log(1 + x) = x - \frac{x^2}{2} + \frac{x^3}{3} - \frac{x^4}{4} + \cdots$$

Let

$$x = \left(\frac{t^2}{2n} + \frac{t^{3k}}{6n^{3/2}} + \cdots \right)$$

Then,

$$\log M_n(t) = n \log(1 + x)$$

$$= n \left(x - \frac{x^2}{2} + \cdots \right)$$

$$= n \left[\left(\frac{t^2}{2n} + \frac{t^{3k}}{6n^{3/2}} + \cdots \right) - \frac{1}{2} \left(\frac{t^2}{2n} + \frac{t^{3k}}{6n^{3/2}} + \cdots \right)^2 + \cdots \right]$$

where the succeeding terms in the expansion involve x^3, x^4, and so on. Multiplying through by n, we see that the first term, $t^2/2$, does not involve n, whereas all other terms have n to a positive power in the denominator. Thus, it can be shown that

$$\lim_{n \to \infty} \log M_n(t) = t^2/2$$

or

$$\lim_{n \to \infty} M_n(t) = e^{t^2/2}$$

which is the moment-generating function for a standard normal random variable. Applying Theorem 7.3, we conclude that Y_n converges in distribution toward a standard normal random variable. ∎

Another way to say that a random variable converges in distribution toward a standard normal is to say that it is asymptotically normal. It is noteworthy that Theorem 7.4 is not the most general form of the central limit theorem. Similar theorems exist for certain cases in which the values of X_i are not identically distributed and in which they are dependent.

The probability distribution that arises from looking at many independent values of $\overline{X}$, for a fixed sample size, selected from the same population is called the *sampling distribution* of $\overline{X}$. The practical importance of the central limit theorem is that, for large n, the sampling distribution of $\overline{X}$ can be closely approximated by a normal distribution. More precisely,

$$P(\overline{X} \leq b) = P\left(\frac{\overline{X} - \mu}{\sigma/\sqrt{n}} \leq \frac{b - \mu}{\sigma/\sqrt{n}}\right)$$

$$\approx P\left(Z \leq \frac{b - \mu}{\sigma/\sqrt{n}}\right)$$

where Z is a standard normal random variable.

We can observe an approximate sampling distribution of $\overline{X}$ by looking at the following results from a computer simulation. Samples of size n were drawn from a population having the probability density function

$$f(x) = \begin{cases} \dfrac{1}{10}e^{-x/10} & \text{for } x > 0 \\ 0 & \text{elsewhere} \end{cases}$$

The sample mean was then computed for each sample. The relative frequency histogram of these mean values for 1000 samples of size $n = 5$ is shown in Figure 7.1. Figures 7.2 and 7.3 show similar results for 1000 samples of size $n = 25$ and $n = 100$, respectively. Although all the relative frequency histograms have roughly a bell shape, the tendency toward a symmetric normal curve is better for the larger n. A smooth curve drawn through the bar graph of Figure 7.3 would be nearly identical to a normal density function with a mean of 10 and a variance of $(10)^2/100 = 1$.

The central limit theorem provides a very useful result for statistical inference, since it enables us to know not only that $\overline{X}$ has mean μ and variance σ^2/n if the population has mean μ and variance σ^2, but also that the probability distribution for $\overline{X}$ is approximately normal. For example, suppose that we wish to find an interval (a, b) such that

$$P(a \leq \overline{X} \leq b) = 0.95$$

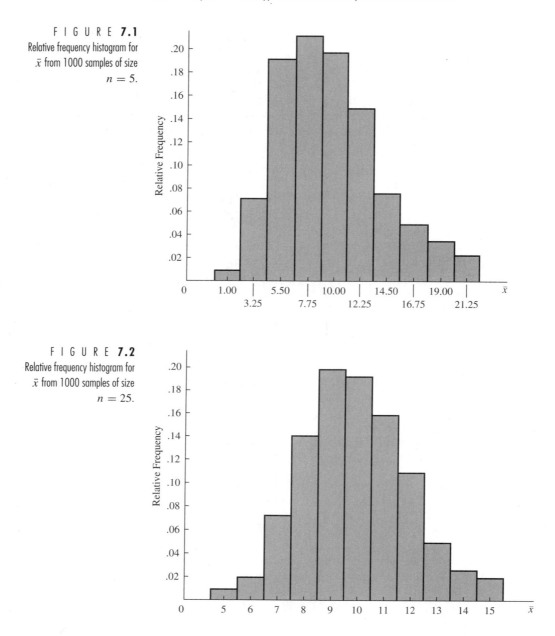

F I G U R E **7.1**

Relative frequency histogram for
$\bar{x}$ from 1000 samples of size
$n = 5$.

F I G U R E **7.2**

Relative frequency histogram for
$\bar{x}$ from 1000 samples of size
$n = 25$.

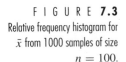

FIGURE **7.3**
Relative frequency histogram for
$\bar{x}$ from 1000 samples of size
$n = 100$.

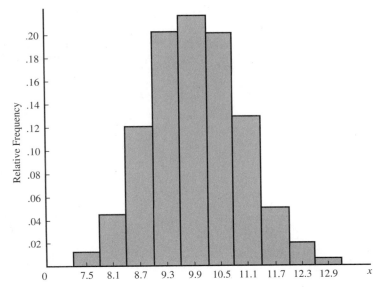

This probability is equivalent to

$$P\left(\frac{a-\mu}{\sigma/\sqrt{n}} \leq \frac{\overline{X}-\mu}{\sigma/\sqrt{n}} \leq \frac{b-\mu}{\sigma/\sqrt{n}}\right) = 0.95$$

for constants μ and σ. Because $(\overline{X} - \mu)/(\sigma/\sqrt{n})$ has approximately a standard normal distribution, the equality can be approximated by

$$P\left(\frac{a-\mu}{\sigma\sqrt{n}} \leq Z \leq \frac{b-\mu}{\sigma/\sqrt{n}}\right) = 0.95$$

where Z has a standard normal distribution. From Table 4 in the Appendix, we know that

$$P(-1.96 \leq Z \leq 1.96) = 0.95$$

and, hence,

$$\frac{a-\mu}{\sigma/\sqrt{n}} = -1.96 \qquad\qquad \frac{b-\mu}{\sigma/\sqrt{n}} = 1.96$$

or

$$a = \mu - \frac{1.96\sigma}{\sqrt{n}} \qquad\qquad b = \mu + \frac{1.96\sigma}{\sqrt{n}}$$

EXAMPLE   **7.6**   The fracture strength of a certain type of glass averages 14 (measured in thousands of pounds per square inch), with a standard deviation of 2.

1   What is the probability that the average fracture strength of 100 pieces of this glass exceeds 14.5?

2   Find an interval that includes the average fracture strength of 100 pieces of this glass, with probability 0.95.

Solution   **1**   The average strength $\overline{X}$ has an approximately normal distribution with $\mu = 14$ and a standard deviation of

$$\frac{\sigma}{\sqrt{n}} = \frac{2}{\sqrt{100}} = 0.2$$

Thus,

$$P(\overline{X} > 14.5) = P\left(\frac{\overline{X} - \mu}{\sigma/\sqrt{n}} > \frac{14.5 - \mu}{\sigma/\sqrt{n}}\right)$$

is approximately equal to

$$P\left(Z > \frac{14.5 - 14}{0.2}\right) = P\left(Z > \frac{0.5}{0.2}\right)$$
$$= P(Z > 2.5)$$
$$= 0.5 - 0.4938$$
$$= 0.0062$$

from Table 4. The probability of seeing an average value (for $n = 100$) that is more than 0.5 units above the population mean is very small in this case.

2   We have seen that

$$P\left[\mu - 1.96\left(\frac{\sigma}{\sqrt{n}}\right) \leq \overline{X} \leq \mu + 1.96\left(\frac{\sigma}{\sqrt{n}}\right)\right] = 0.95$$

for a normally distributed $\overline{X}$. In this problem,

$$\mu - 1.96\left(\frac{\sigma}{\sqrt{n}}\right) = 14 - 1.96\left(\frac{2}{\sqrt{100}}\right) = 13.6$$

and

$$\mu + 1.96\left(\frac{\sigma}{\sqrt{n}}\right) = 14 + 1.96\left(\frac{2}{\sqrt{100}}\right) = 14.4$$

Approximately 95% of sample mean fracture strengths, for samples of size 100, should lie between 13.6 and 14.4.   ▪

E X A M P L E   **7.7**   A certain machine that is used to fill bottles with liquid has been observed over a long period of time, and the variance in the amounts of fill has been found to be approximately $\sigma^2 = 1$ ounce. The mean ounces of fill μ, however, depend on an adjustment that may change from day to day or from operator to operator. If $n = 25$ observations on ounces of fill dispensed are to be taken on a given day (all with the same machine setting), find the probability that the sample mean will be within 0.3 ounce of the true population mean for that setting.

Solution   We shall assume that $n = 25$ is large enough for the sample mean $\overline{X}$ to have approximately a normal distribution. Then

$$P(|\overline{X} - \mu| \leq 0.3) = P[-0.3 \leq (\overline{X} - \mu) \leq 0.3]$$

$$= P\left[-\frac{0.3}{\sigma/\sqrt{n}} \leq \frac{\overline{X} - \mu}{\sigma/\sqrt{n}} \leq \frac{0.3}{\sigma/\sqrt{n}}\right]$$

$$= P\left[-0.3\sqrt{25} \leq \frac{\overline{X} - \mu}{\sigma/\sqrt{n}} \leq 0.3\sqrt{25}\right]$$

$$= P\left[-1.5 \leq \frac{\overline{X} - \mu}{\sigma/\sqrt{n}} \leq 1.5\right]$$

Since $(\overline{X} - \mu)/(\sigma\sqrt{n})$ has approximately a standard normal distribution, the preceding probability is approximately

$$P[-1.5 \leq Z \leq 1.5] = 0.8664 \quad \blacksquare$$

E X A M P L E   **7.8**   Achievement test scores from all high-school seniors in a certain state have a mean and a variance of 60 and 64, respectively. A specific high-school class of $n = 100$ students had a mean score of 58. Is there evidence to suggest that this high-school class is substandard? (Calculate the probability that the sample mean is at most 58 when $n = 100$.)

Solution   Let $\overline{X}$ denote the mean of a random sample of $n = 100$ scores from a population with $\mu = 60$ and $\sigma^2 = 64$. We want to approximate $P(\overline{X} \leq 58)$. We know from Theorem 7.4 that $\sqrt{n}(\overline{X} - \mu)/\sigma$ is approximately a standard normal variable, which we denote by Z. Hence,

$$P(\overline{X} \leq 58) \approx P\left(Z \leq \frac{58 - 60}{\sqrt{64/100}}\right)$$

$$= P(Z \leq -2.5)$$

$$= 0.0062$$

from Table 4 in the Appendix. Because this probability is so small, it is unlikely that the specific class of interest can be regarded as a random sample from a population with $\mu = 60$ and $\sigma^2 = 64$. There is evidence to suggest that this high-school class could be set aside as substandard. $\blacksquare$

We have seen in Chapter 3 that a binomially distributed random variable Y can be written as a sum of independent Bernoulli random variables X_i. Symbolically,

$$Y = \sum_{i=1}^{n} X_i$$

where $X_i = 1$ with probability p and $X_i = 0$ with probability $1 - p$, for $i = 1, \ldots, n$. Y can represent the number of successes in a sample of n trials or measurements, such as the number of thermistors conforming to standards in a sample of n thermistors.

Now the *fraction* of successes in the n trials is

$$\frac{Y}{n} = \frac{1}{n} \sum_{i=1}^{n} X_i = \overline{X}$$

so Y/n is a sample mean. In particular, for large n, we can affirm that Y/n has approximately a normal distribution with a mean of

$$E\left(\frac{Y}{n}\right) = \frac{1}{n} \sum E(X_i) = p$$

and a variance of

$$V(Y/n) = \frac{1}{n^2} \sum_{i=1}^{n} V(X_i)$$
$$= \frac{1}{n^2} \sum_{i=1}^{n} p(1-p)$$
$$= \frac{p(1-p)}{n}$$

The normality follows from the central limit theorem. Because $Y = n\overline{X}$, we know that Y has approximately a normal distribution with a mean of np and a variance of $np(1-p)$. Because binomial probabilities are cumbersome to calculate for a large n, we make extensive use of this normal approximation to the binomial distribution.

Figure 7.4 shows the histogram of a binomial distribution for $n = 20$ and $p = 0.6$. The heights of the bars represent the respective binomial probabilities. For this distribution, the mean is $np = 20(0.6) = 12$, and the variance is $np(1-p) = 20(0.6)(0.4) = 4.8$. Superimposed on the binomial distribution is a normal distribution with mean $\mu = 12$ and variance $\sigma^2 = 4.8$. Notice how closely the normal curve approximates the binomial histogram.

For the situation displayed in Figure 7.4, suppose that we wish to find $P(Y \le 10)$. By the exact binomial probabilities found in Table 2 of the Appendix,

$$P(Y \le 10) = 0.245$$

This value is the sum of the heights of the bars from $y = 0$ up to and including $y = 10$.

Looking at the normal curve in Figure 7.4, we can see that the areas in the bars at $y = 10$ and below are best approximated by the area under the curve to the left of 10.5. The extra 0.5 is added so that the total bar at $y = 10$ is included in the area

FIGURE **7.4**
A binomial distribution
$n = 20$, $p = 0.6$; and a
normal distribution,
$\mu = 12$, $\sigma^2 = 4.8$.

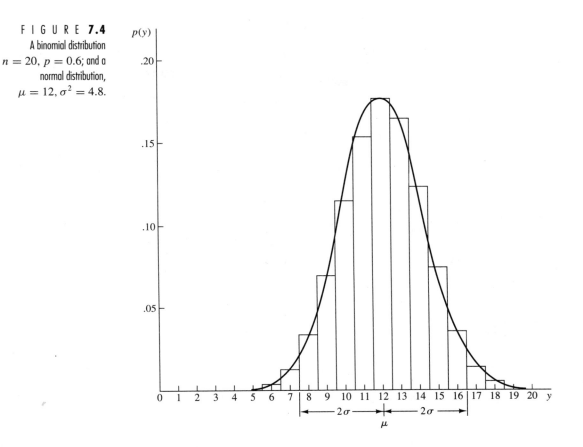

under consideration. Thus, if W represents a normally distributed random variable
with $\mu = 12$ and $\sigma^2 = 4.8$ ($\sigma = 2.2$), then

$$P(Y \leq 10) \approx P(W \leq 10.5)$$
$$= P\left(\frac{W - \mu}{\sigma} \leq \frac{10.5 - 12}{2.2}\right)$$
$$= P(Z \leq -0.68)$$
$$= 0.5 - 0.2517$$
$$= 0.2483$$

from Table 4 in the Appendix. We see that the normal approximation of 0.248 is close
to the exact binomial probability of 0.245. The approximation would even be better
if n were larger.

The normal approximation to the binomial distribution works well even for mod-
erately large n, as long as p is not close to zero or to one. A useful rule of thumb is to
make sure that n is large enough to guarantee that $p \pm 2\sqrt{p(1-p)/n}$ will lie within
the interval $(0, 1)$, before using the normal approximation. Otherwise, the binomial
distribution may be so asymmetric that the symmetric normal distribution cannot
provide a good approximation.

E X A M P L E   **7.9**   Silicon wafers received by a microchip plant are inspected for conformance to specifications. From a large lot of wafers, $n = 100$ are inspected. If the number of nonconformances Y is no more than 12, the lot is accepted. Find the approximate probability of acceptance if the actual proportion of nonconformances in the lot is $p = 0.2$.

Solution   The number of nonconformances Y has a binomial distribution if the lot indeed is large. Before using the normal approximation, we should check to confirm that

$$p \pm 2 \sqrt{\frac{p(1 - p)}{n}} = 0.2 \pm 2 \sqrt{\frac{(0.2)(0.8)}{100}}$$
$$= 0.2 \pm 0.08$$

is entirely within the interval $(0, 1)$, which it is. Thus, the normal approximation should work well.

Therefore, the probability of accepting the lot is

$$P(Y \leq 12) \approx P(W \leq 12.5)$$

where W is a normally distributed random variable with $\mu = np = 20$ and $\sigma = \sqrt{np(1 - p)} = 4$. It follows that

$$P(W \leq 12.5) = P\left(\frac{W - \mu}{\sigma} \leq \frac{12.5 - 20}{4}\right)$$
$$= P(Z \leq -1.88)$$
$$= 0.5 - 0.4699$$
$$= 0.0301$$

There is only a small probability of accepting any lot that has 20% nonconforming wafers.   ▪

E X A M P L E   **7.10**   Candidate A believes that she can win a city election if she can poll at least 55% of the voters in precinct I. She also believes that about 50% of the entire city's voters favor her. If $n = 100$ voters show up to vote at precinct I, what is the probability that candidate A will receive at least 55% of the votes?

Solution   Let X denote the number of voters at precinct I who vote for candidate A. We must approximate $P(X/n \geq 0.55)$, when p, the probability that a randomly selected voter favors candidate A, is 0.5. If we think of the $n = 100$ voters at precinct I as a random sample from the city, then X has a binomial distribution with $p = 0.5$.

Applying Theorem 7.4, we find that

$$P(X/n \geq 0.55) \approx P\left[Z \geq \frac{0.545 - 0.5}{\sqrt{0.5(0.5)/100}}\right]$$
$$= P(Z \geq 0.9)$$

$$= 0.1841$$

from Table 4 in the Appendix.   ■

We can make use of the central limit theorem to generate normally distributed random variables. Recall that the theorem makes the following claim:

> Let $X_1, X_2, \ldots, X_n$ be independent and identically distributed random variables, with $E(X_i) = \mu$ and $V(X_i) = \sigma^2$. Let $Y_n = \sqrt{n}(\overline{X} - \mu)/\sigma$, where $\overline{X} = (\sum X_i)/n$. Then Y_n converges in distribution toward a standard normal random variable as $n \to \infty$.

We can generate normal random variables as follows:

1 Generate n values of x from a specified distribution with known μ and σ^2. Compute the sample mean $\bar{x}_i$.

2 Evaluate $z_i = \sqrt{n}(\bar{x}_i - \mu)/\sigma$. The z_i will be values of a standard normal random variable with a mean of 0 and variance of 1.

3 The normal variable y_i will be given by $y_i = \mu_y + z_i \sigma_y$.

4 Repeat steps 1 through 3 m times, where m is the number of normal random variables desired.

The distribution most often used in step 1 is the uniform distribution on $(0, 1)$. When we use this distribution to evaluate z_i, $\mu = 0.5$ and $\sigma = 1/\sqrt{12}$. To make the algorithm more efficient, let $n = 12$.

Exercises

7.7 Shear strength measurements for spot welds of a certain type have been found to have a standard deviation of approximately 10 psi. If 100 test welds are to be measured, find the approximate probability that the sample mean will be within 1 psi of the true population mean.

7.8 If shear strength measurements have a standard deviation of 10 psi, how many test welds should be used in the sample if the sample mean is to be within 1 psi of the population mean, with a probability of approximately 0.95?

7.9 Soil acidity is measured by a quantity called the pH, which may range from 0 to 14 for soils ranging from extreme alkalinity to extreme acidity. Many soils have an average pH in the more-or-less-neutral 5 to 8 range. A scientist wants to estimate the average pH for a large field from n randomly selected core samples, measuring the pH in each sample. If the scientist selects $n = 40$ samples, find the approximate probability that the sample mean of the 40 pH measurements will be within 0.2 unit of the true average pH for the field.

7.10 Suppose that the scientist in Exercise 7.9 would like the sample mean to be within 0.1 of the true mean, with probability 0.90. How many core samples should he take?

7.11   Resistors of a certain type have resistances that average 200 ohms, with a standard deviation of 10 ohms. Suppose that 25 of these resistors are to be used in a circuit.

   a   Find the probability that the average resistance of the 25 resistors is between 199 and 202 ohms.

   b   Find the probability that the *total* resistance of the 25 resistors does not exceed 5100 ohms. [*Hint:* Notice that

$$P\left(\sum_{i=1}^{n} X_i > a\right) = P(n\overline{X} > a) = P(\overline{X} > a/n)$$

   for the situation described.]

   c   What assumptions are necessary for the answers in parts (a) and (b) to be good approximations?

7.12   One-hour carbon monoxide concentrations in air samples from a large city average 12 ppm, with a standard deviation of 9 ppm. Find the probability that the average concentration in 100 samples selected randomly will exceed 14 ppm.

7.13   Unaltered bitumens, as commonly found in lead–zinc deposits, have atomic hydrogen/carbon ratios that average 1.4/1 with a standard deviation of 0.05. Find the probability that 25 samples of bitumen have an average H/C ratio of less than 1.3/1.

7.14   The downtime per day for a certain computing facility averages 4.0 hours, with a standard deviation of 0.8 hour.

   a   Find the probability that the average daily downtime for a period of 30 days is between 1 and 5 hours.

   b   Find the probability that the *total* downtime for the 30 days is less than 115 hours.

   c   What assumptions must be true for the answers in parts (a) and (b) to be valid approximations?

7.15   The strength of a thread is a random variable with mean 0.5 lb and standard deviation 0.2 lb. Assume that the strength of a rope is the sum of the strengths of the threads in the rope.

   a   Find the probability that a rope consisting of 100 threads will hold 45 lb.

   b   How many threads are needed for a rope to provide 99% assurance that it will hold 50 lb?

7.16   Many bulk products, such as iron ore, coal, and raw sugar, are sampled for quality by a method that requires many small samples to be taken periodically as the material moves along a conveyor belt. The small samples are then aggregated and mixed to form one composite sample. Let Y_i denote the volume of ith small sample from a particular lot; and suppose that $Y_1, \ldots, Y_n$ constitute a random sample, with each Y_i having a mean of μ and a variance of σ^2. The average volume μ of the samples can be set by adjusting the size of the sampling device. Suppose that the variance σ^2 of sampling volumes is known to be approximately 4 for a particular situation (measurements are in cubic inches). The total volume of the composite sample is required to exceed 200 cubic inches with a probability of approximately 0.95 when $n = 50$ small samples are selected. Find a setting for μ that will satisfy the sampling requirements.

7.17   The service times for customers coming through a checkout counter in a retail store are independent random variables, with a mean of 1.5 minutes and a variance of 1.0 minute. Approximate the probability that 100 customers can be serviced in less than 2 hours of total service time.

7.18   Referring to Exercise 7.17, find the number of customers n such that the probability of servicing all n customers in less than 2 hours is approximately 0.1.

7.19   Suppose that $X_1, \ldots, X_{n_1}$ and $Y_1, \ldots, Y_{n_2}$ constitute random samples from populations with means μ_1 and μ_2 and variances σ_1^2 and σ_2^2, respectively. Then the central limit theorem can be extended to show that $\overline{X} - \overline{Y}$ is approximately normally distributed, for large n_1 and n_2, with mean $\mu_1 - \mu_2$ and variance $(\sigma_1^2/n_1 + \sigma_2^2/n_2)$.

Water flow through soil depends on, among other things, the porosity (volume proportion due to voids) of the soil. To compare two types of sandy soil, $n_1 = 50$ measurements are to be taken on the porosity of soil A, and $n_2 = 100$ measurements are to be taken on the porosity of soil B. Assume that $\sigma_1^2 = 0.01$ and $\sigma_2^2 = 0.02$. Find the approximate probability that the difference between the sample means will be within 0.05 unit of the true difference between population means, $\mu_1 - \mu_2$.

7.20 Referring to Exercise 7.19, suppose that samples are to be selected with $n_1 = n_2 = n$. Find the value of n that will allow the difference between the sample means to be within 0.04 unit of $\mu_1 - \mu_2$ with a probability of approximately 0.90.

7.21 An experiment is designed to test whether operator A or operator B gets the job of operating a new machine. Each operator is timed on 50 independent trials involving the performance of a certain task on the machine. If the sample means for the 50 trials differ by more than 1 second, the operator with the smaller mean will get the job. Otherwise, the outcome of the experiment will be considered a tie. If the standard deviations of times for both operators are assumed to be 2 seconds, what is the probability that operator A will get the job on the basis of the experiment, even though both operators have equal ability?

7.22 The median age of residents of the United States is 31 years. If a survey of 100 randomly selected United States residents is taken, find the approximate probability that at least 60 of them will be under 31 years of age.

7.23 A lot acceptance sampling plan for large lots calls for sampling 50 items and accepting the lot if the number of nonconformances is no more than 5. Find the approximate probability of acceptance if the true proportion of nonconformances in the lot is as follows.

  a   10%

  b   20%

  c   30%

7.24 Of the customers who enter a showroom for stereo equipment, only 30% make purchases. If 40 customers enter the showroom tomorrow, find the approximate probability that at least 15 will make purchases.

7.25 The quality of computer disks is measured by the number of missing pulses. For a certain brand of disk, 80% are generally found to contain no missing pulses. If 100 such disks are inspected, find the approximate probability that 15 or fewer contain missing pulses.

7.26 The capacitances of a certain type of capacitor are normally distributed, with a mean of 53 μf and a standard deviation of 2 μf. If 64 such capacitors are to be used in an electronic system, approximate the probability that at least 12 of them will have capacitances below 50 μf.

7.27 The daily water demands for a city pumping station exceed 500,000 gallons with probability of only 0.15. Over a 30-day period find the approximate probability that demand for more than 500,000 gallons per day occurs no more than twice.

7.28 At a specific intersection, vehicles entering from the east are equally likely to turn left, turn right, or proceed straight ahead. If 500 vehicles enter this intersection from the east tomorrow, what are the approximate probabilities of the following events?

  a   150 or fewer turn right.

  b   At least 350 turn.

7.29 Waiting times at a service counter in a pharmacy are exponentially distributed, with a mean of 10 minutes. If 100 customers come to the service counter in a day, approximate the probability that at least half of them must wait for more than 10 minutes.

7.30 A large construction firm has won 60% of the jobs for which it has bid. Suppose this firm bids on 25 jobs next month.

  a   Approximate the probability that it will win at least 20 of these jobs.

  b   Find the exact binomial probability that it will win at least 20 of these jobs. Compare this result to your answer in part (a).

  c   What assumptions must be true for your answers in parts (a) and (b) to be valid?

7.31   An auditor samples 100 of a firm's travel vouchers to determine how many of these vouchers are improperly documented. Find the approximate probability that more than 30% of the sampled vouchers will show up as being improperly documented if, in fact, only 20% of all the firm's vouchers are improperly documented.

7.5   Combination of Convergence in Probability and Convergence in Distribution

We may often be interested in the limiting behavior of the product or quotient of several functions of a set of random variables. The following theorem, which combines convergence in probability with convergence in distribution, applies to the quotient of two functions, X_n and Y_n.

THEOREM   **7.5**

> Suppose that X_n converges in distribution toward a random variable X, and that Y_n converges in probability toward unity. Then X_n / Y_n converges in distribution toward X.
>
>   The proof of Theorem 7.5 is beyond the scope of this text, but we can observe its usefulness in the following example.   ∎

EXAMPLE   **7.11**   Suppose that $X_1, \ldots, X_n$ are independent and identically distributed random variables, with $E(X_i) = \mu$ and $V(X_i) = \sigma^2$. Define S'^2 as

$$S'^2 = \frac{1}{n} \sum_{i=1}^{n} (X_i - \overline{X})^2$$

Show that

$$\sqrt{n} \, \frac{\overline{X} - \mu}{S'}$$

converges in distribution toward a standard normal random variable.

Solution   In Example 7.2, we showed that S'^2 converges in probability toward σ^2. Hence, it follows from Theorem 7.2 parts (c) and (d) that S'^2/σ^2 (and hence, S'/σ) converges in probability toward 1. We also know from Theorem 7.4 that

$$\sqrt{n} \left(\frac{\overline{X} - \mu}{\sigma} \right)$$

converges in distribution toward a standard normal random variable. Therefore,

$$\sqrt{n} \left(\frac{\overline{X} - \mu}{S'} \right) = \sqrt{n} \left(\frac{\overline{X} - \mu}{\sigma} \right) \bigg/ \frac{S'}{\sigma}$$

converges in distribution toward a standard normal random variable, by Theorem 7.5. ∎

7.6 Summary

Distributions of functions of random variables are important in both theoretical and practical work. Often, however, the distribution of a certain function cannot be found—at least not without a great deal of effort. Thus, approximations to the distributions of functions of random variables play a key role in probability theory. To begin with, probability itself can be thought of as a ratio of the number of "successes" (a random variable) to the total number of trials in a random experiment. That this ratio converges, under certain conditions, toward a constant is a fundamental result in probability upon which much theory and many applications are built. This **law of large numbers** is the reason, for example, that opinion polls work, if the sample underlying the poll is truly random.

The average, or mean, of random variables is one of the most commonly used functions. The **central limit theorem** provides a very useful normal approximation to the distribution of such averages. This approximation works well under very general conditions and is one of the most widely used results in all of probability theory.

Supplementary Exercises

7.32 A large industry has an average hourly wage of $4.00 per hour, with a standard deviation of $0.50. A certain ethnic group consisting of 64 workers has an average wage of $3.90 per hour. Is it reasonable to assume that the ethnic group is a random sample of workers from the industry? (Calculate the probability of randomly obtaining a sample man that is less than or equal to $3.90 per hour.)

7.33 An anthropologist wishes to estimate the average height of men for a certain race of people. If the population standard deviation is assumed to be 2.5 inches and if she randomly samples 100 men, find the probability that the difference between the sample mean and the true population mean will not exceed 0.5 inch.

7.34 Suppose that the anthropologist of Exercise 7.33 wants the difference between the sample mean and the population mean to be less than 0.4 inch, with probability 0.95. How many men should she sample to achieve this objective?

7.35 A machine is shut down for repairs if a random sample of 100 items selected from the daily output of the machine reveals at least 15% defectives. (Assume that the daily output is a large number of items.) If the machine, in fact, is producing only 10% defective items, find the probability that it will be shut down on a given day.

7.36 A pollster believes that 20% of the voters in a certain area favor a bond issue. If 64 voters are randomly sampled from the large number of voters in this area, approximate the probability that the sampled fraction of voters favoring the bond issue will not differ from the true fraction by more than 0.06.

7.37 Twenty-five heat lamps are connected in a greenhouse so that, when one lamp fails, another takes over immediately. (Only one lamp is turned on at any time.) The lamps operate independently, each with a mean life of 50 hours and a standard deviation of 4 hours. If the greenhouse is not checked for 1300 hours after the lamp system is turned on, what is the probability that a lamp will be burning at the end of the 1300-hour period?

7.38 Suppose that $X_1, \ldots, X_n$ are independent random variables, each with a mean of μ_1 and a variance of σ_1^2. Suppose, too, that $Y_1, \ldots, Y_n$ are independent random variables, each with a mean of μ_2 and a variance of σ_2^2. Show that, as $n \to \infty$, the random variable

$$\frac{(\overline{X} - \overline{Y}) - (\mu_1 - \mu_2)}{\sqrt{(\sigma_1^2 + \sigma_2^2)/n}}$$

converges in distribution toward a standard normal random variable.

7.39 An experimenter is counting bacteria of two types, A and B, in a certain liquid. Let X denote the number of type A bacteria per cubic centimeter, and let Y denote the number of type B bacteria per cubic centimeter. X and Y are assumed to have Poisson distributions, with means λ_1 and λ_2, respectively. The experimenter plans to observe a number of independent values for X and Y. Identify the function of the observed values of X and Y you would suggest as an estimator of the ratio

$$\frac{\lambda_1}{\lambda_1 + \lambda_2}$$

Why would you suggest this function as the estimator?

7.40 Let Y have a chi-square distribution with n degrees of freedom; that is, Y has the density function

$$f(y) = \begin{cases} \dfrac{1}{2^{n/2}\Gamma(n/2)} y^{(n/2)-1} e^{-y/2} & \text{for } y \geq 0 \\ 0 & \text{elsewhere} \end{cases}$$

Show that the random variable

$$\frac{Y - n}{\sqrt{2n}}$$

is asymptotically standard normal in distribution, as $n \to \infty$.

7.41 A machine in a heavy-equipment factory produces steel rods of length Y, where Y is a normal random variable with a mean μ of 6 inches and a variance σ^2 of 0.2. The cost C of repairing a rod that is not exactly 6 inches in length is proportional to the square of the error and is given (in dollars) by

$$C = 4(Y - \mu)^2$$

If 50 rods with independent lengths are produced in a given day, approximate the probability that the total cost for repairs for that day will exceed $48.

Extended Applications of Probability

8.1 The Poisson Process

In this section, we apply probability theory to modeling random phenomena that change over time—the so-called *stochastic processes*. Thus, a time parameter is introduced, and the random variable $Y(t)$ is regarded as being a function of time. (Later we shall see that the concept of *time* can be generalized to include space.) Examples are endless: $Y(t)$ could represent the size of a biological population at time t; the cost of operating a complex industrial system for t time units; the distance displaced by a particle in time t; or the number of customers waiting to be served at a checkout counter t time units after its opening.

Although we shall not discuss stochastic processes exhaustively, a few elementary concepts are in order. A stochastic process $Y(t)$ is said to have *independent* increments if, for any set of time points $t_0 < t_1 < \cdots < t_n$, the random variables $[Y(t_i) - Y(t_{i-1})]$ and $[Y(t_j) - Y(t_{j-1})]$ are independent for $i \neq j$. (See Section 5.3 for a discussion of independent random variables.) The process is said to have *stationary* independent increments if, in addition, the random variable $[Y(t_2 + h) - Y(t_1 + h)]$ and $[Y(t_2) - Y(t_1)]$ have identical distributions for any $h > 0$. We begin our discussion of specific processes by considering an elementary counting process with wide applicability: the Poisson process.

Let $Y_{(t)}$ denote the number of occurrences of some event in the time interval $(0, t)$, where $t > 0$. This could be the number of accidents at a particular intersection, the number of times a computer breaks down, or any similar count. We derived the Poisson distribution in Section 3.7 by looking at a limiting form of the binomial distribution, but now we derive the Poisson process by working directly from a set of axioms.

The Poisson process is defined as satisfying the following four axioms:

Axiom 1: $P[Y(0) = 0] = 1$.

Axiom 2: $Y(t)$ has stationary independent increments. (The number of occurrences in two nonoverlapping time intervals of the same length are independent and have the same probability distribution.)

Axiom 3: $P\{[Y(t + h) - Y(t)] = 1\} = \lambda h + o(h)$, where $h > 0$, where $o(h)$ is a generic notation for a term such that $o(h)/h \to 0$ as $h \to 0$, and where λ is a constant.

Axiom 4: $P\{[Y(t + h) - Y(t)] > 1\} = o(h)$.

Axiom 3 says that the chance of one occurrence in a small interval of time h is roughly proportional to h; and Axiom 4 states that the chance of more than one such occurrence tends to reach zero faster than h itself does. These two axioms imply that

$$P\{[Y(t + h) - Y(t)] = 0\} = 1 - \lambda h - o(h)$$

The probability distribution can now be derived from the axioms. Consider the interval $(0, t + h)$ partitioned into the two disjoint pieces $(0, t)$ and $(t, t + h)$. Then we can write

$$P[Y(t + h) = k] = \sum_{j=0}^{k} P\{Y(t) = j, \quad [Y(t + h) - Y(t)] = k - j\}$$

because the events inside the right-hand probability are mutually exclusive. Because independent increments are assumed, we have

$$P[Y(t + h) = k] = \sum_{j=0}^{k} P[Y(t) = j]P\{[Y(t + h) - Y(t)] = k - j\}$$

By stationarity and Axiom 1, this becomes

$$P[Y(t + h) = k] = \sum_{j=0}^{k} P[Y(t) = j]P[Y(h) = k - j]$$

To simplify the notation, let $P[Y(t) = j] = P_j(t)$. Then,

$$P_k(t + h) = \sum_{j=0}^{k} P_j(t) P_{k-j}(h)$$

and Axiom 4 reduces this to

$$P_k(t + h) = P_{k-1}(t) P_1(h) + P_k(t) P_0(h) + o(h)$$

Axiom 3 further reduces the equation to

$$P_k(t + h) = P_{k-1}(t)[\lambda h + o(h)] + P_k(t)[1 - \lambda h - o(h)]$$
$$= P_{k-1}(t)(\lambda h) + P_k(t)(1 - \lambda h) + o(h)$$

We can write this equation as

$$\frac{1}{h}[P_k(t+h) - P_k(t)] = \lambda P_{k-1}(t) - \lambda P_k(t) + \frac{o(h)}{h}$$

and on taking limits as $h \to 0$, we have

$$\frac{dP_k(t)}{dt} = \lim_{h \to 0} \frac{1}{h}[P_k(t+h) - P_k(t)] = \lambda P_{k-1}(t) - \lambda P_k(t)$$

To solve this differential equation, we take $k = 0$ to obtain

$$\frac{dP_k(t)}{dt} = P_0'(t) = -\lambda P_0(t)$$

since $P_{k-1}(t)$ will be replaced by zero for $k = 0$. Thus, we have

$$\frac{P_0'(t)}{P_0(t)} = -\lambda$$

or (on integrating both sides)

$$\ln P_0(t) = -\lambda t + c$$

The constant is evaluated by the boundary condition $P_0(0) = P[Y(0) = 0] = 1$, by Axiom 1, which gives $c = 0$. Hence,

$$\ln P_0(t) = -\lambda t$$

or

$$P_0(t) = e^{-\lambda t}$$

By letting $k = 1$ in the original differential equation, we can obtain

$$P_1(t) = \lambda t e^{-\lambda t}$$

and recursively,

$$P_k(t) = \frac{1}{k!}(\lambda t)^k e^{-\lambda t}, \qquad k = 0, 1, 2, \ldots$$

Thus, $Y(t)$ has a Poisson distribution, with mean λt.

The notion of a Poisson process is easily extended to points randomly dispersed in a plane, where the interval $(0, t)$ is replaced by a planar region (quadrat) of area A. Then, under analogous axioms, $Y(A)$—the number of points in a quadrat of area A—has a Poisson distribution with mean λA. Similarly, it can be applied to points randomly dispersed in space when the interval $(0, t)$ is replaced by a cube, a sphere, or some other three-dimensional figure in volume V.

E X A M P L E   **8.1**   Suppose that in a certain area plants are randomly dispersed, with a mean density of 20 per square yard. If a biologist randomly locates 100 sampling quadrats of 2 square yards each in the area, how many of them can be expected to contain no plants?

Solution   Assuming that the plant counts per unit area have a Poisson distribution, with a mean of 20 per square yard, the probability of no plants in a 2-square-yard area is

$$P[Y(2) = 0] = e^{-2\lambda}$$
$$= e^{-2(20)}$$
$$= e^{-40}$$

If the 100 quadrats are randomly located, the expected number of quadrats containing zero plants is

$$100P[Y(2) = 0] = 100e^{-40} \quad \blacksquare$$

8.2   Birth and Death Processes: Biological Applications

In Section 8.1, we were interested in counting the number of occurrences of a single type of event, such as accidents, defects, or plants. Let us extend this idea to counts of two types of events, which we shall call *births* and *deaths*. For illustrative purposes, a *birth* may denote a literal birth in a population of organisms, and similarly for a *death*. More specifically, we shall think of a birth as a cell division that produces two new cells, and we shall think of a death as the removal of a cell from the system.

Birth and death processes are of use in modeling the dynamics of population growth. In human populations, such models may be used to influence decisions on housing programs, food management, natural resource management, and a host of related economic matters. In animal populations, birth and death models are used to predict the size of insect populations and to measure the effectiveness of nutrition or eradication programs.

Let λ denote the birth rate, and let θ denote the death rate of each cell in the population; assume that the probability of birth or death for an individual cell is independent of the size and age of the population. If $Y(t)$ denotes the size of the population at time t, we write

$$P[Y(t) = n] = P_n(t)$$

More specifically, assume that the probability of a birth in a small interval of time h, given that the population size at the start of the interval is n, has the form $n\lambda h + o(h)$. Similarly, the probability of a death is given by $n\theta h + o(h)$. The probability of more than one birth or death in h is on order $o(h)$. For an individual cell (the case $n = 1$), this says that the probability of division in a small interval of time is $\lambda h + o(h)$. A differential equation for $P_n(t)$ is developed as follows. If a population is of size n at time $(t + h)$, it must have been of size $(n - 1), n,$ or $(n + 1)$ at time t. Thus,

$$P_n(t + h) = \lambda(n - 1)h P_{n-1}(t) + [1 - n\lambda h - n\theta h]P_n(t) + \theta(n + 1)h P_{n+1}(t) + o(h)$$

or

$$\frac{1}{h}[P_n(t+h) - P_n(t)] =$$

$$\lambda(n-1)P_{n-1}(t) - n(\lambda+\theta)P_n(t) + \theta(n-1)P_{n+1}(t) + \frac{o(h)}{h}$$

On taking limits as $h \to 0$, we have

$$\frac{dP_n(t)}{dt} = \lambda(n-1)P_{n-1}(t) - n(\lambda+\theta)P_n(t) + \theta(n+1)P_{n+1}(t)$$

A general solution to this equation can be obtained with some difficulty, but let us instead look at some special cases. If $\theta = 0$, which implies that no deaths are taking place in the time of interest, we have a pure birth process exemplified by the differential equation

$$\frac{dP_n(t)}{dt} = \lambda(n-1)P_{n-1}(t) - n\lambda P_n(t)$$

A solution to this equation is

$$P_n(t) = \binom{n-1}{i-1} e^{-\lambda i t}(1 - e^{-\lambda t})^{n-i}$$

where i is the size of the population at time $t = 0$; that is, $P_i(0) = 1$. Notice that this solution is a negative binomial probability distribution. The pure birth process may be a reasonable model of a real population if the time interval is short or if deaths are neglected.

EXAMPLE 8.2   Find the expected size of a population, governed by a pure birth process, at time t if the size was i at time 0.

Solution   The mean of a negative binomial distribution written as

$$\binom{n-1}{r-1} p^r (1-p)^{n-r}$$

is given by r/p. Thus, the mean of the pure birth process is $i/e^{-\lambda t}$ or $ie^{\lambda t}$. If the birth rate is known (at least approximately), this provides a formula for estimating what the expected population size will be t time units in the future.   ∎

Returning to the case where $\theta > 0$, we can show that

$$P_0(t) = \left[\frac{\theta e^{(\lambda-\theta)t} - \theta}{\lambda e^{(\lambda-\theta)t} - \theta}\right]^t$$

(see Baily 1964).

On taking the limit of $P_0(t)$ as $t \to \infty$, we get the probability of ultimate extinction of the population, which turns out to be

$$\lim_{t \to \infty} P_0(t) = \begin{cases} (\theta/\lambda)^t & \text{for } \lambda > \theta \\ 1 & \text{for } \lambda < \theta \\ 1 & \text{for } \lambda = \theta \end{cases}$$

Thus, the population has a chance of persisting indefinitely only if the birth rate λ is larger than the death rate θ; otherwise, it is certain to become extinct. We can find $E[Y(t)]$ for the birth and death process without first finding the probability distribution $P_n(t)$. The method is outlined in Exercise 8.2, and the solution is

$$E[Y(t)] = ie^{(\lambda - \theta)t}$$

8.3 Queues: Engineering Applications

Queuing theory is concerned with probabilistic models that predict the behavior of customers arriving at a certain station and demanding some kind of service. The word *customer* may refer to actual customers at a service counter; or more generally, it may refer to automobiles entering a highway or a service station, telephone calls coming into a switchboard, breakdowns of machines in a factory, or comparable phenomena. Queues are classified according to an input distribution (the distribution of customer arrivals), a service distribution (the distribution of service time per customer), and a queue discipline, which specifies the number of servers and the manner of dispensing service (such as "first come, first served").

Queuing theory forms useful probabilistic models for a wide range of practical problems, including the design of stores and public buildings, the optimal arrangement of workers in a factory, and the planning of highway systems. Consider a system that involves one station dispensing service to customers on a first-come, first-served basis. This could be a store with a single checkout counter or a service station with a single gasoline pump. Suppose that customer arrivals constitute a Poisson process with intensity λ per hour, and that departures of customers from the station constitute an independent Poisson process with intensity θ per hour. (This implies that the service time is an exponential random variable with mean $1/\theta$.) Hence, the probability of a customer arrival in a small interval of time h is $\lambda h + o(h)$, and the probability of a departure is $\theta h + o(h)$, with a probability of more than one arrival (or departure) of $o(h)$. Notice that these probabilities are identical to the ones for births and deaths (discussed in the preceding section), but with the dependence on n suppressed. Thus, if $Y(t)$ denotes the number of customers in the system (being served and waiting to be served) at time t, and if $P_n(t) = P[Y(t) = n]$, then, as in Section 8.2,

$$\frac{dP_n(t)}{dt} = \lambda P_{n-1}(t) - (\lambda + \theta) P_n(t) + \theta P_{n+1}(t)$$

It can be shown that, for a large value of t, this equation has a solution P_n that does not depend on t. Such a solution is called an *equilibrium distribution*. If the solution is free of t, it must satisfy

$$0 = \lambda P_{n-1} - (\lambda + \theta) P_n + \theta P_{n+1}$$

A solution to this equation is given by

$$P_n = \left(1 - \frac{\lambda}{\theta}\right)\left(\frac{\lambda}{\theta}\right)^n, \qquad n = 0, 1, 2, \ldots$$

provided that $\lambda < \theta$. P_n represents the probability of there being n customers in the system at any time t that is far removed from the start of the system.

EXAMPLE 8.3   In a system operating as has just been indicated, find the expected number of customers (including the one being served,) at the station at some time far removed from the start.

Solution   The equilibrium distribution is a version of the geometric distribution given in Chapter 3. If Y has the distribution given by P_n, then $X = Y + 1$ will have the distribution given by

$$P(X = m) = P(Y = m - 1) = \left(1 - \frac{\lambda}{\theta}\right)\left(\frac{\lambda}{\theta}\right)^{m-1}, \qquad m = 1, 2, \ldots$$

From Chapter 3, we know that

$$E(X) = 1 \left/ \left(1 - \frac{\lambda}{\theta}\right)\right. = \frac{\theta}{\theta - \lambda}$$

Hence,

$$\begin{aligned}
E(Y) &= E(X) - 1 \\
&= \frac{\theta}{\theta - \lambda} - 1 \\
&= \frac{\lambda}{\theta - \lambda}
\end{aligned}$$

again assuming that $\lambda < \theta$. ∎

8.4 Arrival Times for the Poisson Process

Let $Y(t)$ denote a Poisson process (see Section 8.1), with mean λ per unit time. It is sometimes of interest to study properties of the actual times at which events have occurred, given that $Y(t)$ is fixed. Suppose that $Y(t) = n$, and let $0 < U_{(1)} < U_{(2)} < \cdots < U_{(n)} < t$ be the actual times that events occur in the interval from 0 to t. Let the joint density function of $U_{(1)}, \ldots, U_{(n)}$ be denoted by $g(u_1, \ldots, u_n)$. If we think of du_i as being a very small interval, then $g(u_1, \ldots, u_n)du_1 \cdots du_n$ is equal to the probability that one event will occur in each of the intervals $(u_i, u_i + du_i)$ and none will occur elsewhere, given that n events occur in $(0, t)$. Thus,

$$g(u_i, \ldots, u_n)du_1 \cdots du_2 = \frac{1}{\dfrac{(\lambda t)^n e^{-\lambda t}}{n!}}[\lambda du_1 e^{-\lambda du_1} \cdots \lambda du_n e^{-\lambda du_n} e^{-\lambda(t-du_1-\cdots-du_n)}]$$

$$= \frac{n!}{t^n}du_1 \cdots du_n$$

It follows that

$$g(u_i, \ldots, u_n) = \frac{n!}{t^n} \qquad u_1 < u_2 < \cdots < u_n$$

or in other words, $U_{(1)}, \ldots, U_{(n)}$ behave as an ordered set of n independent observations from the uniform distribution on $(0, t)$. This implies that the unordered occurrence times $U_1, \ldots, U_n$ are independent uniform random variables.

E X A M P L E   **8.4**   Suppose that telephone calls coming into a switchboard follow a Poisson process with a mean of ten calls per minute. A particularly slow period of 2 minutes' duration yielded only four calls. Find the probability that all four calls came in the first minute.

Solution   Here we are given that $Y(2) = 4$. We can then evaluate the density function of $U_{(4)}$ to be

$$g_4(u) = \left(\frac{4}{2^4}\right)u^3, \qquad 0 \le u \le 2$$

This comes from the fact that $U_{(4)}$ is the largest-order statistic in a sample of $n = 4$ observations from a uniform distribution. Thus, the probability in question is

$$P[U_{(4)} \le 1] = \int_0^1 g_4(u)du$$

$$= \frac{4}{2^4}\int_0^1 u^3 du$$

$$= \frac{1}{2^4}$$

$$= \frac{1}{16} \qquad \blacksquare$$

8.5 Infinite Server Queue

We now consider a queue with an infinite number of servers. This queue could model a telephone system with a very large number of channels or the claims department of an insurance company in which all claims are processed immediately. Customers arrive at random times and keep a server busy for a random length of time. The number of servers is large enough, however, that a customer never has to wait for service to begin. A random variable of much interest in this system is $X(t)$, the number of servers busy at time t.

Let the customer arrivals be a Poisson process with a mean of λ per unit time. The customer arriving at time U_n, measured from the start of the system at $t = 0$, keeps a server busy for a random service time Y_n. The service times $Y_1, Y_2, \ldots$ are assumed to be independent and identically distributed, with distribution function $F(y)$. $X(t)$, the number of servers busy at time t, can then be written as

$$X(t) = \sum_{n=1}^{N(t)} w(t, U_n, Y_n)$$

where $N(t)$, the number of arrivals in $(0, t)$, has a Poisson distribution and

$$w(t, U_n, Y_n) = \begin{cases} 1 & \text{if } 0 < U_n \leq t \leq U_n + Y_n \\ 0 & \text{otherwise} \end{cases}$$

The condition $U_n \leq t \leq U_n + Y_n$ is precisely the condition that a customer arriving at time U_n is still being served at time t. The function $w(t, U_n, Y_n)$ simply keeps track of those customers still receiving service at t. Operations similar to those used in Section 6.7 for the compound Poisson distribution will reveal that $X(t)$ has a Poisson distribution with a mean given by

$$E[X(t)] = E[N(t)]E[w(t, U_n, Y_n)]$$

Because $E[N(t)] = \lambda t$, it remains for us to evaluate

$$E[w(t, U_n, Y_n)] = P(0 \leq U_n \leq t \leq U_n + Y_n) = P(Y_n \geq t - U_n)$$

Recall that U_n is uniformly distributed on $(0, t)$; and thus,

$$P(Y_n \geq t - U_n) = \int_0^t P(Y_n \geq t - u | U_n = u) \frac{du}{t}$$
$$= \frac{1}{t} \int_0^t P(Y_n \geq t - u) du$$
$$= \frac{1}{t} \int_0^t [1 - F(t - u)] du$$

Making the change of variable $t - u = s$, we have

$$P(Y_n \geq t - U_n) = \frac{1}{t} \int_0^t [1 - F(s)] ds$$

or

$$E[X(t)] = \lambda \int_0^t [1 - F(s)] ds$$

E X A M P L E   **8.5**   Suppose that the telephone switchboard of Example 8.4 (which averaged $\lambda = 10$ incoming calls per minute) has a large number of channels. The calls are of random and independent length, but they average 4 minutes each. Approximate the probability that, after the switchboard has been in service for a long time, there will be no busy channels at some specified time t_0.

Solution   Notice that $\int_0^\infty [1 - F(s)]ds = E(Y)$. Thus, for a large value of t_0,

$$E[X(t_0)] = \lambda \int_0^{t_0} [1 - F(s)]ds$$
$$= \lambda E(Y)$$
$$= 10(4)$$
$$= 40$$

Since $X(t)$ has a Poisson distribution,

$$P[X(t_0) = 0] = \exp\left\{-\lambda \int_0^{t_0} [1 - F(s)]ds\right\}$$
$$= \exp(-40) \quad \blacksquare$$

8.6   Renewal Theory: Reliability Applications

Reliability theory deals with probabilistic models that predict the behavior of systems of components, enabling one to formulate optimum maintenance policies, including replacement and repair of components, and to measure operating characteristics of the system, such as the proportion of downtime, the expected life, and the probability of survival beyond a specified age. The "system" could be a complex electronic system (such as a computer), an automobile, a factory, or even a human body. In the remainder of this section, this system will be simplified to a single component that has a random life length but may be replaced before it fails—such as a light bulb in a socket.

For a Poisson process, as described in Section 8.1, the time between any two successive occurrences of the event that is being observed has an exponential distribution, with mean $1/\lambda$, where λ is the expected number of occurrences per unit time (see Section 4.4). A more general stochastic process can be obtained by treating interarrival times as nonnegative random variables, identically distributed but not necessarily exponential.

Suppose that $X_1, X_2, \ldots, X_n$ represent a sequence of independent, identically distributed random variables, each with distribution function $F(x)$ and probability density function $f(x)$. Assume, too, that $E(X_i) = \mu$ and $V(X_i) = \sigma^2 < \infty$. The values of X_i could represent lifetimes of identically constructed electronic components, for example. If a system operates by inserting a new component, letting it burn continuously until it fails, and then repeating the process with an identical new component that is assumed to operate independently, then the occurrences of interest are in-service failures of components; and X_i is the interarrival time between failure number $(i - 1)$ and failure number i. A random variable of interest in such problems is N_t, the number of in-service failures in the time interval $(0, t)$. Notice that

$$N_t = \text{Maximum integer } k, \text{ such that } \sum_{i=1}^{k} X_i \leq t$$

with $N_t = 0$ if $X_1 > t$. Let us now consider the probability distribution of N_t. If X_i is an exponential random variable, then N_t has a Poisson distribution. If X_i is not exponential, however, the distribution of N_t may be difficult or impossible to obtain. Thus we shall investigate the asymptotic behavior of N_t as t tends toward infinity.

Notice that, with $S_r = \sum_{i=1}^{r} X_i$,

$$P(N_t \geq r) = P(S_r \leq t)$$

The limiting probabilities will be unity unless $r \rightarrow \infty$, and some restrictions must be placed on the relationship between r and t. Thus, suppose that $r \rightarrow \infty$ and $t \rightarrow \infty$ in such a way that

$$\frac{t - r\mu}{\sqrt{r\sigma}} \rightarrow c$$

for some constant c. This implies that $(\mu r / t) \rightarrow 1$ as $r \rightarrow \infty$ and $t \rightarrow \infty$. Now the preceding probability equality can be written as

$$P\left[\frac{N_t - (t/\mu)}{t^{1/2} \mu^{-3/2} \sigma} \geq \frac{r - (t/\mu)}{t^{1/2} \mu^{-3/2} \sigma} \right] = P\left(\frac{S_r - r\mu}{r^{1/2} \sigma} \leq \frac{t - r\mu}{r^{1/2} \sigma} \right)$$

We have

$$\frac{r - (t/\mu)}{t^{1/2} \mu^{-3/2} \sigma} = \frac{\mu r - t}{r^{1/2} \sigma} \left(\frac{\mu r}{t} \right)^{1/2}$$

and this quantity tends toward $-c$ as $t \rightarrow \infty$ and $r \rightarrow \infty$. It follows that

$$P\left[\frac{N_t - (t/\mu)}{t^{1/2} \mu^{-3/2} \sigma} \geq -c \right] = P\left(\frac{S_r - r\mu}{r^{1/2} \sigma} \leq c \right)$$

However,

$$\frac{S_r - r\mu}{r^{1/2} \sigma} = \sqrt{r} \left(\frac{\overline{X} - \mu}{\sigma} \right)$$

which has approximately a standard normal distribution for large values of r. Letting $\Phi(x)$ denote the standard normal distribution function, we can conclude that

$$P\left[\frac{N_t - (t/\mu)}{t^{1/2} \mu^{-3/2} \sigma} \geq -c \right] \rightarrow \Phi(c)$$

as $t \rightarrow \infty$ and $r \rightarrow \infty$, or

$$P\left[\frac{N_t - (t/\mu)}{t^{1/2} \mu_{-3/2} \sigma} \leq -c \right] \rightarrow [1 - \Phi(c)] = \Phi(-c)$$

Hence, for large t, the values of N_t can be regarded as approximately normally distributed, with a mean of t/μ and a variance of $t\sigma^2/\mu^3$.

EXAMPLE   **8.6**   A fuse in an electronic system is replaced with an identical fuse upon failure. It is known that these fuses have a mean life of 10 hours, with a variance of 2.5 hours. Beginning with a new fuse, the system is to operate for 400 hours.

1   If the cost of a fuse is $5, find the expected amount to be paid for fuses over the 400-hour period.

2   If 42 replacement fuses are in stock, find the probability that they will all be used in the 400-hour period.

Solution   **1**   No distribution of lifetimes for the fuses is given, so asymptotic results must be employed. (Notice that $t = 400$ is large in comparison to $\mu = 10$.) Letting N_t denote the number of replacements, we have

$$E(N_t) = \frac{t}{\mu} = \frac{400}{10} = 40$$

We would expect to use 40 replacements, plus the one initially placed into the system. Thus, the expected cost of fuses is

$$(5)(41) = \$205$$

2 N_t is approximately normal, with mean $t/\mu = 40$ and variance

$$\frac{t\sigma^2}{\mu^3} = \frac{400(2.5)}{1000} = 1$$

Thus,

$$P(N_t \geq 42) = P\left[\frac{N_t - (t/\mu)}{t^{1/2}\mu^{-3/2}\sigma} \geq \frac{42 - 4}{1}\right]$$
$$= P(Z \geq 2)$$
$$= 0.0228$$

where Z denotes a standard normal random variable.   ■

To minimize the number of in-service failures, components often are replaced at age T or at failure, whichever comes first. Here, *age* refers to length of time in service, and T is a constant. Under this age replacement scheme, the interarrival times are given by

$$Y_i = \min(X_i, T)$$

Consider an age replacement policy in which c_1 represents the cost of replacing a component that has failed in service and c_2 represents the cost of replacing a component (which has not failed) at age T. (Usually, $c_1 > c_2$.) To find the total replacement costs up to time t, we must count the total number of replacements N_t and the number of in-service failures. Let

$$W_i = \begin{cases} 1 & \text{if } X_i < T \\ 0 & \text{if } X_i \geq T \end{cases}$$

Then $\displaystyle\sum_{i=1}^{N_t} W_i$ denotes the total number of in-service failures. Thus, the total replacement cost up to time t is

$$C_t = c_1 \sum_{i=1}^{N_t} W_i + c_2 \left(N_t - \sum_{i=1}^{N_t} W_i \right)$$

The expected replacement cost per unit time is frequently of interest, and this can be approximated after observing that, under the conditions stated in this section,

$$E\left(\sum_{i=1}^{N_t} W_i \right) = E(N_t)E(W_i)$$

Thus,

$$E(C_t) = c_1 E(N_t)E(W_i) + c_2 E(N_t)[1 - E(W_i)]$$

which, for large t, becomes

$$\frac{t}{E(Y_i)} [c_1 P(X_i < T) + c_2 P(X_i \geq T)]$$

It follows that

$$\frac{1}{t} E(C_t) = \frac{1}{E(Y_i)} \{c_1 F(T) + c_2[1 - F(T)]\}$$

where $F(x)$ is still the distribution function of X_i.

An optimum replacement policy can be found by choosing T so as to minimize the expected cost per unit time. This concept is continued in Exercise 8.11.

8.7 Summary

The study of the behavior of random events that occur over time is called *stochastic processes*. A few examples of both the theory and the application of stochastic processes are presented here in order to round out the study of important areas of probability.

Perhaps the most common model for random events occurring over time is the Poisson process—an extension of the Poisson distribution introduced in Chapter 3. The Poisson process is generally considered the standard model for events that occur in a purely random fashion across time.

If two types of items—say, births and deaths—are being counted across time, a reasonable probabilistic model is a little more difficult to develop. Many applications of models for birth and death processes have been developed in the biological sciences. One interesting result is that the pure birth process leads to the negative binomial model for population size.

The number of people in a waiting line, or queue, and the way that number changes with time is another interesting stochastic process. It can be shown that, under equilibrium conditions, the number of people in a queue follows a geometric distribution. In an infinite-server queue, there is no waiting time for customers and the problem of interest switches to the number of servers who are busy at any given moment.

Other details related to stochastic processes include the actual arrival times of the events under study and their interarrival times. For the Poisson process, the arrival times within a fixed time interval are shown to be related to the order statistics from a uniform distribution. Renewal theory provides a useful normal approximation to the distribution of the number of events that occur during a fixed interval of time.

Exercises

8.1 During the workday, telephone calls come into an office at the rate of one call every 3 minutes.

a Find the probability that no more than one call will come into this office during the next 5 minutes. (Are the assumptions for the Poisson process reasonable here?)

b An observer in the office hears two calls come in during the first 2 minutes of a 5-minute visit. Find the probability that no more calls will arrive during the visit.

8.2 For the birth and death processes of Section 8.2, $E[Y(t)] = ie^{(\lambda-\theta)t}$, where i is the population size at $t = 0$. Show this by observing that

$$m(t) = E[Y(t)] = \sum_{n=1}^{\infty} n P_n(t)$$

and

$$m'(t) = \frac{dm(t)}{dt} = \sum_{n=0}^{\infty} n P_n'(t)$$

Now, use the expression for $P_n'(t)$ given in Section 8.2, and evaluate the sum to obtain a differential equation relating $m'(t)$ to $m(t)$. The solution follows.

8.3 Referring to Section 8.3, show that the geometric equilibrium distribution P_n is a solution to the differential equation defining the equilibrium state.

8.4 In a single-server queue, as defined in Section 8.3, let W denote the total waiting time, including her own service time, of a customer entering the queue a long time after the start of operations. (Assume that $\lambda < \theta$.) Show that W has an exponential distribution by writing

$$P[W \le w] = \sum_{n=0}^{\infty} P[W \le w|n \text{ customers in queue}] P_n$$

$$= \sum_{n=0}^{\infty} P[W \leq w | n \text{ customers in queue}] \left(\frac{\lambda}{\theta}\right)^n \left(1 - \frac{\lambda}{\theta}\right)$$

and observing that the conditional distribution of W for fixed n is a gamma distribution with $\alpha = n + 1$ and $\beta = 1/\theta$. (The interchange of $\sum$ and ∞ is permitted here.)

8.5   If $Y(t)$ and $X(t)$ are independent Poisson processes, show that $Y(t) + X(t)$ is also a Poisson process.

8.6   Claims arrive at an insurance company according to a Poisson process, with a mean of 20 per day. The claims are placed in the processing system immediately upon arrival, but the service time per claim is an exponential random variable with a mean of 10 days. On a certain day, the company was completely caught up on its claims service (no claims were left to be processed). Find the probability that, 5 days later, the company was engaged in servicing at least two claims.

8.7   Assume that telephone calls arrive at a switchboard according to a Poisson process, with an average rate of λ per minute. One time period t minutes long is known to have included exactly two calls. Find the probability that the length of time between these calls exceeds d minutes, for some constant $d \leq t$.

8.8   Let $Y(A)$ denote a Poisson process in a plane, as given in Section 8.1; and let R denote the distance from a random point to the nearest point of realization of the process.

a   Show that $U = R^2$ is an exponential random variable with mean $(1/\lambda\pi)$.

b   Suppose that n independent values of R can be obtained by sampling n points. Let W denote the sum of the squares of these values. Show that W has a gamma distribution.

c   Show that $(n - 1)/\pi W$ is an unbiased estimator of λ; that is, $E[(n - 1)/\pi W] = \lambda$.

8.9   Let V denote the volume of a three-dimensional figure, such as a sphere. The axioms of the Poisson process as applied to three-dimensional point processes—such as stars in the heavens or bacteria in water—yield the fact that $Y(V)$, the number of points in a figure of volume V, has a Poisson distribution with mean λV. Moreover, $Y(V_1)$ and $Y(V_2)$ are independent if V_1 and V_2 are not overlapping volumes.

a   If a point is chosen at random in three-dimensional space, show that the distance R to the nearest point in the realization of a Poisson process has the density function

$$f(r) = \begin{cases} 4\lambda\pi r^2 e^{-(4/3)\lambda\pi r^3} & \text{for } r \geq 0 \\ 0 & \text{elsewhere} \end{cases}$$

b   If R is as in part (a), show that $U = R^3$ has an exponential distribution.

8.10   Let Y denote a random variable with distribution function $F(y)$, such that $F(0) = 0$. If $X = \min(Y, T)$ for a constant T, show that

$$E(X) = \int_0^T [1 - F(x)] dx$$

8.11   Referring to Section 8.6, the optimum age replacement interval T_0 is defined as the one that minimizes the expected cost per unit time, as given in Section 8.6.

a   Find T_0, assuming that the components have exponential life lengths with mean θ.

b   Show that a finite T_0 always exists if $r(t)$ is strictly increasing toward infinity. (Notice that this will be the case for Weibull life lengths if $m > 1$.)

Appendix Tables

T A B L E 1   Random numbers

Column Row	1	2	3	4	5	6	7	8	9	10	11	12	13	14
1	10480	15011	01536	02011	81647	91646	69179	14194	62590	36207	20969	99570	91291	90700
2	22368	46573	25595	85393	30995	89198	27982	53402	93965	34095	52666	19174	39615	99505
3	24130	48360	22527	97265	76393	64809	15179	24830	49340	32081	30680	19655	63348	58629
4	42167	93093	06243	61680	07856	16376	39440	53537	71341	57004	00849	74917	97758	16379
5	37570	39975	81837	16656	06121	91782	60468	81305	49684	60672	14110	06927	01263	54613
6	77921	06907	11008	42751	27756	53498	18602	70659	90655	15053	21916	81825	44394	42880
7	99562	72905	56420	69994	98872	31016	71194	18738	44013	48840	63213	21069	10634	12952
8	96301	91977	05463	07972	18876	20922	94595	56869	69014	60045	18425	84903	42508	32307
9	89579	14342	63661	10281	17453	18103	57740	84378	25331	12566	58678	44947	05585	56941
10	85475	36857	53342	53988	53060	59533	38867	62300	08158	17983	16439	11458	18593	64952
11	28918	69578	88231	33276	70997	79936	56865	05859	90106	31595	01547	85590	91610	78188
12	63553	40961	48235	03427	49626	69445	18663	72695	52180	20847	12234	90511	33703	90322
13	09429	93969	52636	92737	88974	33488	36320	17617	30015	08272	84115	27156	30613	74952
14	10365	61129	87529	85689	48237	52267	67689	93394	01511	26358	85104	20285	29975	89868
15	07119	97336	71048	08178	77233	13916	47564	81056	97735	85977	29372	74461	28551	90707
16	51085	12765	51821	51259	77452	16308	60756	92144	49442	53900	70960	63990	75601	40719
17	02368	21382	52404	60268	89368	19885	55322	44819	01188	65255	64835	44919	05944	55157
18	01011	54092	33362	94904	31273	04146	18594	29852	71585	85030	51132	01915	92747	64951
19	52162	53916	46369	58586	23216	14513	83149	98736	23495	64350	94738	17752	35156	35749
20	07056	97628	33787	09998	42698	06691	76988	13602	51851	46104	88916	19509	25625	58104
21	48663	91245	85828	14346	09172	30168	90229	04734	59193	22178	30421	61666	99904	32812
22	54164	58492	22421	74103	47070	25306	76468	26384	58151	06646	21524	15227	96909	44592
23	32639	32363	05597	24200	13363	38005	94342	28728	35806	06912	17012	64161	18296	22851
24	29334	27001	87637	87308	58731	00256	45834	15398	46557	41135	10367	07684	36188	18510
25	02488	33062	28834	07351	19731	92420	60952	61280	50001	67658	32586	86679	50720	94953
26	81525	72295	04839	96423	24878	82651	66566	14778	76797	14780	13300	87074	79666	95725
27	29676	20591	68086	26432	46901	20849	89768	81536	86645	12659	92259	57102	80428	25280
28	00742	57392	39064	66432	84673	40027	32832	61362	98947	96067	64760	64584	96096	98253
29	05366	04213	25669	26422	44407	44048	37937	63904	45766	66134	75470	66520	34693	90449
30	91921	26418	64117	94305	26766	25940	39972	22209	71500	64568	91402	42416	07884	69618
31	00582	04711	87917	77341	42206	35126	74087	99547	81817	42607	43808	76655	62028	76630
32	00725	69884	62797	56170	86324	88072	76222	36086	84637	93161	76038	65855	77919	88006
33	69011	65795	95876	55293	18988	27354	26575	08625	40801	59920	29841	80150	12777	48501

T A B L E **1**   (continued)

Column Row	1	2	3	4	5	6	7	8	9	10	11	12	13	14
34	25976	57948	29888	88604	67917	48708	18912	82271	65424	69774	33611	54262	85963	03547
35	09763	83473	73577	12908	30883	18317	28290	35797	05998	41688	34952	37888	38917	88050
36	91576	42595	27958	30134	04024	86385	29880	99730	55536	84855	29080	09250	79656	73211
37	17955	56349	90999	49127	20044	59931	06115	20542	18059	02008	73708	83517	36103	42791
38	46503	18584	18845	49618	02304	51038	20655	58727	28168	15475	56942	53389	20562	87338
39	92157	89634	94824	78171	84610	82834	09922	25417	44137	48413	25555	21246	35509	20468
40	14577	62765	35605	81263	39667	47358	56873	56307	61607	49518	89656	20103	77490	18062
41	98427	07523	33362	64270	01638	92477	66969	98420	04880	45585	46565	04102	46880	45709
42	34914	63976	88720	82765	34476	17032	87589	40836	32427	70002	70663	88863	77775	69348
43	70060	28277	39475	46473	23219	53416	94970	25832	69975	94884	19661	72828	00102	66794
44	53976	54914	06990	67245	68350	82948	11398	42878	80287	88267	47363	46634	06541	97809
45	76072	29515	40980	07391	58745	25774	22987	80059	39911	96189	41151	14222	60697	59583
46	90725	52210	83974	29992	65831	38857	50490	83765	55657	14361	31720	57375	56228	41546
47	64364	67412	33339	31926	14883	24413	59744	92351	97473	89286	35931	04110	23726	51900
48	08962	00358	31662	25388	61642	34072	81249	35648	56891	69352	48373	45578	78547	81788
49	95012	68379	93526	70765	10592	04542	76463	54328	02349	17247	28865	14777	62730	92277
50	15664	10493	20492	38391	91132	21999	59516	81652	27195	48223	46751	22923	32261	85653
51	16408	81899	04153	53381	79401	21438	83035	92350	36693	31238	59649	91754	72772	02338
52	18629	81953	05520	91962	04739	13092	97662	24822	94730	06496	35090	04822	86774	98289
53	73115	35101	47498	87637	99016	71060	88824	71013	18735	20286	23153	72924	35165	43040
54	57491	16703	23167	49323	45021	33132	12544	41035	80780	45393	44812	12515	98931	91202
55	30405	83946	23792	14422	15059	45799	22716	19792	09983	74353	68668	30429	70735	25499
56	16631	35006	85900	98275	32388	52390	16815	69298	82732	38480	73817	32523	41961	44437
57	96773	20206	42559	78985	05300	22164	24369	54224	35083	19687	11052	91491	60383	19746
58	38935	64202	14349	82674	66523	44133	00697	35552	35970	19124	63318	29686	03387	59846
59	31624	76384	17403	53363	44167	64486	64758	75366	76554	31601	12614	33072	60332	92325
60	78919	19474	23632	27889	47914	02584	37680	20801	72152	39339	34806	08930	85001	87820
61	03931	33309	57047	74211	63445	17361	62825	39908	05607	91284	68833	25570	38818	46920
62	74426	33278	43972	10119	89917	15665	52872	73823	73144	88662	88970	74492	51805	99378
63	09066	00903	20795	95452	92648	45454	09552	88815	16553	51125	79375	97596	16296	66092
64	42238	12426	87025	14267	20979	04508	64535	31355	86064	29472	47689	05974	52468	16834
65	16153	08002	26504	41744	81959	65642	74240	56302	00033	67107	77510	70625	28725	34191
66	21457	40742	29820	96783	29400	21840	15035	34527	33310	06116	95240	15957	16572	06004
67	21581	57802	02050	89728	17937	37621	47075	42080	97403	48626	68995	43805	33386	21597
68	55612	78095	83197	33732	05810	24813	86902	60397	16489	03264	88525	42786	05269	92532
69	44657	66999	99324	51281	84463	60563	79312	93454	68876	25471	93911	25650	12682	73572
70	91340	84979	46949	81973	37949	61023	43997	15263	80644	43942	89203	71795	99533	50501
71	91227	21199	31935	27022	84067	05462	35216	14486	29891	68607	41867	14951	91696	85065
72	50001	38140	66321	19924	72163	09538	12151	06878	91903	18749	34405	56087	82790	70925
73	65390	05224	72958	28609	81406	39147	25549	48542	42627	45233	57202	94617	23772	07896
74	27504	96131	83944	41575	10573	08619	64482	73923	36152	05184	94142	25299	84387	34925
75	37169	94851	39117	89632	00959	16487	65536	49071	39782	17095	02330	74301	00275	48280
76	11508	70225	51111	38351	19444	66499	71945	05422	13442	78675	84081	66938	93654	59894
77	37449	30362	06694	54690	04052	53115	62757	95348	78662	11163	81651	50245	34971	52924
78	46515	70331	85922	38329	57015	15765	97161	17869	45349	61796	66345	81073	49106	79860

T A B L E **2**   Binomial probabilities

Tabulated values are $\sum_{x=0}^{k} p(x)$. (Computations are rounded at the third decimal place.)

(a) $n = 5$

k	p .01	.05	.10	.20	.30	.40	.50	.60	.70	.80	.90	.95	.99
0	.951	.774	.590	.328	.168	.078	.031	.010	.002	.000	.000	.000	.000
1	.999	.977	.919	.737	.528	.337	.188	.087	.031	.007	.000	.000	.000
2	1.000	.999	.991	.942	.837	.683	.500	.317	.163	.058	.009	.001	.000
3	1.000	1.000	1.000	.993	.969	.913	.812	.663	.472	.263	.081	.023	.001
4	1.000	1.000	1.000	1.000	.998	.990	.969	.922	.832	.672	.410	.226	.049

(b) $n = 10$

k	p .01	.05	.10	.20	.30	.40	.50	.60	.70	.80	.90	.95	.99
0	.904	.599	.349	.107	.028	.006	.001	.000	.000	.000	.000	.000	.000
1	.996	.914	.736	.376	.149	.046	.011	.002	.000	.000	.000	.000	.000
2	1.000	.988	.930	.678	.383	.167	.055	.012	.002	.000	.000	.000	.000
3	1.000	.999	.987	.879	.650	.382	.172	.055	.011	.001	.000	.000	.000
4	1.000	1.000	.998	.967	.850	.633	.377	.166	.047	.006	.000	.000	.000
5	1.000	1.000	1.000	.994	.953	.834	.623	.367	.150	.033	.002	.000	.000
6	1.000	1.000	1.000	.999	.989	.945	.828	.618	.350	.121	.013	.001	.000
7	1.000	1.000	1.000	1.000	.998	.988	.945	.833	.617	.322	.070	.012	.000
8	1.000	1.000	1.000	1.000	1.000	.998	.989	.954	.851	.624	.264	.086	.004
9	1.000	1.000	1.000	1.000	1.000	1.000	.999	.994	.972	.893	.651	.401	.096

T A B L E **2**   (continued)

(c) $n = 15$

k	p .01	.05	.10	.20	.30	.40	.50	.60	.70	.80	.90	.95	.99
0	.860	.463	.206	.035	.005	.000	.000	.000	.000	.000	.000	.000	.000
1	.990	.829	.549	.167	.035	.005	.000	.000	.000	.000	.000	.000	.000
2	1.000	.964	.816	.398	.127	.027	.004	.000	.000	.000	.000	.000	.000
3	1.000	.995	.944	.648	.297	.091	.018	.002	.000	.000	.000	.000	.000
4	1.000	.999	.987	.836	.515	.217	.059	.009	.001	.000	.000	.000	.000
5	1.000	1.000	.998	.939	.722	.403	.151	.034	.004	.000	.000	.000	.000
6	1.000	1.000	1.000	.982	.869	.610	.304	.095	.015	.001	.000	.000	.000
7	1.000	1.000	1.000	.996	.950	.787	.500	.213	.050	.004	.000	.000	.000
8	1.000	1.000	1.000	.999	.985	.905	.696	.390	.131	.018	.000	.000	.000
9	1.000	1.000	1.000	1.000	.996	.966	.849	.597	.278	.061	.002	.000	.000
10	1.000	1.000	1.000	1.000	.999	.991	.941	.783	.485	.164	.013	.001	.000
11	1.000	1.000	1.000	1.000	1.000	.998	.982	.909	.703	.352	.056	.005	.000
12	1.000	1.000	1.000	1.000	1.000	1.000	.996	.973	.873	.602	.184	.036	.000
13	1.000	1.000	1.000	1.000	1.000	1.000	1.000	.995	.965	.833	.451	.171	.010
14	1.000	1.000	1.000	1.000	1.000	1.000	1.000	1.000	.995	.965	.794	.537	.140

(d) $n = 20$

k	p .01	.05	.10	.20	.30	.40	.50	.60	.70	.80	.90	.95	.99
0	.818	.358	.122	.002	.001	.000	.000	.000	.000	.000	.000	.000	.000
1	.983	.736	.392	.069	.008	.001	.000	.000	.000	.000	.000	.000	.000
2	.999	.925	.677	.206	.035	.004	.000	.000	.000	.000	.000	.000	.000
3	1.000	.984	.867	.411	.107	.016	.001	.000	.000	.000	.000	.000	.000
4	1.000	.997	.957	.630	.238	.051	.006	.000	.000	.000	.000	.000	.000
5	1.000	1.000	.989	.804	.416	.126	.021	.002	.000	.000	.000	.000	.000
6	1.000	1.000	.998	.913	.608	.250	.058	.006	.000	.000	.000	.000	.000
7	1.000	1.000	1.000	.968	.772	.416	.132	.021	.001	.000	.000	.000	.000
8	1.000	1.000	1.000	.990	.887	.596	.252	.057	.005	.000	.000	.000	.000
9	1.000	1.000	1.000	.997	.952	.755	.412	.128	.017	.001	.000	.000	.000
10	1.000	1.000	1.000	.999	.983	.872	.588	.245	.048	.003	.000	.000	.000
11	1.000	1.000	1.000	1.000	.995	.943	.748	.404	.113	.010	.000	.000	.000
12	1.000	1.000	1.000	1.000	.999	.979	.868	.584	.228	.032	.000	.000	.000
13	1.000	1.000	1.000	1.000	1.000	.994	.942	.750	.392	.087	.002	.000	.000
14	1.000	1.000	1.000	1.000	1.000	.998	.979	.874	.584	.196	.011	.000	.000
15	1.000	1.000	1.000	1.000	1.000	1.000	.994	.949	.762	.370	.043	.003	.000
16	1.000	1.000	1.000	1.000	1.000	1.000	.999	.984	.893	.589	.133	.016	.000
17	1.000	1.000	1.000	1.000	1.000	1.000	1.000	.996	.965	.794	.323	.075	.001
18	1.000	1.000	1.000	1.000	1.000	1.000	1.000	.999	.992	.931	.608	.264	.017
19	1.000	1.000	1.000	1.000	1.000	1.000	1.000	1.000	.999	.988	.878	.642	.182

T A B L E **2**   (continued)

(e) $n = 25$

k	p .01	.05	.10	.20	.30	.40	.50	.60	.70	.80	.90	.95	.99
0	.778	.277	.072	.004	.000	.000	.000	.000	.000	.000	.000	.000	.000
1	.974	.642	.271	.027	.002	.000	.000	.000	.000	.000	.000	.000	.000
2	.998	.873	.537	.098	.009	.000	.000	.000	.000	.000	.000	.000	.000
3	1.000	.966	.764	.234	.033	.002	.000	.000	.000	.000	.000	.000	.000
4	1.000	.993	.902	.421	.090	.009	.000	.000	.000	.000	.000	.000	.000
5	1.000	.999	.967	.617	.193	.029	.002	.000	.000	.000	.000	.000	.000
6	1.000	1.000	.991	.780	.341	.074	.007	.000	.000	.000	.000	.000	.000
7	1.000	1.000	.998	.891	.512	.154	.022	.001	.000	.000	.000	.000	.000
8	1.000	1.000	1.000	.953	.677	.274	.054	.004	.000	.000	.000	.000	.000
9	1.000	1.000	1.000	.983	.811	.425	.115	.013	.000	.000	.000	.000	.000
10	1.000	1.000	1.000	.994	.902	.586	.212	.034	.002	.000	.000	.000	.000
11	1.000	1.000	1.000	.998	.956	.732	.345	.078	.006	.000	.000	.000	.000
12	1.000	1.000	1.000	1.000	.983	.846	.500	.154	.017	.000	.000	.000	.000
13	1.000	1.000	1.000	1.000	.994	.922	.655	.268	.044	.002	.000	.000	.000
14	1.000	1.000	1.000	1.000	.998	.966	.788	.414	.098	.006	.000	.000	.000
15	1.000	1.000	1.000	1.000	1.000	.987	.885	.575	.189	.017	.000	.000	.000
16	1.000	1.000	1.000	1.000	1.000	.996	.946	.726	.323	.047	.000	.000	.000
17	1.000	1.000	1.000	1.000	1.000	.999	.978	.846	.488	.109	.002	.000	.000
18	1.000	1.000	1.000	1.000	1.000	1.000	.993	.926	.659	.220	.009	.000	.000
19	1.000	1.000	1.000	1.000	1.000	1.000	.998	.971	.807	.383	.033	.001	.000
20	1.000	1.000	1.000	1.000	1.000	1.000	1.000	.991	.910	.579	.098	.007	.000
21	1.000	1.000	1.000	1.000	1.000	1.000	1.000	.998	.967	.766	.236	.034	.000
22	1.000	1.000	1.000	1.000	1.000	1.000	1.000	1.000	.991	.902	.463	.127	.002
23	1.000	1.000	1.000	1.000	1.000	1.000	1.000	1.000	.998	.973	.729	.358	.026
24	1.000	1.000	1.000	1.000	1.000	1.000	1.000	1.000	1.000	.996	.928	.723	.222

T A B L E **3**
Poisson distribution function

$$F(x, \lambda) = \sum_{k=0}^{x} e^{-\lambda} \frac{\lambda^k}{k!}$$

λ	x 0	1	2	3	4	5	6	7	8	9
.02	.980	1.000								
.04	.961	.999	1.000							
.06	.942	.998	1.000							
.08	.923	.997	1.000							
.10	.905	.995	1.000							
.15	.861	.990	.999	1.000						
.20	.819	.982	.999	1.000						
.25	.779	.974	.998	1.000						
.30	.741	.963	.996	1.000						
.35	.705	.951	.994	1.000						
.40	.670	.938	.992	.999	1.000					
.45	.638	.925	.989	.999	1.000					
.50	.607	.910	.986	.998	1.000					
.55	.577	.894	.982	.998	1.000					
.60	.549	.878	.977	.997	1.000					
.65	.522	.861	.972	.996	.999	1.000				
.70	.497	.844	.966	.994	.999	1.000				
.75	.472	.827	.959	.993	.999	1.000				
.80	.449	.809	.953	.991	.999	1.000				
.85	.427	.791	.945	.989	.998	1.000				
.90	.407	.772	.937	.987	.998	1.000				
.95	.387	.754	.929	.981	.997	1.000				
1.00	.368	.736	.920	.981	.996	.999	1.000			
1.1	.333	.699	.900	.974	.995	.999	1.000			
1.2	.301	.663	.879	.966	.992	.998	1.000			
1.3	.273	.627	.857	.957	.989	.998	1.000			
1.4	.247	.592	.833	.946	.986	.997	.999	1.000		
1.5	.223	.558	.809	.934	.981	.996	.999	1.000		
1.6	.202	.525	.783	.921	.976	.994	.999	1.000		
1.7	.183	.493	.757	.907	.970	.992	.998	1.000		
1.8	.165	.463	.731	.891	.964	.990	.997	.999	1.000	
1.9	.150	.434	.704	.875	.956	.987	.997	.999	1.000	
2.0	.135	.406	.677	.857	.947	.983	.995	.999	1.000	
2.2	.111	.355	.623	.819	.928	.975	.993	.998	1.000	
2.4	.091	.308	.570	.779	.904	.964	.988	.997	.999	1.000
2.6	.074	.267	.518	.736	.877	.951	.983	.995	.999	1.000
2.8	.061	.231	.469	.692	.848	.935	.976	.992	.998	.999
3.0	.050	.199	.423	.647	.815	.916	.966	.988	.996	.999

λ	x 0	1	2	3	4	5	6	7	8	9
3.2	.041	.171	.380	.603	.781	.895	.955	.983	.994	.998
3.4	.033	.147	.340	.558	.744	.871	.942	.977	.992	.997
3.6	.027	.126	.303	.515	.706	.844	.927	.969	.988	.996
3.8	.022	.107	.269	.473	.668	.816	.909	.960	.984	.994
4.0	.018	.092	.238	.433	.629	.785	.889	.949	.979	.992
4.2	.015	.078	.210	.395	.590	.753	.867	.936	.972	.989
4.4	.012	.066	.185	.359	.551	.720	.844	.921	.964	.985
4.6	.010	.056	.163	.326	.513	.686	.818	.905	.955	.980
4.8	.008	.048	.143	.294	.476	.651	.791	.887	.944	.975
5.0	.007	.040	.125	.265	.440	.616	.762	.867	.932	.968
5.2	.006	.034	.109	.238	.406	.581	.732	.845	.918	.960
5.4	.005	.029	.095	.213	.373	.546	.702	.822	.903	.951
5.6	.004	.024	.082	.191	.342	.512	.670	.797	.886	.941
5.8	.003	.021	.072	.170	.313	.478	.638	.771	.867	.929
6.0	.002	.017	.062	.151	.285	.446	.606	.744	.847	.916
6.2	.002	.015	.054	.134	.259	.414	.574	.716	.826	.902
6.4	.002	.012	.046	.119	.235	.384	.542	.687	.803	.886
6.6	.001	.010	.040	.105	.213	.355	.511	.658	.780	.869
6.8	.001	.009	.034	.093	.192	.327	.480	.628	.755	.850
7.0	.001	.007	.030	.082	.173	.301	.450	.599	.729	.830
7.2	.001	.006	.025	.072	.156	.276	.420	.569	.703	.810
7.4	.001	.005	.022	.063	.140	.253	.392	.539	.676	.788
7.6	.001	.004	.019	.055	.125	.231	.365	.510	.648	.765
7.8	.000	.004	.016	.048	.112	.210	.338	.481	.620	.741
8.0	.000	.003	.014	.042	.100	.191	.313	.453	.593	.717
8.5	.000	.002	.009	.030	.074	.150	.256	.386	.523	.653
9.0	.000	.001	.006	.021	.055	.116	.207	.324	.456	.587
9.5	.000	.001	.004	.015	.040	.089	.165	.269	.392	.522
10.0	.000	.000	.003	.010	.029	.067	.130	.220	.333	.458
10.5	.000	.000	.002	.007	.021	.050	.102	.179	.279	.397
11.0	.000	.000	.001	.005	.015	.038	.079	.143	.232	.341
11.5	.000	.000	.001	.003	.011	.028	.060	.114	.191	.289
12.0	.000	.000	.001	.002	.008	.020	.046	.090	.155	.242
12.5	.000	.000	.000	.002	.005	.015	.035	.070	.125	.201
13.0	.000	.000	.000	.001	.004	.011	.026	.054	.100	.166
13.5	.000	.000	.000	.001	.003	.008	.019	.041	.079	.135
14.0	.000	.000	.000	.000	.002	.006	.014	.032	.062	.109
14.5	.000	.000	.000	.000	.001	.004	.010	.024	.048	.088
15.0	.000	.000	.000	.000	.001	.003	.008	.018	.037	.070

T A B L E **3**
(continued)

λ	x 10	11	12	13	14	15	16	17	18	19
2.8	1.000									
3.0	1.000									
3.2	1.000									
3.4	.999	1.000								
3.6	.999	1.000								
3.8	.998	.999	1.000							
4.0	.997	.999	1.000							
4.2	.996	.999	1.000							
4.4	.994	.998	.999	1.000						
4.6	.992	.997	.999	1.000						
4.8	.990	.996	.999	1.000						
5.0	.986	.995	.998	.999	1.000					
5.2	.982	.993	.997	.999	1.000					
5.4	.977	.990	.996	.999	1.000					
5.6	.972	.988	.995	.998	.999	1.000				
5.8	.965	.984	.993	.997	.999	1.000				
6.0	.957	.980	.991	.996	.999	.999	1.000			
6.2	.949	.975	.989	.995	.998	.999	1.000			
6.4	.939	.969	.986	.994	.997	.999	1.000			
6.6	.927	.963	.982	.992	.997	.999	.999	1.000		
6.8	.915	.955	.978	.990	.996	.998	.999	1.000		
7.0	.901	.947	.973	.987	.994	.998	.999	1.000		
7.2	.887	.937	.967	.984	.993	.997	.999	.999	1.000	
7.4	.871	.926	.961	.980	.991	.996	.998	.999	1.000	
7.6	.854	.915	.954	.976	.989	.995	.998	.999	1.000	
7.8	.835	.902	.945	.971	.986	.993	.997	.999	1.000	
8.0	.816	.888	.936	.966	.983	.992	.996	.998	.999	1.000
8.5	.763	.849	.909	.949	.973	.986	.993	.997	.999	.999
9.0	.706	.803	.876	.926	.959	.978	.989	.995	.998	.999
9.5	.645	.752	.836	.898	.940	.967	.982	.991	.996	.998
10.0	.583	.697	.792	.864	.917	.951	.973	.986	.993	.997
10.5	.521	.639	.742	.825	.888	.932	.960	.978	.988	.994
11.0	.460	.579	.689	.781	.854	.907	.944	.968	.982	.991
11.5	.402	.520	.633	.733	.815	.878	.924	.954	.974	.986
12.0	.347	.462	.576	.682	.772	.844	.899	.937	.963	.979
12.5	.297	.406	.519	.628	.725	.806	.869	.916	.948	.969
13.0	.252	.353	.463	.573	.675	.764	.835	.890	.930	.957
13.5	.211	.304	.409	.518	.623	.718	.798	.861	.908	.942
14.0	.176	.260	.358	.464	.570	.669	.756	.827	.883	.923
14.5	.145	.220	.311	.413	.518	.619	.711	.790	.853	.901
15.0	.118	.185	.268	.363	.466	.568	.664	.749	.819	.875

T A B L E **3**   (continued)

λ	x 20	21	22	23	24	25	26	27	28	29
8.5	1.000									
9.0	1.000									
9.5	.999	1.000								
10.0	.998	.999	1.000							
10.5	.997	.999	.999	1.000						
11.0	.995	.998	.999	1.000						
11.5	.992	.996	.998	.999	1.000					
12.0	.988	.994	.997	.999	.999	1.000				
12.5	.983	.991	.995	.998	.999	.999	1.000			
13.0	.975	.986	.992	.996	.998	.999	1.000			
13.5	.965	.980	.989	.994	.997	.998	.999	1.000		
14.0	.952	.971	.983	.991	.995	.997	.999	.999	1.000	
14.5	.936	.960	.976	.986	.992	.996	.998	.999	.999	1.000
15.0	.917	.947	.967	.981	.989	.994	.997	.998	.999	1.000

λ	x 4	5	6	7	8	9	10	11	12	13
16	.000	.001	.004	.010	.022	.043	.077	.127	.193	.275
17	.000	.001	.002	.005	.013	.026	.049	.085	.135	.201
18	.000	.000	.001	.003	.007	.015	.030	.055	.092	.143
19	.000	.000	.001	.002	.004	.009	.018	.035	.061	.098
20	.000	.000	.000	.001	.002	.005	.011	.021	.039	.066
21	.000	.000	.000	.000	.001	.003	.006	.013	.025	.043
22	.000	.000	.000	.000	.001	.002	.004	.008	.015	.028
23	.000	.000	.000	.000	.000	.001	.002	.004	.009	.017
24	.000	.000	.000	.000	.000	.000	.001	.003	.005	.011
25	.000	.000	.000	.000	.000	.000	.001	.001	.003	.006

TABLE **3**
(continued)

λ	14	15	16	17	18	19	20	21	22	23
16	.368	.467	.566	.659	.742	.812	.868	.911	.942	.963
17	.281	.371	.468	.564	.655	.736	.805	.861	.905	.937
18	.208	.287	.375	.469	.562	.651	.731	.799	.855	.899
19	.150	.215	.292	.378	.469	.561	.647	.725	.793	.849
20	.105	.157	.221	.297	.381	.470	.559	.644	.721	.787
21	.072	.111	.163	.227	.302	.384	.471	.558	.640	.716
22	.048	.077	.117	.169	.232	.306	.387	.472	.556	.637
23	.031	.052	.082	.123	.175	.238	.310	.389	.472	.555
24	.020	.034	.056	.087	.128	.180	.243	.314	.392	.473
25	.012	.022	.038	.060	.092	.134	.185	.274	.318	.394

λ	24	25	26	27	28	29	30	31	32	33
16	.978	.987	.993	.996	.998	.999	.999	1.000		
17	.959	.975	.985	.991	.995	.997	.999	.999	1.000	
18	.932	.955	.972	.983	.990	.994	.997	.998	.999	1.000
19	.893	.927	.951	.969	.980	.988	.993	.996	.998	.999
20	.843	.888	.922	.948	.966	.978	.987	.992	.995	.997
21	.782	.838	.883	.917	.944	.963	.976	.985	.991	.994
22	.712	.777	.832	.877	.913	.940	.959	.973	.983	.989
23	.635	.708	.772	.827	.873	.908	.936	.956	.971	.981
24	.554	.632	.704	.768	.823	.868	.904	.932	.953	.969
25	.473	.553	.629	.700	.763	.818	.863	.900	.929	.950

λ	34	35	36	37	38	39	40	41	42	43
19	.999	1.000								
20	.999	.999	1.000							
21	.997	.998	.999	.999	1.000					
22	.994	.996	.998	.999	.999	1.000				
23	.988	.993	.996	.997	.999	.999	1.000			
24	.979	.987	.992	.995	.997	.998	.999	.999	1.000	
25	.966	.978	.985	.991	.995	.997	.998	.999	.999	1.000

Source: Reprinted by permission from E. C. Molina, *Poisson's Exponential Binomial Limit* (Princeton, N.J.: D. Van Nostrand Company, 1947).

TABLE **4**
Normal curve areas

z	.00	.01	.02	.03	.04	.05	.06	.07	.08	.09
.0	.0000	.0040	.0080	.0120	.0160	.0199	.0239	.0279	.0319	.0359
.1	.0398	.0438	.0478	.0517	.0557	.0596	.0636	.0675	.0174	.0753
.2	.0793	.0832	.0871	.0910	.0948	.0987	.1026	.1064	.1103	.1141
.3	.1179	.1217	.1255	.1293	.1331	.1368	.1406	.1443	.1480	.1517
.4	.1554	.1591	.1628	.1664	.1700	.1736	.1772	.1808	.1844	.1879
.5	.1915	.1950	.1985	.2019	.2054	.2088	.2123	.2157	.2190	.2224
.6	.2257	.2291	.2324	.2357	.2389	.2422	.2454	.2486	.2517	.2549
.7	.2580	.2611	.2642	.2673	.2704	.2734	.2764	.2794	.2823	.2852
.8	.2881	.2910	.2939	.2967	.2995	.3023	.3051	.3078	.3106	.3133
.9	.3159	.3186	.3212	.3238	.3264	.3289	.3315	.3340	.3365	.3389
1.0	.3413	.3438	.3461	.3485	.3508	.3531	.3554	.3577	.3599	.3621
1.1	.3643	.3665	.3686	.3708	.3729	.3749	.3770	.3790	.3810	.3830
1.2	.3849	.3869	.3888	.3907	.3925	.3944	.3962	.3980	.3997	.4015
1.3	.4032	.4049	.4066	.4082	.4099	.4115	.4131	.4147	.4162	.4177
1.4	.4192	.4207	.4222	.4236	.4251	.4265	.4279	.4292	.4306	.4319
1.5	.4332	.4345	.4357	.4370	.4382	.4394	.4406	.4418	.4429	.4441
1.6	.4452	.4463	.4474	.4484	.4495	.4505	.4515	.4525	.4535	.4545
1.7	.4554	.4564	.4573	.4582	.4591	.4599	.4608	.4616	.4625	.4633
1.8	.4641	.4649	.4656	.4664	.4671	.4678	.4686	.4693	.4699	.4706
1.9	.4713	.4719	.4726	.4732	.4738	.4744	.4750	.4756	.4761	.4767
2.0	.4772	.4778	.4783	.4788	.4793	.4798	.4803	.4808	.4812	.4817
2.1	.4821	.4826	.4830	.4834	.4838	.4842	.4846	.4850	.4854	.4857
2.2	.4861	.4864	.4868	.4871	.4875	.4878	.4881	.4884	.4887	.4890
2.3	.4893	.4896	.4898	.4901	.4904	.4906	.4909	.4911	.4913	.4916
2.4	.4918	.4920	.4922	.4925	.4927	.4929	.4931	.4932	.4934	.4936
2.5	.4938	.4940	.4941	.4943	.4945	.4946	.4948	.4949	.4951	.4952
2.6	.4953	.4955	.4956	.4957	.4959	.4960	.4961	.4962	.4963	.4964
2.7	.4965	.4966	.4967	.4968	.4969	.4970	.4971	.4972	.4973	.4974
2.8	.4974	.4975	.4976	.4977	.4977	.4978	.4979	.4979	.4980	.4981
2.9	.4981	.4982	.4982	.4983	.4984	.4984	.4985	.4985	.4986	.4986
3.0	.4987	.4987	.4987	.4988	.4988	.4989	.4989	.4989	.4990	.4990

Source: Abridged from Table I of A. Hald. *Statistical Tables and Formulas* (New York: John Wiley & Sons, 1952). Reproduced by permission of A. Hald and the publisher.

Notes On Computer Simulations

The simulations presented in the computer activities sections were written in the Turbo Pascal program language. Instead of the actual Pascal code, brief descriptions are given for how each simulation was obtained.

CHAPTER 3

Page 131
1 Use the algorithm for generating a binomial random variable to obtain the variable X, which is binomial with parameters n_1, p_1, and to obtain the variable Y, which is binomial with parameters n_2, p_2.
2 Let $W = X + Y$.
3 Repeat steps 1 and 2 for the desired number of simulated values for W.
4 Compute the sample mean and the sample standard deviation for W.
5 Construct a histogram for all values of W simulated.

Page 132
1 Declare the number of distinct coupons desired, denoted as N.
2 Generate a random number R, which is $U(0, 1)$.
3 Take the integer part of $N * R + 1$.
4 Repeat steps 2 and 3 until each distinct integer $1, 2, \ldots, N$ has appeared at least once.
5 Let $X = $ Number of integers generated to satisfy step 4.
6 Repeat steps 2 through 5 for the desired number of simulated values for X.
7 Compute the sample mean and the sample standard deviation for X.
8 Construct a histogram for all values of X simulated.

CHAPTER 4

Page 213
1 Use the algorithm for generating Poisson random variables to obtain X (which is a Poisson variable with parameter λ_1) and to obtain Y (which is a Poisson variable with a parameter of λ_2).
2 Let $W = X / Y$.
3 Repeat steps 1 and 2 for the desired number of simulated values for W.
4 Compute the sample mean and the sample standard deviation for W.
5 Construct a histogram for the simulated values of W.

Pages 213–216

1 Use the algorithm for generating exponential random variables to obtain X (which is an exponential variable with mean θ_x) and to obtain Y (which is an exponential variable with a mean of θ_y).

2 Let $W = \max\{X, Y\}$.

3 Repeat steps 1 and 2 for the desired number of simulated values for W.

4 Compute the sample mean and the sample standard deviation for W.

5 Construct a histogram for the simulated values of W.

Pages 214, 216–217

1 Use one of the algorithms for simulating normal random variables to obtain X (which is a normal variable with parameters μ_x and σ_x^2) and to obtain Y (which is a normal variable with parameters μ_y and σ_y^2).

2 Let $W = X/(X + Y)$.

3 Repeat steps 1 and 2 for the desired numbers of simulated values for W.

4 Compute the sample mean and the sample standard deviation for W.

5 Construct a histogram for the simulated values of W.

References

Baily, N. T. J. 1964. *The Elements of Stochastic Processes with Applications to the Natural Sciences*. New York: John Wiley & Sons.

Barlow, R. E., and F. Proschan. 1965. *Mathematical Theory of Reliability*. New York: John Wiley & Sons.

Casella, G., and R. L. Berger. 1990. *Statistical Inference*. Belmont, CA: Wadsworth.

Cohen, A. C., B. J. Whitten, and Y. Ding. 1984. Modified Moment Estimation for the Three-Parameter Weibull Distribution. *Journal of Quality Technology* 16(3): 164–68.

Feller, W. 1968. *An Introduction to Probability Theory and Its Applications*, 3d ed., vol. 1. New York: John Wiley & Sons.

Ferrari, P., E. Cascetta, A. Nuzzolo, P. Treglia, and P. Olivotto. 1984. A Behavioral Approach to the Measurement of Motorway Circulation, Comfort and Safety. *Transportation Research*, 18A: 46–51.

Goranson, U. G., and J. Hall. 1980. Airworthiness of Long-Life Jet Transport Structures. *Aeronautical Journal* (November): 374–85.

Kamerad, D. B. 1983. The 55MPH Speed Limit: Costs, Benefits, and Implied Trade-Offs. *Transportation Research* 17A(1): 51–64.

Karlin, S. 1968. *A First Course in Stochastic Processes*. New York: Academic Press.

Kemeny, J. G., A. Schleifer, J. L. Snell, and G. L. Thompson. 1962. *Finite Mathematics with Business Applications*. Englewood Cliffs, N.J.: Prentice-Hall.

Kennedy, W. J., Jr., and J. E. Gentle. 1980. *Statistical Computing*. New York: Marcel Dekker.

Larsen, R. J., and M. L. Marx. 1985. *An Introduction to Probability and Its Applications*. Englewood Cliffs, N.J.: Prentice-Hall.

Meyer, P. L. 1970. *Introductory Probability and Statistical Applications*, 2d ed. Reading, Mass.: Addison-Wesley.

Mosteller, F., R. E. K. Rourke, and G. B. Thomas. 1970. *Probability with Statistical Applications*, 2d ed. Reading, Mass.: Addison-Wesley.

Nelson, W. 1967. The Truncated Normal Distribution with Applications to Component Sorting. *Industrial Quality Control* (November 1967): 261–68.

Parzen, E. 1964. *Modern Probability Theory and Its Applications*. New York: John Wiley & Sons.

Perruzzi, J. J., and E. J. Hilliard. 1984. Modeling Time-Delay Measurement Errors Using a Generalized Beta Density Function. *Journal of the Acoustical Society of America* 75(1): 197–201.

Pyke, R. 1965. Spacings. *Journal of the Royal Statistical Society* B(27): 395–436.

Ross, S. 1988. *A First Course in Probability*, 3d ed. New York: Macmillan.

Sullivan, B. E. 1984. Some Observations on the Present and Future Performance of Surface Public Transportation in the U.S. and Canada. *Transportation Research* 18A(2).

Yang, M. C. K., and D. H. Robinson. 1986. *Understanding and Learning Statistics by Computer*. Philadelphia: World Scientific Publishing.

Zamurs, J. 1984. Assessing the Effect of Transportation Control Strategies on Carbon Monoxide Concentrations. *Air Pollution Control Association Journal* 34(6): 637–42.

Zimmels, Y. 1983. Theory of Hindered Sedimentation of Polydisperse Mixtures. *AIChE Journal* 29(4): 669–76.

Answers to Selected Problems

Chapter 2

2.1 **a** 3   **b** 13   **c** 7   **d** 9

2.2 **a** $JD, JS, DM, DN, MN, JM, JN, DS, MS, SN$

     b $A = \{JD, JM, JS, JN, DM, DS, DN\}$

     c $B = \{JM, JS, JN, DM, DS, DN\}$

     d $\overline{A} = \{MS, MN, SN\}$

     e $\overline{A} = \{MS, MN, SN\}$

       $AB = B$

       $A \cup B = A$

       $\overline{AB} = \{JD, MS, MN, SN\}$

2.5 **a** $L = $ Left turn, $R = $ Right turn, $S = $ Straight

     b $P(L) = P(R) = P(S) = 1/3$ **c** 2/3

2.6 **a** 0.4   **b** 0.9   **c** 0.4   **d** 0.1

     e 0.5 .9   **f** 0.1   **g** 0.1

2.7 **a** 1/3   **b** 6/15   **c** 19/48   **d** 2/3

2.8 **a** 0.43   **b** 0.05   **c** 0.59   **d** 0.74

     e 0.91

2.9 **a** 0.08   **b** 0.16   **c** 0.14   **d** 0.84

2.11 **a** $SS, SR, SL, RS, RR, RL, LS, LR, LL$

     b 5/9   **c** 5/9

2.12 **a** (I, I), (I, II), (II, I), (II, II)   **b** $P[(I, I)] = 1/4$

                                      $P[(I, I)] + P[(II, II)] = 1/2$

2.13 **a** (I, I), (I, II), (I, III), (II, I), (II, II), (II, III), (III, I), (III, II), (III, III)

     b 1/3   **c** 5/9

2.14 3/10

2.15 **a** 42   **b** 21

2.16 1/5

2.17 5040

2.18 **a** 168   **b** 1/8

2.20 **a** 3/5 **b** 2/5 **c** 3/10

2.21 **a** 0.09375 **b** 0.140625 **c** 0.578125 **d** 0.5625

2.22 **a** 0.48 **b** 0.04 **c** 0.48

2.23 **a** 24 **b** 1/2

2.24 **a** 1680 **b** 1/12

2.25 **a** $\frac{12}{60}$; no **b** $\frac{4}{60}$; perhaps **c** $\frac{1}{60}$; yes

2.26 (4, 1) (4, 1) **2.27** 5/8

2.28 **a** 2/3 **b** 1/9 **c** 1/3

2.29 **a** 0.4979 **b** 0.5144 **c** 0.7443 **d** 0.4399
  e 0.6894

2.30 **a** 0.16 **b** $1 - 0.16 = 0.84$ **c** 0.96 **d** no

2.31 **a** 0.10 **b** 0.03 **c** 0.06 **d** 0.018

2.32 **a** 0.6316 **b** 0.3684 **c** 0.6522

2.33 **a** 0.8053 **b** 0.9947 **c** 0.8096

2.34 **a** 1/10 **b** 7/10 **c** 1/7

2.35 **a** 0.7225 **b** 0.19

2.36 **a** 0.999 **b** 0.9009

2.37 series: 0.81 parallel: 0.99

2.38 0.40 **2.39** 0.8235 **2.40** 0.0833

2.41 **a** 3/5 **b** 2/5 **c** yes **d** yes

2.44 0.5073

2.45 **a** $1/k$ **b** n/m **c** $1 - \dfrac{n}{mk}$ **d** $\dfrac{m-n}{mk-n}$ **e** $\dfrac{m}{mk-n}$

2.46 $\dfrac{mp}{1 + (m-1)p}$

2.50 **a** 24/40 **b** 24/60 **c** yes **d** no **e** 1

2.51 **a**
$$HHHH \quad THHH$$
$$HHHT \quad THHT$$
$$HHTH \quad THTH$$
$$HHTT \quad THTT$$
$$HTHH \quad TTHH$$
$$HTHT \quad TTHT$$
$$HTTH \quad TTTH$$
$$HTTT \quad TTTT$$

  b $A = \{HHHT, HHTH, HTHH, THHH\}$ **c** 1/4

2.52 **a**
$$N_1N_2 \quad N_2N_3 \quad N_3N_4 \quad N_4D_1 \quad D_1D_2$$
$$N_1N_3 \quad N_2N_4 \quad N_3D_1 \quad N_4D_2 \quad D_1D_3$$
$$N_1N_4 \quad N_2D_1 \quad N_3D_2 \quad N_4D_3 \quad D_2D_3$$
$$N_1D_1 \quad N_2D_2 \quad N_3D_3 \quad N_1D_2 \quad N_2D_3$$
$$N_1D_3$$

  b $A = \{N_1N_2, N_1N_3, N_1N_4, N_2N_3, N_2N_4, N_3N_4\}$ **c** 2/7 **d** .5625

2.53 **a**

Year	Desired Probability
1973	0.07
1974	0.32
1975	0.29

b

Year	Desired Probability
1973	0.23
1974	0.43
1975	0.39

c

Year	Desired Probability
1973	0.59
1974	0.75
1975	0.65

2.54 **a** 36   **b** 1/6

2.56 **a** 0.57   **b** 0.18   **c** 0.9   **d** 0.3158

2.57 120   **2.58** 9,000,000   **2.59** 720

2.60 18   **2.61** 40,320

2.62 **a** 0.0362   **b** 0.000495, 0.001981

2.63 **a** 0.216   **b** 0.936   **c** 0.648

2.64 **a** 1/8   **b** 1/64   **c** no

2.65 1/16   **2.66** 0.5952

2.67 **a** 5/16   **b** $27(1/2)^{10}$

2.68 **a** $(1/4)^7$ **b** $9(1/4)^3$

2.69 no   **2.70** 0.8704   **2.71** 0.0625

2.74 no   **2.75** 1/2   **2.76** 1/7

2.78 A

2.79 **a** 0.00892   **b** 0.9890

Chapter 3

3.1

X	p(X)
0	1/30
1	3/10
2	1/2
3	1/6

3.2 **a**

Y	p(Y)
0	1/16
1	1/4
2	3/8
3	1/4
4	1/16

b no

3.3

X	p(X)
0	0.2585
1	0.4419
2	0.2518
3	0.0478

(assumes independence)

3.4

X	p(X)
0	1/8
1	3/8
2	3/8
3	1/8

3.5 **a**

X	p(X)
0	0.0053
1	0.0575
2	0.2331
3	0.4201
4	0.2840

b 0.9947

3.6

X	p(X)
0	0.1296
1	0.3456
2	0.3456
3	0.1536
4	0.0256

3.7

x	$p(x)$
0	8/27
1	12/27
2	6/27
3	1/27

x	$p(y)$
0	2744/3375
1	588/3375
2	42/3375
3	1/3375

$x + y$	$p(x + y)$
0	0.24090
1	0.41295
2	0.26179
3	0.07445
4	0.00935
5	0.00053
6	0.00003

3.8 **a**

x	$p(x)$
0	0.49
1	0.28
2	0.18
3	0.04
4	0.01

b 0.51

3.9 **a**

x	$p(x)$
0	1/6
1	2/3
2	1/6

b

x	$p(x)$
0	1/2
1	1/2

c $p(0) = 1$

3.10

y	$p(y)$
0	0.019
1	0.252
2	0.729

3.11 **a** 0, 2/3 **b** 0, 12/5

3.12 **a** $\mu = 2.601$ **b** median $= 2$
$\sigma = 1.414$

3.13 $\mu = 37, \sigma = 11$, approximately

3.14 median ≈ 31 mean ≈ 41 standard deviation ≈ 11

3.15 $\mu = 0.4$ $\sigma = 0.6633$

3.16 $\mu = 0.2$ $\sigma^2 = 0.18$

3.17 $60,000 for firm I $120,000 for both firms

3.18 $\mu = 1$ $\sigma^2 = 1/2$

3.19 **a** (1.4702, 6.5298) **b** yes **3.20** 41

3.21 **a** (84.1886, 115.8114) **b** no

3.22 **a** $\mu = \$200$ **b** $360
$\sigma = \$80$

3.23 **a** 0.1536 **b** 0.1808 **c** 0.9728
d 0.8 **e** 0.64

3.24 **a** 0.250 **b** 0.057 **c** 0.180

3.25 **a** 0.537 **b** 0.098

3.26 **a** 0.109 **b** 0.999 **c** 0.589

3.27 **a** 16 **b** 3.2

3.28 **a** 0.672 **b** 0.672 **3.29** 8

3.30 **a** 1/16 **b** 1/4 **3.31** 0.5931

3.32 **a** independence **b** 0.7379

3.33 **a** 0.99 **b** 0.9999 **3.34** 2

3.35 **a** 0.1536 **b** 0.9728

3.36 **a** $400,000 **b** $474,342 **3.37** 3.96

3.38 **a** 22   **b** 45   **3.39** 840

3.40 **a** 0.9   **b** $P(Y > 4 | Y > 2) = P(Y > 2) = (1 - p)^2$

3.41 **a** 0.648   **b** 1   **3.42** 0.09

3.43 **a** 0.04374   **b** 0.99144

3.44 0.1

3.45 **a** $\mu = 10/9$ **b** $\mu = 10/3$
$\sigma^2 = 0.1235$ $\sigma^2 = 0.3704$

3.46 0.0645

3.47 $\mu = 150$ $\sigma^2 = 4500$ no

3.48 **a** 0.09877   **b** 0.14815

3.49 **a** 0.128   **b** 0.03072

3.50 $\mu = 15$ $\sigma^2 = 60$

3.51 **a** 0.06561   **b** $\mu = 4.4444$
$\sigma^2 = 0.4938$

3.52 **a**

Try	Probability
1	0.4
2	0.24
3	0.144

     b 0.1728

3.53 **a** 3/16   **b** 3/16   **c** 1/8   **d** 1/2

3.54 **a** 0.0902   **b** 0.143   **c** 0.857   **d** 0.2407

3.55 **a** 0.0183   **b** 0.908   **c** 0.997

3.56 **a** 0.905   **b** 0.005   **c** 0.819

3.57 **a** 0.467   **b** 0.188

3.58 **a** 0.022   **b** 0.866

3.59 **a** 0.8187   **b** 0.5488   **3.60** 0.353

3.61 **a** 0.140   **b** 0.042   **c** 0.997

3.62 $\mu = 80$ $\sigma^2 = 800$ no

3.63 **a** 1.44×10^{-5} **b** 1.44×10^{-5}

3.64 **a** 0.9817   **b** $1 - e^{-12}$

3.65 $\mu = 320$ $\sigma = 56.5685$

3.66 **a** 0.9997   **b** 2

3.67 $E[Y(Y - 1)] = \lambda^2$ **3.68** 47.5 hours

3.69 **a** 0.001   **b** 0.000017

3.70 **a** 4/7   **b** 6/7   **c** 1/3   **d** 4/7

3.71 1/42

3.72 $\mu = 100$ $\sigma^2 = 5000/3$ $P(\text{Cost} < \$222.48) = 8/9$

3.73 **a** 4/5   **b** 1/5

3.74 1/30; probably not random

3.75   **a**

y	p(y)
0	7/15
1	7/15
2	1/15

b

y	p(y)
0	1/6
1	1/2
2	3/10
3	1/30

3.76   **a**   1/14   **b**   13/14   **c**   1/14

3.77   3/14

3.78   **a**   41/42   **b**   11/42   **c**   1/42   **d**   ∅

3.79   **a**   9/14   **b**   13/14   **c**   1

3.80

y	p(y)
0	1/15
1	8/15
2	6/15

3.81   **a**   1   **b**   1   **c**   18/19   **d**   49/57   **e**   728/969

3.82   **a**   1   **b**   1   **c**   1   **d**   113/114   **e**   938/969

3.83   10/21   **3.84** $M(t) = [pt + 1 - p]$

3.89 $P(t) = [pt + (1 - p)]^n$ **3.91**   0.75

3.94 $E(\text{Duration}) = 3$ probability that A wins = 1/4

3.97 $P(Y = 4) = 0.32805$ $P(Y \geq 1) = 0.99999$

3.98   **a**   0.9606   **b**   0.9994

3.99   **a**   1   **b**   0.5905   **c**   0.1681   **d**   0.03125   **e**   0

3.100 (a, b, c)

p	$p(Y \leq 0)$	$p(Y \leq 1)$	$p(Y \leq 2)$
0	1.0000	1.0000	1.0000
0.05	0.5987	0.9139	0.9885
0.10	0.3487	0.7361	0.9298
0.30	0.0283	0.1493	0.3828
0.50	0.0010	0.0107	0.0597
1.00	0.0000	0.0000	0.0000

3.101   **a** $n = 25, a = 5$ **b** $n = 25, a = 5$

3.102 $E(C) = \$80$ $\sqrt{V(C)} = \$9.49$ ($\$61.02, \$98.98$)

3.103   **a**   0.758   **b** $\mu = 12, \sigma = \sqrt{12}$ **c**   (5, 19)

3.104   **a**   0.083   **b**   0.895

3.105   **a**   5   **b**   0.007   **c**   0.384

3.106   **a** $ke^{\lambda t}$ **b**   3.2974

3.107   0.993   **3.108**   0.04096   **3.110**   0.837   **3.111**   0.18522

3.112   **a**   0.081   **b**   0.81

3.113 $P(Y = 5) = 0.01536$ $P(Y \geq 5) = 0.0256$

3.114   **a**   0.019   **b**   0.1745

3.115 $E(Y) = 900$ $V(Y) = 90$ $P(81.026 < Y < 918.974) \geq 0.75$

3.116

	$p(y)$	
y	(a) Binomial	(b) Poisson
0	0.358	0.368
1	0.378	0.368
2	0.189	0.184
3	0.059	0.061
4	0.013	0.015

3.117 **a**

y	$p(y)$
0	$(2/3)^4$
1	$2(2/3)^4$
2	$(2/3)^3$
3	$1/3(2/3)^3$
4	$(1/3)^4$

b 1/9   **c** 4/3   **d** 8/9

3.119 **a** 0.1192   **b** 0.117
yes

3.120 \$149.09   **3.121** 3

3.122 **a** $\dfrac{N}{K}(1 + K[1 - (0.95)^K])$ **b** 5   **c** $(0.5738)N$

3.123 Number of combinations = 6,760,000 $E(\text{Winnings}) = \$0.0311$ no

Chapter 4

4.2 **a** 6   **b** 0.648   **c** 0.3929

d $F(b) = \begin{cases} 0 & \text{for } b < 0 \\ 3b^2 - 2b^3 & \text{for } 0 \le b \le 1 \\ 1 & \text{for } b > 1 \end{cases}$

4.3 **a** 0.84375   **b** 4

4.4 **a** $F(x) = \begin{cases} 0 & \text{for } x < 0 \\ \dfrac{x^3}{256}(16 - 3x) & \text{for } 0 \le x \le 4 \\ 1 & \text{for } x > 4 \end{cases}$

b 11/16   **c** 67/256   **d** 3.4298

4.5 **b** $F(x) = \begin{cases} 0 & \text{for } x < 5 \\ \dfrac{(x - 7)^3}{8} + 1 & \text{for } 5 \le x \le 7 \\ 1 & \text{for } x > 7 \end{cases}$

c 7/8   **d** 37/56

4.6 **a** 1/2   **b** 1/4

4.7 **a** 3/4   **b** 4/5   **c** 1

d $F(x) = \begin{cases} 0 & \text{for } x < 0 \\ x^2 & \text{for } 0 \le x \le 1 \\ 1 & \text{for } x > 1 \end{cases}$

4.8 **a** 0.08392   **b** 0.99046

4.9 $E(X) = 60$ $V(X) = 1/3$

4.10 **a** $E(X) = 2/3$ **b** $E(Y) = 220/3$ **c** $(-20.9476, 167.6142)$
$V(X) = 1/18$ $V(Y) = 20{,}000/9$

4.11 4

4.12 **a** $E(X) = 2.4$ **b** $E(\text{Weekly costs}) = 480$ **c** yes; $P(Y > 600) = 0.2617$
$V(X) = 0.64$ $V(\text{Weekly costs}) = 25{,}600$

4.13 **a** $E(X) = 5.5$ **b** $(4.7254, 6.2746)$ **c** yes; $P(X < 5.5) = 0.5781$
$V(x) = 0.15$

4.14 0.7368

4.15 **a** $F(x) = \begin{cases} 0 & \text{for } x < a \\ \dfrac{x-a}{b-a} & \text{for } a \le x \le b \\ 1 & \text{for } x > b \end{cases}$

     b $\dfrac{b-c}{b-a}$ **c** $\dfrac{b-d}{b-c}$

4.16 **a** 2/5   **b** 1/5   **c** $22,500

4.17 **a** 1/20   **b** 1/20   **c** 1/2

4.18 0.2   **4.19** 3/4   **4.20** 2/5

4.21 **a** 1/8   **b** 1/8   **c** 1/4

4.22 **a** 1/5   **b** $E(X) = 0$
                         $V(X) = 1/1200$

4.23 **a** 2/7   **b** $E(X) = 0.015$
                         $V(X) = 49/120,000$

4.24 $E(\text{Volume}) = 2.0 \times 10^{-5}$ $V(\text{Volume}) = 3.4795 \times 10^{-10}$

4.25 1/6   **4.26** 1/4

4.27 **a** 1/2   **b** 1/4   **4.28** 3/8

4.29 **a** $E(X) = 60$ **b** 4
           $V(X) = 100/3$

4.31 **a** 0.2865   **b** 0.1481   **4.32** 0.7355

4.33 **a** 0.1353   **b** 460.52 cfs

4.34 **a** $\mu = \frac{1}{2}, \sigma^2 = \frac{1}{4}$ **b** 0.9975

4.35 **a** $E(C) = 1100$ **b** no; $P(C > 2000) = 0.14$
           $V(C) = 2,920,000$

4.36 **a** 0.5057   **b** 1936

4.37 **a** 0.3679   **b** 0.6065

4.38 0.2636

4.39 **a** 0.6065   **b** 0.7788   **c** no

4.40 **a** 0.08208   **b** 0.02732

4.41 **a** 0.2865   **b** 0.5091

4.42 **a** 0.5353   **b** 0.5353

4.43 **a** 0.3653   **b** 0.1903   **c** 50.66 minutes

4.44 $E(\text{Area}) = 200\pi$ $V(\text{Area}) = 200,000\pi^2$

4.45 **a** $E(X) = 3.2$ **b** (0, 8.26)
           $V(X) = 6.4$

4.46 **a** $E(X) = 30,000$ **b** no; $P(X > 35,000) < .06$
           $V(X) = 1,500,000$

4.47 **a** $E(L) = 276$ **b** (0, 930.963)
           $V(L) = 47,664$

4.48 **a** $E(Y) = 1$ **b** $E(Y) = 3/2$
           $V(Y) = 1/2$ $V(Y) = 3/4$

           $f(y) = \begin{cases} 4ye^{-2y} & \text{for } y > 0 \\ 0 & \text{for } y \le 0 \end{cases}$ $f(y) = \begin{cases} 4y^2e^{-2y} & \text{for } y > 0 \\ 0 & \text{for } y \le 0 \end{cases}$

4.49 **a** $E(Y) = 20$ **b** $E(A) = 10$
        $V(Y) = 200$ $V(A) = 50$

4.50 **a** $E(Y) = 140$ **b** 206.93
        $V(Y) = 280$

$$f(y) = \begin{cases} \dfrac{1}{2^{70}(69!)} y^{69} e^{-y/2} & \text{for } y > 0 \\ 0 & \text{for } y \le 0 \end{cases}$$

4.51 **a** $E(Y) = 240$ **b** $(0, 809.21)$
        $\sqrt{V(Y)} = 189.74$

4.52 $E(Y) = 60$
     $V(Y) = 900$

$$f(y) = \begin{cases} \dfrac{1}{303,750} y^3 e^{-y/15} & \text{for } y > 0 \\ 0 & \text{for } y \le 0 \end{cases}$$

4.53 $E(Y) = 9.6$
     $V(Y) = 30.72$

$$f(x) = \begin{cases} \dfrac{1}{65.536} y^2 e^{-y/3.2} & \text{for } y > 0 \\ 0 & \text{for } y \le 0 \end{cases}$$

4.54 $$f(x) = \begin{cases} \dfrac{1}{4} x e^{-x/2} & \text{for } y > 0 \\ 0 & \text{for } x \le 0 \end{cases}$$

4.55 **a** 0.3849   **b** 0.3159   **c** 0.3227
    d 0.1586   **e** 0.9178

4.56 **a** 0   **b** 1.15   **c** 1.19
    d −0.30   **e** 1.645   **f** 1.96

4.57 0.0062   **4.58** $425.60   **4.59** 0.0730   **4.60** $\mu = 1$

4.61 **a** 0.9544   **b** 0.8297

4.62 0.5859

4.63 $P(|X| > 5) = 0.6170$ $P(|X| > 10) = 0.3174$

4.64 **a** 0.1498   **b** 0.0224

4.65 **a** 0.0062   **b** 225.6 hours

4.66 **a** 0.0062   **b** 1171 hours

4.67 13.67 ounces   **4.68** 0.5102

4.69 **a** yes   **b** yes   **c** no, no   **d** 4

4.70 **a** no   **b** $\bar{x} = 0.032$
               $s = 0.017$

4.73 **a** 60   **b** $E(X) = 4/7$
               $V(X) = 3/98$

4.75 **a** $E(C) = 17.33$ **b** $(6.387, 28.280)$
        $V(C) = 29.96$

4.76 **a** 0.8208   **b** $E(V) = 4.7$
               $V(V) = 0.01$

4.77 $E(X) = 2/3$ corresponds to angle of $240°$

4.78 **a** 0.75   **b** $E(X) = 1.3$
               $\sqrt{V(X)} = \sqrt{1/18}$

4.79   **a** $E(X) = 1/2$ **b** $E(X) = 1/2$ **c** $E(X) = 1/2$ **d** a
 $V(X) = 1/28$ $V(X) = 1/20$ $V(X) = 1/12$

4.80   **a**   7/8   **b**   0.002128

4.81   **a**   0.6321   **b** $\sqrt{\pi}$

4.82   **a**   0.5547   **b**   0.6966

4.83   0.6576   **4.84**   0.03091

4.85   0.06573   **4.86**   0.09813

4.87   **a**   0.08209   **b**   0.01855   **c** $\dfrac{\sqrt{10\pi}}{2}$, $10\left(1 - \dfrac{\pi}{4}\right)$

4.88 $\hat{\gamma} = 6.22$ $\ln \hat{\theta} = 43.52$

4.89   42.9193

4.90   **a** $2\sqrt{\dfrac{2KT}{m\pi}}$ $\dfrac{3}{2}KT$

4.91 $R(t) = e^{-t_2/\theta}$, $t_2 > 0$ **4.92** $R_3(t) = e^{-nt/\theta}$, θ/n

4.93 a **4.94** $n = 3$ in parallel

4.96 $E(X^2) = 2\theta^2$ **4.97** $M_z(t) = e^{t^2/2}$

4.98 $M_{z^2}(t) = (1 - 2t)^{-1/2}$;   **4.99**   0.4 lb
 gamma (1/2, 2)

4.100   **d**   0.483

4.101   **b** $1 - e^{-2}$ **c** $\dfrac{2}{3}e^{-4/3} + \dfrac{1}{3}e^{-1}$

4.102   **a**   1/2   **b** $F(y) = \begin{cases} 0 & \text{for } y < 0 \\ y^2/4 & \text{for } 0 \le y \le 2 \\ 1 & \text{for } y > 2 \end{cases}$

 **c**   3/4   **d**   3/4

4.103   **a** $-3/8$ **b** $F(y) = \begin{cases} 0 & \text{for } y < 0 \\ \dfrac{y^2}{2} - \dfrac{y^3}{8} & \text{for } 0 \le y \le 2 \\ 1 & \text{for } y > 2 \end{cases}$

 **c** $F(-1) = 0$ **d**   7/64   **e** $E(Y) = 7/6$
 $F(0) = 0$ $V(Y) = 43/180$
 $F(1) = 3/8$

4.104   **a**   6/5   **b** $F(y) = \begin{cases} 0 & \text{for } y \le -1 \\ \dfrac{1}{5}(y + 1) & \text{for } -1 < y \le 0 \\ \dfrac{1}{5}(1 + y + 3y^2) & \text{for } 0 < y \le 1 \\ 1 & \text{for } y > 1 \end{cases}$

 **c** $F(-1) = 0$ **d**   1/4   **e** $E(Y) = 2/5$
 $F(0) = 1/5$ $V(Y) = 41/150$
 $F(1) = 1$

4.105   0.1151   **4.106**   15.87%   **4.107**   0.0015

4.108   0.073   **4.109**   0.3155

4.110   **a** $E(Y) = c/4$ **b** $c/(2 - t)^2$ **c** $c = 4$
 $V(Y) = c(6 - c)/16$

4.111 $E(X^k) = \dfrac{\Gamma(\alpha + \beta)\Gamma(\alpha + k)}{\Gamma(\alpha + \beta + k)\Gamma(\alpha)}$ **4.112** 2.3833×10^{-7}

4.113 **a** 0.9975

b $f(y) = \begin{cases} \dfrac{1}{60,000} y^3 e^{-y/10} & \text{for } y > 0 \\ 0 & \text{for } y \le 0 \end{cases}$

4.114 **a** 105

b $E(X) = 0.375$
$V(X) = 0.02604$

4.115 $1 - e^{-4}$ **4.116** $\dfrac{29}{8} e^{-3/2}$ **4.117** $53.58

4.118 $113.33 **4.119** 0.736 **4.120** 0.875

4.121 **a** R has a Weibull distribution, $\gamma = 2, \theta = \dfrac{1}{\lambda \pi}$ **b** $1/(2\sqrt{\lambda})$

4.122 **a** 0.0045 **b** 0.9726

4.123 **a** $E(X) = e^{11} \times 10^{-2} g$ **b** $(0, 3{,}570{,}245)$ **c** 0.3156
$V(X) = (e^{38} - e^{22}) \times 10^{-4} g^2$

4.124 $M(t) = 1/(1 - t^2)$ $E(Y) = 0$

4.125 **a** $F_X(x) = \begin{cases} 0 & \text{for } x \le 0 \\ 1 - e^{-x/100} & \text{for } 0 < x < 200 \\ 1 & \text{for } x \ge 200 \end{cases}$ **b** 86.4665

4.130 0.04999

Chapter 5

5.1 **a, b**

		x_1		
		0	1	2
	0	1/9	2/9	1/9
x_2	1	2/9	2/9	0
	2	1/9	0	0
	$p(x_1)$	4/9	4/9	1/9

c 1/2

5.2 **b** 0.08 **c** 0.12

5.3 **a**

		x_1	
		0	1
	0	0.063	0.078
	1	0.101	0.056
x_2	2	0.163	0.065
	3	0.169	0.055
	4	0.193	0.057

b

	x_1	
	0	1
	0.45	0.55
	0.64	0.36
	0.71	0.29
	0.75	0.25
	0.77	0.23

c

	0	0.092	0.250
	1	0.146	0.179
x_2	2	0.236	0.210
	3	0.245	0.178
	4	0.280	0.185

5.4 **a** $k = 1$ **b** 2/3

5.5 **a** $f_1(x_1) = 1, \quad 0 \le x_1 \le 1$ **b** 0.5
  c yes

5.6 **a** $f(x_2) = 2(1 - x_2), \quad 0 \le x_2 \le 1$ **b** 0.64
  c no **d** 1/2

5.7 **a** 7/8 **b** 1/2 **c** 2/3

5.8 **a** $f_1(x_1) = 2(1 - x_1), \quad 0 \le x_1 \le 1$
  $f_2(x_2) = 2(1 - x_2), \quad 0 \le x_2 \le 1$
  b no **c** 2/3 **c** no

5.9 **a** 21/64 **b** 1/3

5.10 11/32

5.11   **a**   yes   **b** $\dfrac{3}{2e}$

5.12   7/32   **5.13**   11/36   **5.14**   1/4

5.15   **a**   1/4   **b**   23/144

5.16   1/4

5.17   **a** $E(X_i) = 0.24$ **b** $C_{0V}(X_1, X_2) = -0.0576$ **c** $E(Y) = E(X_1 + X_2) = 0.98$
$V(X_i) = 0.1824$ $V(Y) = 1.07$
$E(X_2) = 0.74$
$V(X_2) = 0.7724$

5.18   **a** $E(Y) = 1$ **b**   (0.1835, 1.8165)
$V(Y) = 1/6$

5.19   **a** $E(Y) = 2/3$ **b**   (1/3, 1)
$V(Y) = 1/8$

5.20   **a** $E(Y) = 61$ **b**   no; $P(Y > 75) \le 0.1653$
$V(Y) = 20$

5.21   **a** $\dfrac{1}{e} - \dfrac{2}{e^2}$ **b**   1/2   **c** $1/e$

d $f_1(y_1) = y_1 e^{-y_1}, \quad y_1 > 0$ $f_2(y_2) = e^{-y_2}, \quad y_2 > 0$

5.22   1/3

5.23   **a**   1   **b**   1   **c**   no; $P(Y_1 - Y_2 > 2) = e^{-2}$

5.24   1/2   **5.26**   0.08953   **5.27**   66,960

5.28   **a**   0.04594   **b**   0.2262

5.29   0.07776   **5.30**   0.09352

5.31   **a**   4/27   **b**   1/27   **c**   4/9

5.32   0.40951   **5.33** $\mu = 2.5, \sigma^2 = 4.875$

5.34   **a**   0.2759   **b**   0.80313

5.35   **a**   0.08575   **b**   0.7627   **c** $E(Y_1) = 40$
$V(Y_1) = 24$

5.36 $M_Y(t) = \left(\dfrac{pe^t}{1 - (1 - p)e^t} \right)^r$ **5.37**   0.05213

5.39   **a**   0.0793   **b**   0.2389

5.40   0.0228   **5.41** $\alpha\beta, \alpha\beta(1 + \beta)$ **5.42** $\lambda p, \lambda p$

5.44   **a**   4   **b** $f_i(x_i) = 2x_i, \qquad 0 \le x_i \le 1$
$i = 1, 2$

c $F(x_1, x_2) \le \begin{cases} 0 & \text{for } x_1 < 0 \text{ or } x_2 < 0 \\ x_1^2 x_2^2 & \text{for } 0 \le x_i \le 1, \quad i = 1, 2 \\ x_1^2 & \text{for } 0 \le x_1 \le 1 \text{ and } x_2 > 1 \\ x_2^2 & \text{for } 0 \le x_2 \le 1 \text{ and } x_1 > 1 \\ 1 & \text{for } x_1 > 1 \text{ and } x_2 > 1 \end{cases}$

d   9/64   **e**   1/4

5.45   **a** $f_1(x_1) = 3x_1^2, \qquad 0 \le x_1 \le 1$ **b**   23/64
$f_2(x_2) = \dfrac{3}{2}(1 - x_2^2), \qquad 0 \le x_2 \le 1$ **c**   0

5.46   **a, b**

			x_2			
		0	1	2	3	$p(x_1)$
	0	0	1/28	1/14	1/84	10/84
	1	1/21	2/7	1/7	0	10/21
x_1	2	1/7	3/14	0	0	5/14
	3	1/21	0	0	0	1/21
	$p(x_2)$	5/21	15/28	3/14	1/84	

c   9/16

5.47 $f(x_1 | X_2 = x_2) = 2x_1,$ $0 \le x_1 \le 1$ yes

5.48   **a** $f(x_1 | x_2) = \dfrac{2x_1}{1 - x_2^2},$ $0 \le x_2 \le x_1 \le 1$

   b $f(x_2 | x_1) = 1/x_1,$ $0 \le x_2 \le x_1 \le 1$

   c $f(x_1 | x_2) \ne f_1(x_1)$ **d**   5/12

5.49   **a** $f(x_1, x_2) = 1/x_1,$ $0 \le x_2 \le x_1 \le 1$

   b   1/2   **c** $\ln 2 / \ln 4$

5.50   **a** $f(x_1) = \dfrac{2}{\pi} \sqrt{1 - x^2},$ $|x_1| \le 1$ **b**   1/2

5.51   **a** $f(x_i) = x_i + \dfrac{1}{2},$ $0 \le x_i \le 1, \ i = 1, 2$ **b**   no

   c $f(x_1 | x_2) = \dfrac{x_1 + x_2}{x_2 + \frac{1}{2}},$ $0 \le x_i \le 1, \ i = 1, 2$

5.52   **a**   1   **b** $f(x_1) = x_1/2$ $0 \le x_1 \le 2$
      $f(x_2) = 2(1 - x_2),$ $0 \le x_2 \le 1$

   c $f(x_1 | x_2) = \dfrac{1}{2(1 - x_2)},$ **d** $f(x_2 | x_1) = \dfrac{2}{x_1},$
      $0 \le 2x_2 \le x_1 \le 1$ $0 \le 2x_2 \le x_1 \le 1$

   e   1/2   **f**   8/9

5.53   **a** $f(x_2) = 2(1 - x_2),$ **b** $f(x_1) = 1 - |x_1|,$ **c**   1/4
      $0 \le x_2 \le 1$ $|x_1| \le 1$

5.54   **a**   2/3   **b**   1/18   **c**   0

5.55   3/160

5.56   **a**   −0.3333   **b, c** $E(X_1 + X_2) = 7/3$
            $V(X_1 + X_2) = 7/18$

5.57   **a**   −1/144   **b**   7/12   **c**   1.0764

5.58   **a**   2   **b**   2/3

5.59   1/4   **5.60** $f(x) = \left(\dfrac{1}{2}\right)^{(x+1)},$ $x = 0, 1, 2, \ldots$

5.61 $f(x_1, x_2, x_3) = \left(\dfrac{1}{\theta}\right)^3 e^{(-1/\theta)(x_1 + x_2 + x_3)}$ $\theta, x_i > 0,$ $i = 1, 2, 3$

5.62 $E(G) = 42$ **5.63** $E(Y) = np$
      $V(G) = 26$ $V(Y) = np(1 - p)$
      no

5.64   **a** $x_1/2$ **b, c**   3/8

5.65   **a, b**   1

5.66   3/8   **5.67**   300

5.69 **a** $(p_1 e^{t_1} + p_2 e^{t_2} + p_3 e^{t_3})^n$ **b** $-np_1 p_2$

5.73 $-\sqrt{\dfrac{p_1 p_2}{(1 - p_1)(1 - p_2)}}$

5.75 **a** $f(x_1, x_2) = \dfrac{1}{9} e^{-(x_1 + x_2)/3}$, $x_1 > 0 \ x_2 > 0$ **b** 0.0446

5.76 **a** 0.0868 **b** $1 - e^{-5}$

Chapter 6

6.1 **a** $f_{u_1}(u) = \dfrac{1}{2}(1 - u)$, $|u| \le 1$ **b** $f_{u_2}(u) = \dfrac{1}{2}(1 + u)$, $|u| \le 1$

   c $f_{u_3}(u) = \dfrac{1 - \sqrt{u}}{\sqrt{u}}$, $0 < u \le 1$

6.2 **a** $f_{u_1}(u) = \dfrac{u^2}{18}$, $|u| \le 3$ **b** $f_{u_2}(u) = \dfrac{3}{2}(3 - u)^2$, $2 \le u \le 4$

   c $f_{u_3}(u) = \dfrac{3\sqrt{u}}{2}$, $0 \le u \le 1$

6.3 **a** $f_u(u) = \begin{cases} \dfrac{u + 4}{100} & \text{for } -4 \le u \le 6 \\ \dfrac{1}{10} & \text{for } 6 < u \le 11 \\ 0 & \text{otherwise} \end{cases}$ **b, c** 67/12

6.4 $f_u(u) = \dfrac{1}{8\sqrt{2(u - 3)}}$, $5 \le u \le 53$

6.5 **a** $f_u(u) = \begin{cases} u & \text{for } 0 \le u \le 1 \\ 2 - u & \text{for } 1 < u \le 2 \\ 0 & \text{otherwise} \end{cases}$ **b** 1

6.6 **a** $f_u(u) = e^{-u}$ $u > 0$ **b** $E(U) = 1$
   $V(U) = 1$

6.7 **a** $f_u(u) = 2u$, $0 \le u \le 1$ **b, c** 2/3

6.8 $f_u(u) = 1$, $0 \le u \le 1$

6.9 **a** $f_{u_1}(u) = \dfrac{1 - u}{2}$, $|u| \le 1$ **b** $f_{u_2}(u) = \dfrac{1 + u}{2}$, $|u| \le 1$

   c $f_{u_3}(u) = \dfrac{1 - \sqrt{u}}{\sqrt{u}}$, $0 < u \le 1$

6.10 **a** $f_{u_1}(x) = \dfrac{u^2}{18}$, $|x| \le 3$ **b** $f_{u_2}(u) = \dfrac{3}{2}(3 - u)^2$, $2 \le u \le 4$

   c $f_{u_3}(u) = \dfrac{3\sqrt{u}}{2}$, $0 \le u \le 1$

6.11 $f_u(u) = 18u(1 - u \pm u \ln u)$

6.12 **a** $f_u(u) = \dfrac{1}{\theta} e^{-u/\theta}$, $u > 0$ **b** $E(u) = \theta$
   $V(u) = \theta^2$

6.13 **a** $f_u(u) = \dfrac{1}{\alpha} e^{-u/\alpha}$, $u > 0$ **b** $\Gamma\left(\dfrac{k}{m} + 1\right)\alpha^{k/m}$

6.14 **a** binomial; (kn, p) **b** binomial; $\left(\sum_{i=1}^{k} n_i, p\right)$ **c** not binomial

6.15 **a** Poisson $(n\lambda)$ **b** Poisson $\left(\sum_{i=1}^{k} \lambda_i\right)$ **c** no

6.16 **a** gamma $\left(\sum_{i=1}^{k} \lambda_i, \beta \right)$ **b** no

6.17 normal $(\mu, \sigma^2/n)$

6.19 **a** σ^2 **b** $2\sigma^4/(n-1)$ **c** $\sigma \sqrt{\dfrac{2}{n-1}} \dfrac{\Gamma(n/2)}{\Gamma((n-1)/2)}$

6.20 **a** $\dfrac{1}{4} e^{-t/4}$ **b** $\dfrac{u}{2} e^{-u^2/4}$

6.21 **a** $f_u(u) = 2(1-u), \quad 0 \le u \le 1$ **b** $f_u(u) = 2u, \quad 0 \le u \le 1$

6.22 $g_1(x) = e^{-(x-4)}, \quad x \ge 4$

6.23 **a** $g_1(x) = n e^{-n(x-\theta)}, \quad x > \theta$ **b** $E(X_{(1)}) = \dfrac{1}{n} + \theta$

6.24 **a** $n x^{n-1}/\theta^n, \quad 0 < x < \theta$ **b** $\dfrac{n}{n+1} \theta$

  c $\left[\dfrac{n}{n+2} - \left(\dfrac{n}{n+1} \right)^2 \right] \theta^2$ **d** $b = \dfrac{n+1}{n}$

  e $\dfrac{n(n-1)}{\theta^n} r^{n-2}(\theta - r), 0 < r < \theta$ **f** $\left(\dfrac{n-1}{n+1} \right) \theta$

6.25 **a** $\theta_1 + \dfrac{1}{n+1}(\theta_2 - \theta_1)$ **b** $\theta_2 - \dfrac{1}{n+1}(\theta_2 - \theta_1)$ **c** $\dfrac{n+1}{n-1}(x_{(n)} - x_{(1)})$

6.26 $\dfrac{1}{\Gamma(1/2)} e^{-\frac{1}{4}r^2}, \quad r > 0$

6.30 **a** $f_{u_1}(u) = \dfrac{1}{2\sqrt{u}}, \quad 0 < u \le 1$ **b** $f_{u_2}(u) = \begin{cases} 1/2 & \text{for } o \le u \le 1 \\ \dfrac{1}{2u^2} & \text{for } u > 1 \\ 0 & \text{otherwise} \end{cases}$

  c $f_{u_3}(u) = u e^{-u}, \quad u > 0$ **d** $f_{u_4}(u) = -\ln u, \quad 0 < u \le 1$

6.31 **a** $p_u(u) = \dfrac{e^{-(\lambda_1 + \lambda_2)} (\lambda_1 + \lambda_2)^u}{u!}, \quad u = 0, 1, 2, \dots$

  b $P(X_1 = u | X_2 + X_1 = m) = \binom{m}{u} \left(\dfrac{\lambda_1}{\lambda_1 + \lambda_2} \right)^u \left(\dfrac{\lambda_2}{\lambda_1 + \lambda_2} \right)^{m-u}, \quad u = 0, 1, 2, \dots$

6.32 **a, b** $E(U_1) = -1/3$ $E(U_2) = 1/3$ $E(U_3) = 1/6$

6.33 $f_u(u) = 1, \quad 0 \le u \le 1$

6.34 **a** $f_u(u) = 2(1-u), \quad 0 \le u \le 1$ **b** $1/3$ **c** $1/18$

6.35 $f_u(u) = \dfrac{1}{\Gamma(4)} u^3 e^{-u}, \quad u > 0$

6.36 $f_u(u) = 1, \quad 0 \le u \le 1$ **6.37** $3/4$

6.38 $f_u(u) = \dfrac{1}{\pi(1+u^2)}, \quad -\infty < u < \infty$ **6.39** $f_u(u) = \begin{cases} \dfrac{1}{4\sqrt{u}} & \text{for } |u| \le 1 \\ \dfrac{1}{8\sqrt{u}} & \text{for } 1 < u \le 9 \end{cases}$

6.40 $[1 - F(x)]^3 [1 + F(x)]$

6.41 $f_R(u) = n(n-1) u^{n-2}(1-u), \quad 0 < u < 1$

6.44 $f_u(u) = u e^{-u^2/2}, \quad u > 0$

6.46 $f_u(u) = \dfrac{1}{2\pi} \left(\dfrac{3u}{4\pi} \right)^{-1/3}, \quad 0 \le u \le \dfrac{4}{3}\pi$

6.47 $f_E(u) = 4\sqrt{\dfrac{2}{\pi}} \left(\dfrac{b}{m} \right)^{3/2} \sqrt{u} e^{-2bu/m}, \quad u > 0$

6.48 Poisson, with mean $\theta\lambda$

Chapter 7

7.1 $\mu = 3/4$ **7.2** $\mu = ab$ **7.3a** $\mu = \theta/2$ **7.4**   no

7.5a $1 - 2e^{-1}$ **7.6**   0.0668   **7.7**   0.6826   **7.8**   385

7.9   0.9090   **7.10**   153

7.11   **a**   0.5328   **b**   0.9772   **c**   independence and random sampling

7.12   0.0132   **7.13**   0

7.14   **a**   1   **b**   0.1230   **c**   independence

7.15   **a**   0.9938   **b**   110

7.16   4.4653   **7.17**   0.0013   **7.18**   88   **7.19**   0.9876

7.20   51   **7.21**   0.0062   **7.22**   0.0287

7.23   **a**   0.5948   **b**   0.0559   **c**   0.0017

7.24   0.3936   **7.25**   0.1292   **7.26**   0.0   **7.27**   0.1539

7.28   **a**   0.0630   **b**   0.0630

7.29   0.0041

7.30   **a**   0.0329   **b**   0.029   **c**   independence

7.31   0.0043   **7.32**   0.0548   **7.33**   0.9544   **7.34**   151

7.35   0.0668   **7.36**   0.7698   **7.37**   0.0071   **7.39** $\overline{X}/(\overline{X} + \overline{Y})$

7.41   .1587

Chapter 8

8.1   **a**   0.5037   **b**   0.3679

8.6 $1 - \exp[-200(1 - e^{-0.5})][1 + 200(1 - e^{-0.5})]$

8.7 $\left(1 - \dfrac{d}{t}\right)^2$ **8.11a** $T_0 \to \infty$